RESIDENTIAL INTEGRATION SERIES

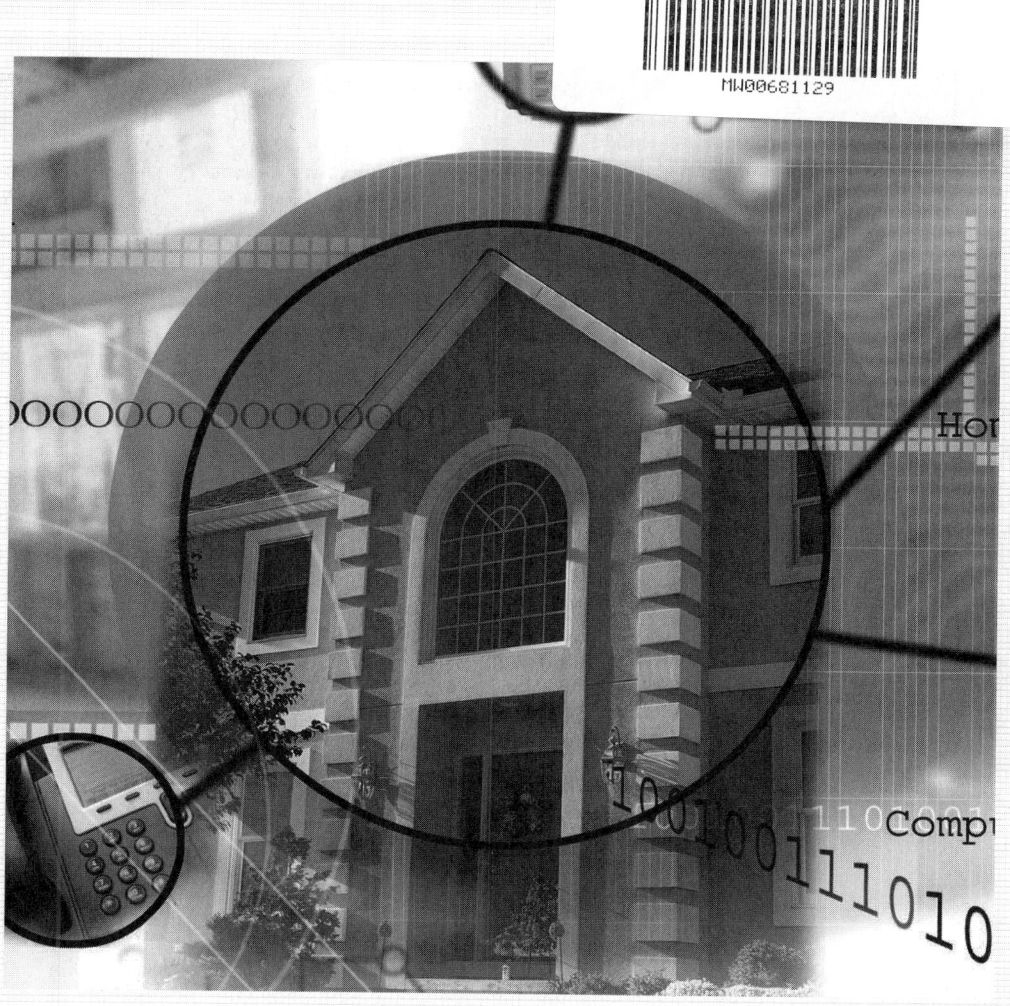

Residential Integrator's Project Management

TODD B. ADAMS

GWENN WILSON, Project Manager Quantum Integrations

THOMSON DELMAR LEARNING

Australia Canada Mexico Singapore Spain United Kingdom United States

Residential Integrator's Project Management
Todd. B. Adams and Gwenn Wilson

Vice President, Technology and Trades ABU:
David Garza

Director of Learning Solutions:
Sandy Clark

Executive Editor:
Stephen Helba

Product Manager:
Sharon Chambliss

Senior Editorial Assistant:
Dawn Daugherty

Marketing Director:
Deborah S. Yarnell

Channel Manager:
Dennis Williams

Marketing Coordinator:
Stacey Wiktorek

Curriculum Manager:
Elizabeth Sugg

Director of Production:
Patty Stephan

Production Manager:
Larry Main

Content Project Manager:
Nicole Stagg

Technology Project Manager:
Kevin Smith

Technology Project Specialist:
Linda Verde

Printed in the United States of America
1 2 3 4 5 XX 08 07 06

For more information contact Thomson Delmar Learning
Executive Woods
5 Maxwell Drive, PO Box 8007,
Clifton Park, NY 12065-8007

Or find us on the World Wide Web at
www.delmarlearning.com

ISBN: 1-4180-1411-7

NOTICE TO THE READER

Publisher does not warrant or guarantee any of the products described herein or perform any independent analysis in connection with any of the product information contained herein. Publisher does not assume, and expressly disclaims, any obligation to obtain and include information other than that provided to it by the manufacturer.

The reader is expressly warned to consider and adopt all safety precautions that might be indicated by the activities herein and to avoid all potential hazards. By following the instructions contained herein, the reader willingly assumes all risks in connection with such instructions.

The publisher makes no representation or warranties of any kind, including but not limited to, the warranties of fitness for particular purpose or merchantability, nor are any such representations implied with respect to the material set forth herein, and the publisher takes no responsibility with respect to such material. The publisher shall not be liable for any special, consequential, or exemplary damages resulting, in whole or part, from the readers' use of, or reliance upon, this material.

Contents

RESIDENTIAL INTEGRATION SERIES

Series Preface

You've heard these catch phrases many times. Home theaters. High-Definition Television (HDTV). Distributed audio and digital video. Lighting and home control systems. Broadband Internet access and wireless networking. What do all of these terms mean? Simply stated, it's cutting-edge technology that is changing how we live—and will live—in our homes.

Why is this technology emerging? For one thing, the need for technical content and information is growing. But perhaps more importantly, people want the convenience and time- and cost-saving benefits of controlling their homes electronically as a way to spend more quality time with their families.

The residential integration industry is enjoying unprecedented growth in the installation of these new technologies. Due to the high demand, residential integration companies are seeking qualified individuals who understand the traditional fields such as new home construction, along with newer disciplines such as structured low-voltage wiring and data systems. Now is the time to join the bandwagon!

Recognizing the need for educational resources in this field, Thomson Delmar Learning is excited to present a four textbook suite, called the *Residential Integration Series*, which addresses this exploding industry. These four texts encompass many of the aspects of the residential integration industry to include the basics of the business and how to get started in it, customer service skills, project management, and finally a text on the information required to prepare for the various certifications that are currently offered within the industry.

More specifically, these textbooks are:

Residential Integrator's Basics is the foundational text for this industry. Here you'll find comprehensive information on computer networks, communications, home automation, cabling, wiring, and other topics. The final chapter addresses the Home of the Future.

Residential Integrator's Customer Relations pinpoints the types of customer service skills a residential integrator needs to be successful in this industry. These include working with both internal and external clients, working in teams (very important in this field), handling difficult client relationships, communications skills, training the client on both technology and equipment, and ensuring client satisfaction.

Residential Integrator's Project Management is based on *A Guide to the Project Management Body of Knowledge* (PMBOK), which is a widely accepted work published by the Project Management Institute (**www.pmi.org**). This text is divided into four sections, each covering a major phase of a residential integration project: The Foundation, Defining the Project, Planning the Project, and Executing, Monitoring/Controlling, and Closing the Project. This text is accompanied by a CD-ROM that contains project management templates specific to the industry that can be adapted to the needs of a particular company.

Residential Integrator's Certification provides in-depth coverage of the information required to prepare for the CompTIA HTI+ exam and the CEDIA Designer Classification Series Exam. This includes low-voltage cabling, high-voltage wiring, computer networking, audio/video, security systems, home controls, and other industry related topics. Two appendices contain exam objectives for both exams.

The textbooks are pedagogically rich with chapter objectives, critical thinking questions, study tips, and other suggested activities, along with chapter summaries and review questions to help you learn and retain the material.

Build Your Perfect Course Solution

It's your course, so why compromise? Now you can create a text that exactly matches your syllabus using **Thomson Custom Solutions** online book-building application, **TextChoice.** TextChoice allows you to easily browse and select content from leading Thomson textbooks and custom collections—even include your own content—to create a text that is tailor-fit to the way you teach. Visit TextChoice at **www.textchoice.com** and learn how Thomson Custom Solutions can help you teach your course, your way.

Preface

The material in this textbook and the accompanying CD is intended for those in the residential integration industry who want to become a project manager or improve their skills as a project manager. Becoming a successful project manager involves more than delivering and setting up home electronic systems. You must learn what equipment to select, how it interacts with other equipment, how best to install it, how to estimate its cost, and how to anticipate your customer's response to it. You also need to learn who your customers are, what their values are, and how to meet and exceed their expectations. An effective project manager accomplishes all this and sticks to the company's mission statement and core beliefs in making day-to-day decisions.

This text represents many hours of research and years of experience in the industry. It is based on A *Guide to the Project Management Body of Knowledge* (PMBOK® Guide). Examples from the residential integration industry and a case study are used to relate the information in PMBOK to the industry.

The textbook is divided into four sections with a total of 18 chapters. Section I is The Foundation, Section II is Defining the Project, Section III is Planning the Project, and Section IV is Executing, Monitoring/Controlling, and Closing the Project. In addition, there are three appendices—Pulling It All Together—ABC Integrations, PMP Certification, and Trade Organizations. The CD includes computer files pertinent to the text.

In the textbook, we develop a case study company, ABC Integrations (a fictitious company), that is used to provide examples for further understanding the concepts developed within the chapters. We also develop a project management plan for ABC Integrations. ABC Integrations' business plan and other ABC Integrations' materials are included on the CD.

Section Descriptions

Section I of the text lays the foundation for studying project management. It introduces basic terminology used in project management, explains the project life cycle, and describes typical members of the project team.

Section II defines the project. To begin, it reviews basic cost accounting. It explains what information is needed before the project begins and shows how to gather that information. It defines the scope statement and explains what to include in it. This section shows how to develop a work breakdown structure (WBS) and how to use the WBS in cost estimating. Cost estimating is explained in the final chapter of this section.

Section III focuses on planning the project. It discusses the scope management plans, and explains how and why to develop plans to manage time, risks, procurement, communications, quality, and cost in a project.

Section IV describes how to execute a project, how to manage and control a project, and how and why to verify a project. Finally, this section explains how to close a project and retain the information for future projects.

Features

Chapter Objectives. Each chapter begins with a detailed list of the concepts to be accomplished within that chapter. This list provides the student with a quick reference to the chapter's contents and is a useful study aid.

Illustrations and Tables. Illustrations are provided to help the student visualize important concepts and procedures presented in the chapter.

Chapter Summaries. Each chapter is followed by a summary of the concepts presented in the chapter. These summaries provide a helpful recap and review of the chapter.

Key Terms. All key terms introduced in bold face type in the chapter are listed at the end of the chapter along with the definitions. This provides students with an opportunity to check their understanding of all the terms presented.

Review Questions. End-of-chapter review questions reinforce the ideas introduced in each chapter. These questions ensure the student has mastered the concepts presented in the chapter.

Supplemental Material

The following supplemental material is available with the text.

Faculty Guide with CD-ROM. The printed faculty guide offers a complete teaching package. Components for each chapter include chapter objectives, key terms, classroom discussion questions, learning activities, teaching tips, chapter summaries, answers to the end-of-chapter review questions, and an outline of the PowerPoint® presentation.

CD-ROM. Each faculty guide has an accompanying CD-ROM that includes a PowerPoint® presentation, a computerized test bank, sample syllabi, and suggestions for additional resources.

About the Authors

Todd B. Adams received his B.S.E.E. from Northeastern University. He is a CEDIA Certified Instructor as well as a CEDIA Certified Electronic System's Designer. Mr. Adams' recent industry awards include CEDIA Volunteer of the Year 2002, and CEDIA Lifestyle Award 2002.

Gwenn Wilson, Project Manager Quantum Integrations
Quantum Integrations, Inc. is a privately held company providing curriculum, instruction, and faculty development solutions for the education and training market. We are dedicated to assisting our customers in effectively providing the highest quality education to an increasingly diverse and challenging student/participant population. We do this through design of innovative new academic programs for colleges and universities, customized curriculum, faculty development courses and materials, and custom faculty development workshops. For more information about Quantum Integrations, please see our website at www.qicurriculum.com.

Project Management Foundation

Introduction

Developing the necessary skills to manage simple and complex projects efficiently in the residential integration industry is similar to developing the project itself. First, you need to lay a firm foundation before you can move forward. In this first section, we lay the groundwork for future levels of skill, knowledge, and understanding. We provide the information that every project manager needs to know to tackle this challenging and rewarding career. In the following three sections of the textbook, we will focus on the processes that a project manager needs to develop and create a case study company to illustrate those processes. But first we introduce project management in its most elemental terms.

In Chapter 1, we start with the basics. We define the term project, explore the role of the project manager, explain the need for a project management plan, discuss project and facility constraints that affect a project, and provide the details of what project management is. In Chapter 2, we focus on the life cycle of a residential integration project. We go over the typical life cycle and review the common characteristics in the life cycle of each project. We list and explain the five phases of a project—initiating the project, planning the project, executing the project, monitoring and controlling the project, and closing the project. Then we elaborate on each life cycle characteristic.

We list the key functions in defining the project and explain each function in detail. These functions include lead generation, qualification, and contract; project scope; cost estimates; schedules; documentation; project monitoring; customer preferences; and project close. We discuss the importance of planning the many details in a project. We explain the five common phases in executing a project. And finally, we discuss closing the project, which includes system testing, customer training, final documentation, project audit, and turnover from project management to the service department.

In Chapter 3, we introduce the project team and provide typical job descriptions for each member. We look at how much authority a project manager has and how the type of organization affects that authority. We also spend some time on the customer, explaining who is the residential integrator's customer, what customer service is, and how to ensure customer satisfaction.

Once we have established the foundation for studying about project management in section one, we move onto sections two, three, and four, where we explain how to create the major processes that must be developed to manage a project successfully.

Introduction to Project Management

OBJECTIVES

After studying this chapter, you should be able to:

- Define the term "project" and list its key components.
- Discuss the role of the project manager.
- Identify project constraints and explain how they affect a project.
- List and define facility constraints.
- Define project management.

OUTLINE

Introduction

What Is a Project?

Facility Constraints

What Is Project Management?

Introduction

A career in project management can be as rewarding as it is challenging. It requires skill, knowledge, and understanding to achieve success. Project management is an art and a science, requiring a high degree of professionalism. The first three chapters in this textbook lay the foundation for understanding the basics in project management in the residential integration industry. A project manager has the ability to manage time, scope, cost, and quality throughout the project. He knows how to pay attention to project constraints and facility constraints in planning and executing a project. A relatively young industry, residential integration already has developed certain standards and protocols based on the experience of others in the field, who have managed a wide range of projects from whole-house audio systems

PMBOK

In this textbook we will discuss project management in the residential integration industry based on *A Guide to the Project Management Body of Knowledge* (PMBOK® Guide). The Project Management Institute (PMI), a trade organization in the project management industry, produces *The PMBOK® Guide*. It represents the sum of knowledge within the profession of project management. It includes proven traditional practices that are widely applied as general project management methodologies across all industries. In this textbook, we take these generally accepted practices and apply them to the residential integration industry. Visit **www.pmi.org** for more information.

and home theaters to total home integration. To begin to understand how to manage projects, we first must understand what a project is and what all projects have in common.

What Is a Project?

So, what exactly is a **project?** A friend may say, "I'm going to clean the garage this weekend, it's going to be a real project." Your boss comes to you on that clear, sunny Friday afternoon and says, "I have a small project for you that you should be able to finish by the close of business." NASA announces, "We have decided to take on the project of sending mankind to Mars."

What do all these projects have in common? They all have these essential elements:

- A project is temporary.
- A project has progressive elaboration.
- Each project has unique characteristics.
- A project requires resources.
- A project has a sponsor.

Let's take a closer look at each one of these elements.

Critical Thinking Question

Think of a homework assignment or a task at home or work you've recently completed.

- How would you relate it to the essential elements of a project?

A Project Is Temporary

A project is temporary in nature; that means that it has a clearly defined start and finish. This, however, does not necessarily mean the timeline is short; a project may take several years to complete. Let's

OUR CASE STUDY

ABC Integrations

Throughout this textbook, we refer to a single company, ABC Integrations, as our case study. In order to demonstrate how a project management plan is built for the residential integration industry, we use ABC Integrations as our model and we create project management documents that support the needs of that company.

The first document we create is Document PM01, which is a project charter used to define the key technologies and standards for a structured cabling system typically installed by ABC Integrations. This document is created and explained in Chapter 9 The Scope Management Plan.

Note that all templates and documents referenced within this text are included on the CD-ROM that accompanies this textbook and are listed by the document number.

put this in terms of the residential integration industry. The typical 2,500 square foot home is built in less than nine months, but homes that are greater than 10,000 square feet can take several years to complete due to the size and scope of the building design. Nevertheless, the project is still temporary.

It should be noted that the temporary nature of the project does not generally extend to the outcome of the project. In fact, projects are undertaken to create a lasting outcome. For example, it may take as little as a week to install a home theater in a family room, but the system will last for many years before needing replacement or upgrades.

The temporary nature of a project also applies to other areas of the venture. The opportunity to obtain the project exists only during a short procurement window. If our organization in our case study, ABC Integrations, does not install a home theater in a timely manner, the company could be let go and soon another company could take on the project. Another consideration is that the project team for each project is assembled as a temporary unit. That is, the builder, carpenter, electrician, and installation team of the organization work together for a limited period of time to accomplish the project. Once the installation is completed, the members of the team individually move on to other projects that have been awarded to the organization. Figure 1–1 shows the temporary nature of a project as it relates to the installation of a home theater.

Study/Career Tip

Consider making old-fashioned flash cards with the definitions of the bolded terms in this and other chapters in this text. Or, create an on-line glossary that you can reference as a study tool. Remember, while some of these terms are common in every day use, you will need to be able to apply them within the context of residential integration.

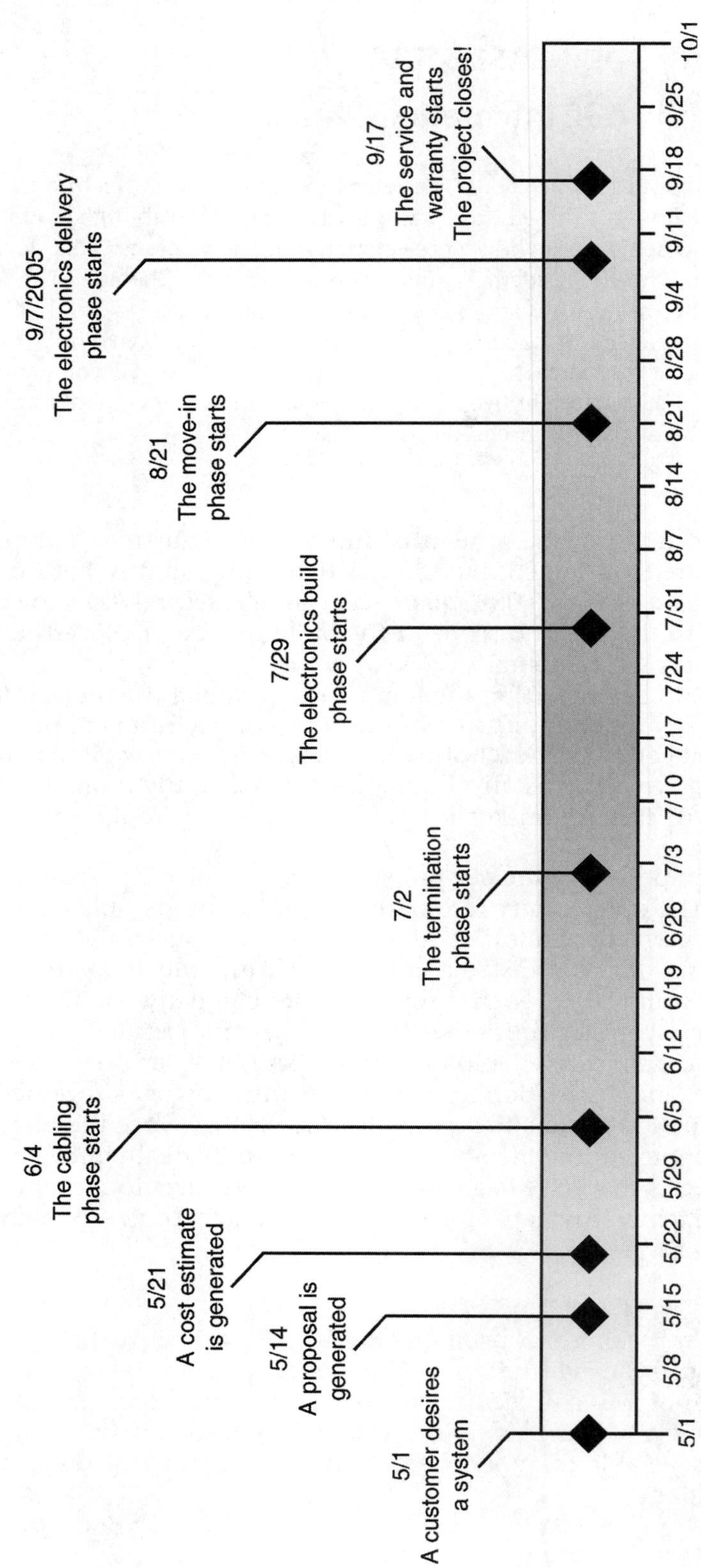

FIGURE 1–1 Temporary Nature of a Project

A Project Has Progressive Elaboration

The term **progressive elaboration** means developing the project in steps and continuing in increments. A project consists of a well-defined collection of small tasks or activities and **events,** with a project schedule, indicating the preferred sequence of execution for these tasks and events. Examples of *tasks* are installing cables, speakers, keypads, and touchscreens. An example of an *event* is an electrician completing his work at the residence. That event triggers the next task. The painters finish their work (event), which triggers the installation of the speakers (task). Note that, within the project management profession, the terms **task** and **activity** are used interchangeably.

Critical Thinking Questions

- Which are tasks?
- Which are events?

Walking the dog
Getting the dog's leash
Buttering a slice of toast
Eating the toast
Watching the local news
Changing the channel
Brushing your teeth
Sleeping

Those progressive tasks and events are carefully integrated within the **scope** of work. The customer not only will want to know about the project **deliverables** (the physical items brought to the home for the project), but she also will want to know in what order the various phases of the project will be carried out.

In a residential integration project, there are distinct phases of installation. These phases progress from one to the other in an organized process. Let's look at an ABC Integrations list of events and tasks as they apply to the installation of a single wallplate, (see Figure 1–2). A wallplate is the connection point for connecting the electronics in a room to the house-wide integration system. Wallplates generally are located on the wall near the floor.

We discuss the progression of a project in greater detail in Chapter 2, The Residential Integration Project Life Cycle.

Each Project Has Unique Characteristics

A successful project concludes with the creation of an end product or deliverable (something that can be or will be delivered to the customer). A project is a unique, one-time undertaking; it will not be executed again exactly the same way, by the same people, or within the same environment. This uniqueness is the art and science of the work performed by the **project manager** (read more about the role of the project manager later in this chapter). In the residential integration industry, every home system that is installed is unique by its very nature. For example, let's say ABC Integrations has two homes that order the same set of equipment for home theater systems.

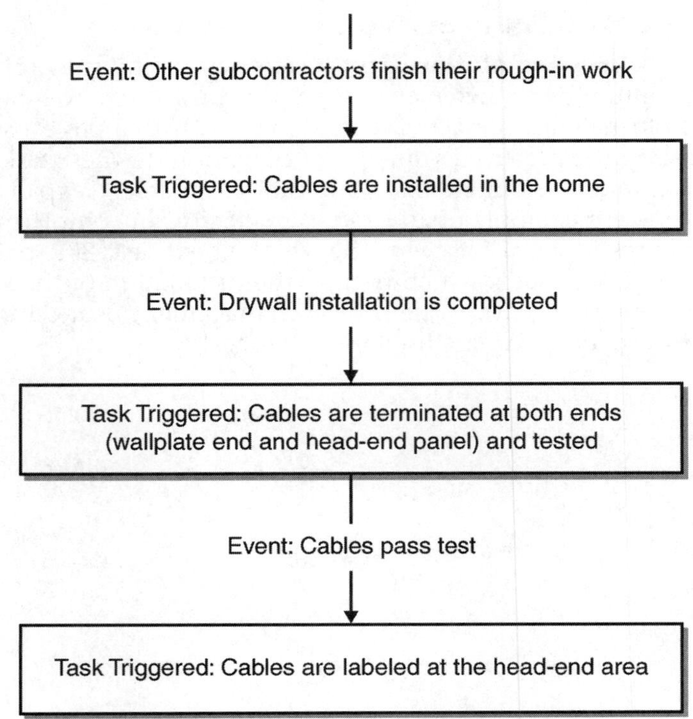

FIGURE 1–2 ABC Integrations' List of Tasks and Events for a Wallplate

Even though they are receiving the same equipment, each installation outcome will be unto its own. The placement of the speakers will depend on the room layout. The equipment in one ABC Integrations project needs to be installed in a custom cabinet, requiring cabinet documentation and a ventilation system to keep the system cool. The other ABC Integrations project is installed in an equipment closet. Other unique characteristics of the two projects include:

- Different contractors doing the installation
- Customized programming of the touchscreen-based remote
- Documentation requirements
- Site location and conditions
- Customer preferences
- Costs.

It is impossible to describe a successive project precisely the same way as the previous project because of the many variables in each project. The project objectives, deliverables, resources, timeline, and costs are different with each project, which is executed within different homes with different styles and dimensions. And it is one main reason why a residential integration project manager's job is so challenging.

Despite its uniqueness, many comparisons can be drawn from project to project to help create a smooth process that can be used in subsequent projects. Some similarities in ABC Integrations' projects include:

- Cable requirements
- Identical equipment
- Residential environment
- Non-technical, family-based customer.

Critical Thinking Questions

- Why would it be important to track the similarities among residential integration projects?
- How will this information benefit the project manager when embarking on a new installation project?

Study/Career Tip

Think about the way you study. You may not study at the same time of day (although that's a very good habit to get into), in the same room, reviewing the same subject. But whether you recognize it or not, you do have a pattern of how you study, such as rereading the chapter, studying review questions, making notes in the margins of the text, and so forth. Just like a residential integration project, your study habits follow an established, proven set of steps that get you from start to finish. It's a project, no matter how you look at it.

A Project Requires Resources

Clearly, a project cannot come to a successful conclusion without resources being put toward it. **Resources** can be people, materials, assets, software, and other items you might need to be ready to use on the job (see Figure 1–3). The project cannot be carried out without these things. For example, when installing a home theater system for the customer, the following is a list of some resources used during the installation process:

Examples of people as resources:

- Installer: 20 hours of installation time
- Project manager: 5 hours of project management time

Examples of materials as resources:

- 125 feet of speaker cable
- 1 set of home theater speakers
- 1 Surround Sound receiver
- 1 DVD player

FIGURE 1–3 An Electrician Would Be Considered on Asset. (Photo courtesy of constructionphotographs.com)

Examples of assets as resources:

- 1 installation truck
- Assorted hand tools
- Assorted power tools

In addition to the internal resources listed above, a number of external resources are employed to carry out the project. A few of those resources are listed below:

Examples of external people as resources:

- Electrician
- General contractor
- Satellite installation subcontractor

Example of external assets as resources:

- Rented equipment, such as a category 5e cable certifier or a staging, power generator
- Other high-value tools required to complete the work

It's important that the project manager track the allocation and use of all resources at the installation site. While the same person might fill different resource requirements, a diligent project manager lists each task separately. For example, if one person within the organization fills the role of the installer *and* the lead installer, both of these tasks and their allotted times are listed as separate resources. This is done so that, in the event a second person is added, the work can be easily divided between functional responsibilities.

Regardless of which resources are utilized for a project, they must be carefully monitored. There is never an endless supply of any resource. The misuse of resources costs the company money, time, and quality. It is a residential integration installer's job to make sure that the resources are not only accounted for on the job, but that they are used in the most economical and timely way.

- Even if you are not the project manager on a job, why do you need to think like one when it comes to the allocation and use of resources?

Critical Thinking Question

A Project Has a Sponsor

All projects have interested parties called **stakeholders.** In the case of residential integration, they can be the builder, the electrician, the roofer, other subcontractors, and the custom installer or home integrator. Typically, the primary stakeholder is the customer, who is also known as the project **sponsor.** In our case study, Fred Smith, the customer (or homeowner), is the one who ultimately bears the financial burden for the project, so he has a vested interest in making sure the project is completed according to his wishes. By signing the scope statement and cost estimate (contract), Fred Smith indicates the project deliverables and sets expectations for the project outcomes.

In many industries, such as manufacturing, a customer is viewed in terms of demographics or as a small statistical piece in the study of a larger group of the population. An individual customer is viewed almost as an afterthought. Not true in the custom installation and residential integration industry. Here the customer sits at the center of the project by driving the project's breadth and constraints. In the automobile industry, for example, cars are designed, manufactured, and brought to market prior to customer involvement. In the residential integration industry, the customer creates the need for the end product, and as a result the product is not designed until there is a specific customer need.

As the sponsor, the customer holds the purse strings of the project.

- As a project manager, how do you go about helping the customer define his need for the end product?

Critical Thinking Question

A Project Has a Manager

Project managers in all industries manage projects. They follow the project from start to finish, and they are in charge of how a project flows. They make sure a project stays on schedule and on budget. They assign tasks and responsibilities to supporting players, who will complete the actual project. And they assume the overall responsibility for the successful completion of the project, regardless of what obstacles might get in the way. Early in your career, you aren't likely to be called upon to be a project manager. However, understanding the skill set required to be a successful project manager is key to your personal success in the industry. In later chapters of this text, we explore the role and responsibilities of a project manager in detail.

Study/Career Tip

Think of yourself in terms of being your own project manager in your daily life. You have the ultimate responsibility for many things at home, at school, and at work, and if they are not accomplished correctly, there are consequences. Those consequences may be deemed minor in the grand scale of things, but the successful management of tasks and events in your daily life will help to reinforce the skills you'll need to be a successful project manager in the residential integration industry.

A Project Has a Plan

All project managers need a project management plan, regardless of the industry. It's the blueprint for success and serves as an audit trail of the many tasks and events that were accomplished in order to complete the project. In addition, project management plans guide the project manager in "lessons learned"—what worked, what didn't, and where improvements can be made. Throughout this textbook, we will be building the ABC Integrations project management plan. Each chapter will expand on prior chapters to build a complete suite of project management templates and forms for such a plan. It is the purpose of the textbook to teach the fundamentals required for you, the project manager, to build your own suite of processes and solutions for your organization. Examples of the tools you'll use include information gathering forms, the project charter, the scope statement template, project memos, and a project-tracking template. By the end of this textbook, you will be well versed in the intricacies of a project management plan and will be armed with the tools and templates you'll need to develop one for your organization.

A Project Has Constraints

Projects are *always* evaluated and measured against the three **constraints—cost,** scope, and time. Project **quality** is affected by the balancing of these three factors.

These three project constraints are interrelated trade-offs, pulling in different directions throughout the project. If the project timeline is accelerated, then overtime charges increase and, as a result, costs will increase. If the customer (sponsor) were to request a lower cost from what was originally estimated, this request would result in a change in the scope of work. It might also impact the time it takes to complete the installation. If additional features are added to the system, again, this would change the scope of the project, which would in turn add to the project cost.

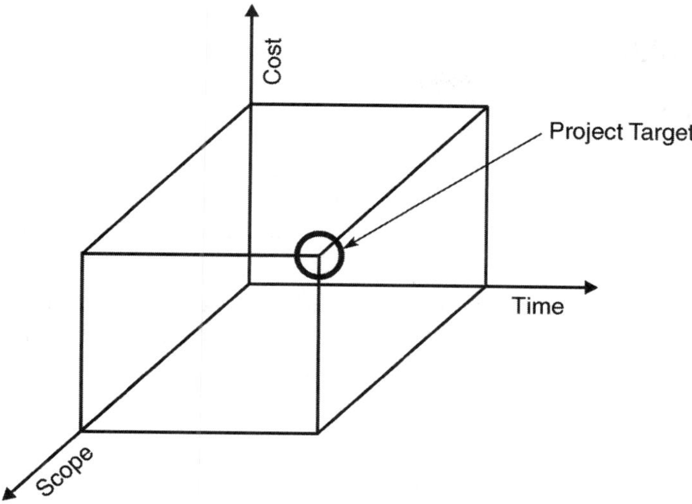

FIGURE 1–4 The Triple Constraints of Project Management

Let's take a look at an example of project constraints. A customer receives a quote for a home theater with a price of $50,000. The customer requests that the project not exceed $46,000. In order to facilitate the cost constraint, the projector that was originally chosen is replaced with a less expensive model. The original projector included a model with higher resolution and higher light output than the subsequent model. While the *cost* constraint is met, the quality of the video is lessened due to the decrease in resolution and light output. The project *time* does not change, as both projectors require the same amount of installation time. The project *scope* has changed because the requirement for resolution and light output of the projector has been decreased.

It is important to not confuse negotiation with cost constraints. The customer might be asking for a discount on the overall price of the system, and if the residential integrator has done a thorough job of pricing the system, the equipment should already be at market price. By simply lowering the price, without changing the scope, there exists a risk of low or no profitability on the project. See Figure 1–4 for a graphical representation of the triple constraints of project management.

Critical Thinking Questions

Think of a project you recently completed, whether at home, school, or work.

- What project constraints did you encounter?
- How did you mitigate those constraints in order to complete the project?

Cost

The cost of a home integration project represents all deliverables; labor costs associated with cable installation; termination, programming,

PROJECT MANAGEMENT PLAN

ABC Integrations Scope Statement PM05

You can choose wallplates with our standard colors of white, ivory, almond, or black, or you can select wallplates with one of the special finishes that are available.

- **Hanging Telephone Wallplate** The hanging telephone wallplate allows connections for two (2) telephone lines.
- **Telephone and Networking Wallplate** The telephone and networking wallplate allows connections to four (4) telephone lines and a home network.
- **Multimedia Wallplate** The multimedia wallplate allows connection to cable television, satellite, four (4) telephone lines, and a home network.

and electronics, cabinetry; building equipment racks; and any manufactured integration systems. The cost also includes more than just the expense of the work itself. The project cost must include consideration for the home integrator's facility; advertising and daily; business operating expenses, including salaries beyond labor costs; benefits; and administrative expenses.

As we learned before, the customer imposes the project budget on the project manager, as well as the time constraint for completion. The budget is highly useful in defining project limits, and is an inherent financial constraint. It would be nice to have deep pockets for every project that you engage in, but the reality for most companies, organizations, and individuals is that budgetary limits must be set. And this is just as well because without a budget, there are no defined limits on the project. There is no end point; the project would never be completed. We explore cost in Chapter 4, Understanding Cost Accounting, and cost estimation in Chapter 8, Cost Estimating.

Scope

In the residential integration industry, the project scope is detailed in a document called the scope statement (also known as scope of work, see document PM05). It is written in language understandable to the customer. No techie vocabulary here. No light emitting diodes (LEDs) or capacitors. The scope statement with customer-centric text clearly states each aspect of the work to be accomplished. Typically, this document is prepared by the sales representative. Here's how ABC Integrations's scope of work, or scope statement, looks for the ordering of wallplates as part of a structured cabling system.

Customers Fred and Wilma Smith review the scope statement, which includes a choice of many wallplates, and select the ones that they want included in their home, thus creating the scope of work for wallplates. Fred and Wilma continue their selections throughout the entire scope statement process, creating the scope of work for the whole project. The scope statement is fully explored in Chapter 6, The Project Scope Statement.

Study/Career Tips

There's an unscientific expression used in many industries that rely on scope statement (or scope of work) documents to conduct business. It's known as "scope creep." What this means is that the customer and/or other involved parties change their mind about products and services, or the availability of such products and services may become an issue. It is remarkably easy for scope creep to hit any project, thus impacting time, costs, and deliverables. At the first sign of scope creep, contact your supervisor to report the requested variances from the established scope statement document. Don't just change the document on the fly, or worse, not document changes at all. In the long run discussion and documentation will save all parties aggravation and will help to ensure the good business reputation of your company.

Time

Time spans the scheduled start and stop times for a subtask, a task, a phase, or an entire project. Although projects often are completed beyond the timeframe initially allotted, the timeframe is important because it helps to structure the work—all the tasks and events that must be completed within the project. A project without a schedule or predetermined timeline is like a football field without an end zone. You may have the ball and crank out a lot of yardage, but you'll never score until you cross the goal line. Plus, you might run out the clock before you put any points up on the scoreboard!

When building a new home there are two milestones along the build process that are the most time-sensitive in terms of any type of residential integration being installed prior to occupation. The dates of these events are scheduled by the general contractor, and all subcontractors involved with building the home have to complete their work in order for these two activities to occur. The two milestones in question are Sheetrock™ installation and the customer's Certificate of Occupancy (CO).

Critical Thinking Question

- Why do you think that Sheetrock™ installation and the Certificate of Occupancy are the most impactful events when it comes to managing time during a residential integration installation project?

Critical Thinking Question

- As an installer who is caught in the middle of all the action, how do you deal with this situation?

The trick in dealing with proper time management is to recognize these crucial events up front and to perform the lion's share of work prior to these two critical periods of construction.

Critical Thinking Questions

We've all heard the expression "that's of high quality."

- What does that mean to you?
- Is it simply an expression that's used when talking about goods such as wool sweaters and fine china?
- How would you equate the term "high quality" to a service such as being served in a restaurant?
- How do you measure the quality of a good versus the quality of a service? Is it the same?

Quality

The culmination of time, scope, and cost is quality. What is quality? The International Organization of Standards (ISO) defines quality as "The totality of characteristics of an entity that bear on its ability to satisfy stated or implied needs."

A little confusing or vague? Well, think of quality as the ability to exceed customer expectations. This is true for the delivery of both goods *and* services. If the customer has a set of expectations for the deliverable(s), whatever those may be, then by exceeding those expectations, you are helping to ensure that your customer is receiving the quality he expects and deserves. After all, he's paying for it!

In the case of home integration installations, high quality projects are those that are delivered on budget, on time, and hit the deliverables as specified in the scope statement. The dollar value of the project does not necessarily indicate its quality.

What is a good measure of quality? There is no better way to provide a high quality product than meeting budgets. For example, a customer purchases a home theater for $10,000 and may view this theater as high quality because it was installed exactly to his wishes, on time, and within budget, (see Figure 1–5). Another customer may view a $100,000 theater as not being high quality if the theater system does not perform as specified in the scope statement, or the installation is continually delayed, or the project is over the initial budget.

The customer can see the equipment within the system is of exceptional quality, but that the service offered to install it did not meet the customer's expectations. While meeting the project budget cannot guarantee quality in the eyes of the customer, it does go a long way in ensuring customer satisfaction.

Facility Constraints

In addition to project constraints, there are facility constraints that the project manager must consider. They are:

- Human Resource Management
- Communications Management

FIGURE 1–5 An Example of a Home Theater

- Risk Management
- Procurement Management

Much of the knowledge used to manage residential integration projects is specific to the industry. However, a basic knowledge of general management skills is also needed. These management skills include human resources, **communications,** business risks, and procurement. We have grouped these elements of expertise into what is known in the industry as facility constraints. Each of these elements has a direct impact on the overall management of the project.

Critical Thinking Questions

- Why do you think it is important for project managers to be well versed in the area of human resource management?
- What does this have to do with installing cable in someone's home?

Human Resource Management

Human resource management can be divided into the following categories:

- *Employment law:* This includes civil rights, the Equal Employment Opportunity Commission (EEOC), Affirmative Action, pregnancy, age and disability legislation, and many other employee protection laws.

- *Effective recruitment:* This includes recruiting methods, key selection tools, adverse impact calculations, and interview biases.
- *Basics of Compensation and Benefits:* This includes base pay structure, incentives, differentials and increases, job analysis and documentation, methods for job evaluation, and commonly offered benefits.
- *Orienting and training employees:* This includes meeting corporate training objectives and providing effective adult learning, both on the job and off the job.
- *Ensuring quality performance:* This includes common appraisal methods and errors, legal concepts surrounding the disciplinary process, and guidelines for conducting disciplinary meetings.

It should be noted that it is not the intention of this textbook to address Human Resource Management. However, project managers need to have a broad understanding of these topics as they can impact employee on-the-job performance, morale, and the ability to recruit, hire, train, and retain excellent workers. There is no greater litmus test about who you are as a company than the employees who are sent into the field to perform quality work, job after job.

Study/Career Tips

It's a known statistic that if a customer is dissatisfied with the customer service they receive, for any reason, they will tell ten people about why they are dissatisfied and who made them that way. Conversely, a satisfied customer rarely tells one or more people. Who do you want your customers talking to, and what about?

Communications Management

Communications management includes creating and utilizing a **communication management plan,** distribution of project information to those with a need to know, **performance reports,** and managing the project stakeholders, including the customer.

These practices interact with each other as well as the other elements of general management. Each one of these aspects of communications management can involve just the project manager or the entire project team during the various phases of the project. They can utilize a wide range of communication vehicles, such as memos, faxes, e-mail, verbal communication, weekly status meetings, and even a corporate Internet or intranet site.

Study/Career Tip

It's fine and well to have good communications among project team members. But don't forget a vital communications link that should never be broken—communication with the customer. Interaction with the sponsor, daily as need be, is critical. Don't leave the customer guessing, or worse, assuming. This can cause project delays, flared tempers, and a hit to your bottom line—profitability.

Risk Management

Risk management is the process of identifying, analyzing, and responding to risks throughout the project. Risk management is frequently overlooked by the novice residential integrator who, because of a lack of industry experience, does not fully understand all of the possible pitfalls during the project life cycle. A few of the risks encountered during the project life cycle are listed below.

- **Ambiguity in Scope** A poorly written scope statement that is unclear in each project deliverable can create the risk that deliverables will not be delivered properly or fall short of customer expectations. For example, if the telephone and networking wallplate states that it includes two category 5e cables, but does not state the purpose for each cable, the purpose of the cables might be misconstrued by the customer, resulting in poor customer satisfaction.
- **Plan and Specification Discrepancies** Often, house plans and other documentation are provided by the customer to the sales representative to prepare the scope statement. In some cases, there are discrepancies between various plans provided by the customer. For example, one plan might call for two category 5e cables to each telephone and networking wallplate, and another document might call for three category 5e cables.
- **Extended Warranty** The length of time of the extended warranty extends the risk on the project. A 90-day warranty inherently has less risk than a five-year warranty.
- **Special Test Equipment/Tools** A home theater project might require the use of a special audio analyzer. Acquiring the tool, learning its proper usage, and training an employee on its use all add to the risk of the project.

We cover project risks in more detail in subsequent chapters. You can think of risk management as an insurance policy—one that lessens potential project delays and costs. Apply a pro-active process to identifying risks before they happen, and formulate an action plan based upon those potential risks. In this way, many possible pitfalls can be bypassed, and others can be minimized by anticipating and identifying the risks.

Critical Thinking Question

The saying goes: "The customer is always right."

- How do you manage this risk when you are at the home doing an installation?

Procurement

Procurement means purchasing goods and/or services from a source that is outside the company, commonly referred to as an outside source. Procurement also is known as purchasing in many companies within the residential integration industry, and typically, the term is limited to the acquisition of goods. For example, the Purchasing Department buys all the office supplies needed by the staff. It buys all the materials needed for a project—the nuts and bolts, equipment, cables, and any other goods needed to complete a project. In many residential integration companies, acquiring services is handled in a different manner than the purchasing of goods. Sometimes purchased services are outsourced. Outsourcing is the term used to define the procurement of services. Chapter 12, Procurement Management Planning, focuses on both purchasing to acquire goods and outsourcing to acquire services in the residential integration industry. Each aspect of procurement is discussed in separate sections, although there is some overlap because the principles of procurement apply to both.

Critical Thinking Questions

- What's a "change order?"
- Why are change orders so important in an industry such as residential integration?
- Why should you always request a signature on a change order by the customer before leaving the premise?

What Is Project Management?

Project management is the application of knowledge, skills, techniques, and tools to meet project constraints while working to control facility constraints (see Figure 1–6). Project management is accomplished through the application and integration of initiating, defining, planning, implementing, and closing the project. Project management involves:

- Identifying project requirements
- Establishing clear and achievable objectives

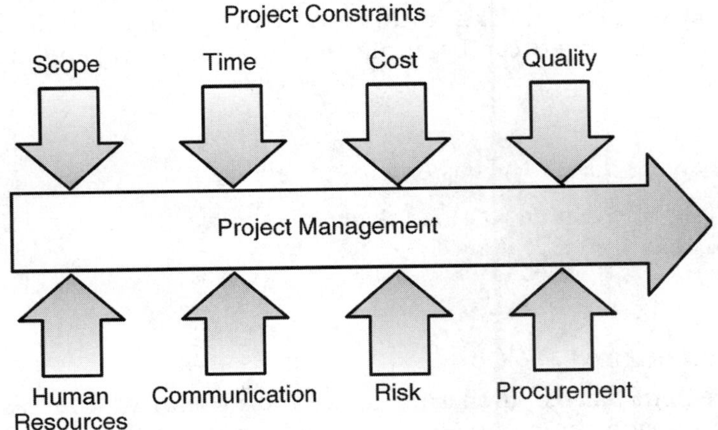

FIGURE 1–6 What Is Project Management?

PROJECT MANAGEMENT PLAN
ABC Integrations Site Survey PM02

SITE SURVEY

PROPOSAL NUMBER: ____________ DRIVING DISTANCE: ____________
PROPOSAL DUE BY: ____________ CUSTOMER NAME: ____________
SITE ADDRESS: ____________
PROJECT SCOPE: ____________

BASEMENT

CEILING HEIGHT: ____________ TYPE: ❑ OPEN ❑ DROP ❑ BOARD & PLASTER ❑ ____
SQUARE FEET: ____________ FINISH: ❑ YES ❑ NO
CONSTRUCTION: ❑ NEW ❑ RETROFIT
ROOM NAMES:

- Balancing project constraints
- Adapting *procedures and personnel* from project to project.

Identifying Project Requirements

Identifying a project's requirements is at the heart of the scope statement. What are the expectations of the customer? What is required to achieve those expectations? As part of our case study, we develop a set of survey forms designed to identify the various requirements of the project (see document PM02). These forms are used by the sales representative to get the right information from the customer to create realistic expectations and to set the project objectives.

What is the relationship between the project manager and the sales representative? The sales representative sells new projects and works with the customer, the primary stakeholder. The project manager fulfills the requirements of the scope statement. As a result, there is a balance between the two in setting expectations with the customer. If the sales representative over promises, then the project manager is doomed to fail, and if the project manager is too conservative in what he permits to be sold, the company could not maintain it place in the market place. We discuss this further in Chapter 3, The Project Team.

Establishing Clear and Achievable Objectives

Again, it's all about the scope statement. By setting detailed **objectives** in writing with the customer, it is far easier to manage the project. A customer might want all white wallplates used throughout his home.

That's certainly an achievable objective. But if all ivory wallplates are delivered and there is no documentation available to the installer indicating the color of the wallplates to be used, then the wrong wallplates could be installed. Depending on how many wallplates there are, it could become a nightmare taking out the wrong color wallplates and reinstalling the right color wallplates. And let's not forget the customer satisfaction factor here, not to mention the perceived quality of work.

Balancing Project Constraints

The ability to successfully balance project constraints comes from experience and is more an art than a science. The ability to manage time, scope, cost, and quality throughout the project is what makes a project manager succeed or fail. In residential integration, a project manager may have as many as fifty projects in progress at any given time. Each of these projects comes with its own budget, timeline, and project deliverables, not to mention all the management issues just discussed. The project manager must balance each aspect of every project to ensure all areas are not left without audit. Pay too much attention to the deliverables, and ignore the timeline, and the project could run past its agreed upon deadline for completion. Additional costs could be the result.

Summary

In Chapter 1 we examined the term "project" as defined by the Project Management Institute (PMI) and its standards document, *The Guide to Project Management Body of Knowledge,* the role of the project manager, the constraints that can affect each project, and other areas a project manager must consider during the course of a job. All of these topics were discussed in relationship to the residential integration industry. Additionally, we provided a definition of project management and laid the groundwork for more in-depth discussions about various aspects of project management in subsequent chapters of this text. Finally, we introduced the case study "ABC Integrations" that we will follow throughout the chapters. A series of templates and documents that will guide us through the case study can be found on the CD-ROM that accompanies this text.

Important points in this chapter are:

- A project has certain essential elements. It's temporary. It has progressive elaboration. It has unique characteristics. It requires resources and a sponsor.
- Project managers assume overall responsibility for the successful completion of a project and monitor all aspects of the project from start to finish.
- Projects always are evaluated and measured against the three constraints—cost, scope, and time. The balancing of these three factors affects the project quality.
- Time is important because it helps to structure the work. Two milestones in the building process are the most time-sensitive. The two milestones, as scheduled by the general builder, are Sheetrock™ installation and the customer's Certificate of Occupancy.
- The culmination of time, scope, and cost is quality. High quality projects are those that are delivered on budget, on time, and hit the deliverables as specified in the scope statement.

- The project manager must consider facility constraints—human resource management, communications management, risk management, and procurement management—at all times throughout the life cycle of the project.
- Project management is the application of knowledge, skills, techniques, and tools to meet project constraints. Project management is responsible for initiating, planning, implementing, and closing the project. Project management identifies project requirements, establishes objectives, balances project constraints, and adapts plans from project to project.

Key Terms

Activity A component or work performed during the course of a project.

Communication management plan The document that describes the communications needs and expectations for the project, how and in what format information will be communicated, when and where each communication will be made, and who is responsible for providing each type of communication. A communication management plan can be formal or informal, highly detailed or broadly framed, based on the requirements of the project stakeholders. The communications management plan is contained in, or is a subsidiary plan of the project management plan.

Communications A process through which information is exchanged among persons using a common system of symbols, signs, and behaviors.

Constraint The state, quality, or sense of being restricted to a given course of action or inaction. An applicable restriction or limitation, either internal or external to the project, that will affect the performance of the project or a process. For example, a schedule constraint is any limitation or restraint placed on the project schedule that affects when an activity can be scheduled and is usually in the form of fixed imposed dates. A cost constraint is any limitation or restraint placed on the project budget such as funds available over time. A project resource constraint is any limitation or restraint placed on resource usage, such as what resource skills or disciplines are available and the amount of a given resource available during a specified period.

Cost The monetary value or price of a project activity or component that includes the monetary work of the resources required to perform and complete the activity or component, or to produce the component. A specific cost can be composed of a combination of cost components including direct labor hours, other direct costs, indirect labor hours, other indirect costs, and purchase price. However, in the earned value management methodology, in some instances, the term cost can represent only labor hours with conversion to monetary worth.

Deliverables Any unique and verifiable product, result, or capability to perform a service that must be produced to complete a process, phase, or project. Often used more narrowly in reference to an external deliverable, which is a deliverable that is subject to approval by the project sponsor or customer.

Event Something that happens, an occurrence, an outcome.

Objective Something toward which work is to be directed, a strategic position to be attained, or a purpose to be achieved, a result to be obtained, a product to be produced, or a service to be performed.

Performance reports Documents and presentations that provide organized and summarized work performance information, earned value management parameters and calculations, and the analyses of project work progress and status. Common formats for performance reports include bar charts, S-curves, histograms, tables, and project schedule network diagrams showing current schedule status.

Progressive elaboration Continuously improving and detailing a plan as more details and specific information and more accurate estimates become available as the project progresses, and thereby producing more accurate and complete plans that result from the successive iterations of the planning process.

Project A temporary endeavor undertaken to create a unique product, service, or result.

Project manager The person assigned by the performing organization to achieve the project objectives.

Project management The application of knowledge, skills, tools, and techniques to project activities in order to meet or exceed stakeholders' needs and expectations from a project.

Quality The degree to which a set of inherent characteristics fulfills requirements.

Resources Skilled human resources (specific disciplines either individually or in crews or teams), equipment, services, supplies, commodities, materials, budgets, or funds.

Scope The sum of all products, services, and results to be provided as a project.

Sponsor The person or group that provides the financial resources, in cash or in kind, for the project.

Stakeholders Persons and organizations such as customers, sponsors, performing organization and the public, that are actively involved in the project, or whose interest may be positively or negatively affected by execution or completion of the project. They may also exert influence over the project and its deliverables.

Task A term for work whose meaning and placement within a structured plan for project work varies from the application area, industry, and brand of project management software.

Review Questions

1. What are the five essential elements of a project?
2. Why is it important to differentiate between events and tasks on the job?
3. Give three examples of resources that may be used during an installation.
4. Why is it important for the project manager to identify and monitor resources during the course of a project?
5. Generally speaking, who is the project sponsor? What is the sponsor's primary role during the project?
6. Three project constraints are scope, time, and cost. Provide a brief description of each.
7. Why is it important for project managers to have an overall sense of human resource management?
8. What is a definition of risk management?

The Residential Integration Project Life Cycle

OBJECTIVES

After studying this chapter, you should be able to:

- Explain a typical residential integration project life cycle.
- List and explain the common characteristics of a residential integration life cycle.
- Discuss how certain key functions define a project.
- Explain the process of planning a project.
- Define the important issues surrounding each implementation phase of a residential integration project.
- List and explain the typical implementation phases of a home integration project.
- Define the roles of the project manager and the project team during the project life cycle.
- Explain how to close a project.

OUTLINE

Introduction

The project life cycle encompasses the whole project from the minute a sales representative closes the deal and has a signed contract to the final minute when a service contract is signed or the project is accepted by the customer. Typically, the project life cycle does not include locating potential customers, time spent selling to the customer, or contact with the customer after the project is completed. A sales representative usually maintains contact with a customer long after the project is complete, potentially to sell future projects and upgrades or gain referrals. In this chapter, we discuss the project life cycle from the signing of the contract to the closing of the project when the customer is trained and the documentation is handed over to the customer. We also discuss the pre-project and post-project areas to provide a complete picture of managing a project.

The Project Life Cycle

In general, residential integration projects are divided into phases to provide better management control. This is typical of most well-managed projects in many industries. When these project phases are taken collectively, they are known as the **project life cycle.** This life cycle connects the beginning of the project to the end, using sequential phases and subphases that are small subsets of each phase. Although there is no single way to define or schedule the phases of a project life cycle, it is important that each organization standardize how it approaches the life cycle process. Each residential integration business, large or small, can draw on industry standards to formulate its own set of phases and segments or subphases. For the purposes of this discussion, we have identified the following phases for a residential integration project life cycle. They are:

- Initiating the Project
- Planning the Project
- Executing the Project
- Monitoring and Controlling the Project
- Closing the Project

Figure 2–1 shows a project life cycle for our case study company, ABC Integrations. In the case study example, a customer hired ABC Integrations on May 1 to install an integrated system, and subsequently the project is progressing through many phases and subphases, until it is closed formally on September 17. As you'll note, each phase has its own life cycle with its own distinct start and end times. For example, the termination phase starts on July 2 and is scheduled to complete by July 5.

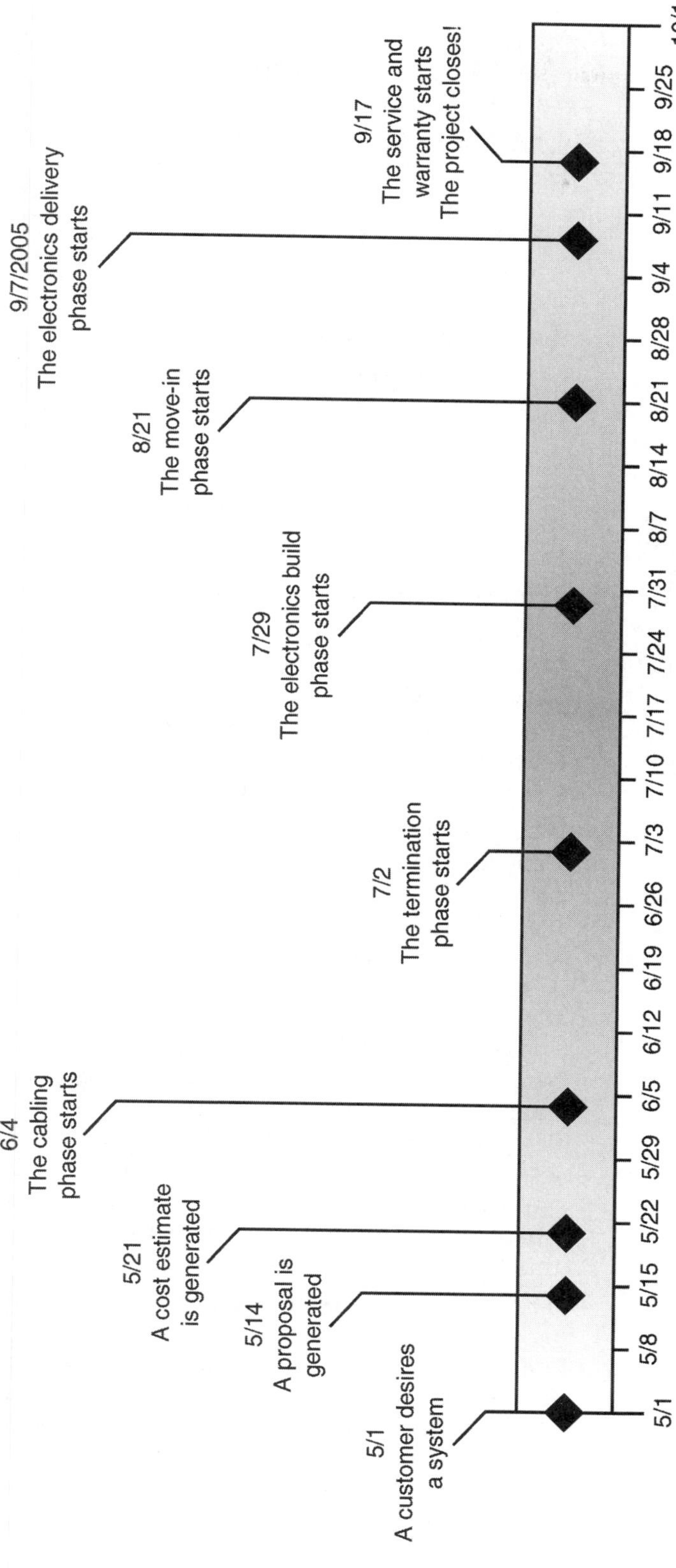

FIGURE 2–1 Residential Integration Life Cycle for ABC Integrations—Timeline View

Study/Career Tip

Conduct your own research on the project management life cycle for industry in general. Visit the Project Management Institute at **www.pmi.org** and type in "project life cycle" in the Search box, or visit **www.google.com** or a similar search engine and do the same.

Critical Thinking Questions

- After conducting a search on "project life cycle" on the Internet, what commonalities do you see among the research you conducted?
- How can you apply these broad methodologies to the residential integration industry?
- How does this affect you as a project manager?

Common Characteristics of the Project Life Cycle

Every project life cycle shares a number of common characteristics. Some of those characteristics are:

- **Phases generally are sequential.** For example, in the second phase of our case study life cycle, see Figure 2–2, we terminate the cables that are installed in the first phase. It should be noted that there might be instances where the cables in one section of the home are terminated, while other sections of the home have yet to be cabled. As we explained in Chapter 1, the practice of overlapping phases is quite common, rather than the exception.
- **Cost and staffing levels are low at the start and end of the project.** To illustrate, prior to starting any work the project goes through an initiating phase, which involves the sales representative and cost engineer, who both develop the scope statement (see document PM05) and cost estimate (see document PM05). At the end of the project, the project manager is the member of the staff who is left to close the project. In the middle phases of the project, various members of the staff complete the work of the project deliverables, which includes the expensive consumer electronics being installed. For ABC Integrations, the relative work hours are shown in Figure 2–3. In this example, the cabling requires the greatest number of project hours, and each subsequent phase requires fewer and fewer hours. The last phase, the equipment delivery phase, requires little time because much of the work has been completed in previous phases.
- **The level of uncertainty is the highest at the start of the project.** As the scope statement and cost estimate are developed, many assumptions about site conditions, technology, and other factors exist. As the project progresses, those assumptions play out and the project becomes progressively detailed. As a result, the risk in the project is lessened through the project life cycle. For example, in a retrofit (existing home) installation, the sales team makes assumptions about the number of cabling hours that will be needed to complete the work. However, when the cabling has been

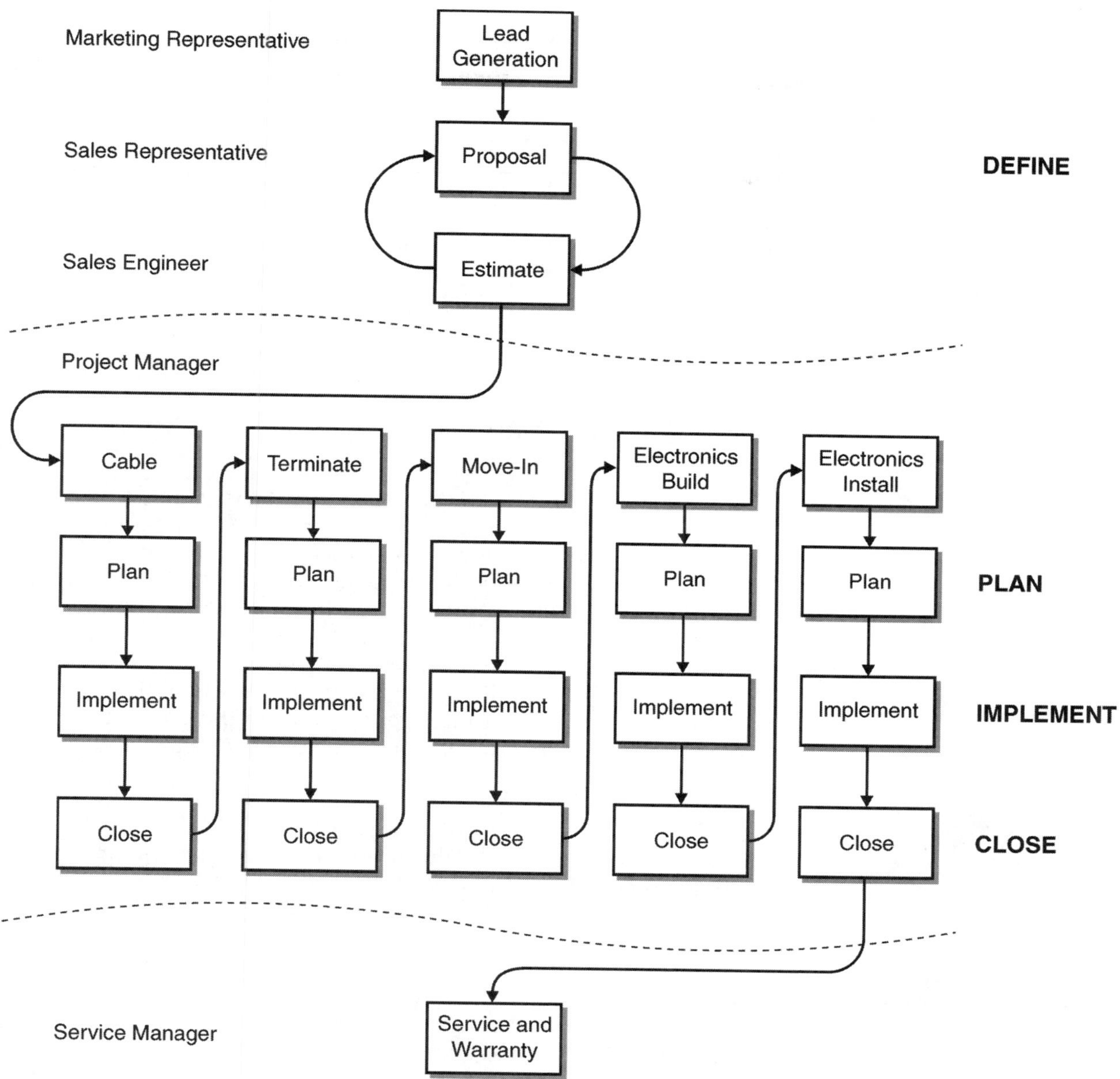

FIGURE 2–2 Residential Integration Project Life Cycle

completed, the number of hours is no longer assumed, it is known, thus minimizing the risk associated with wages for the retrofit installation. The rest of the project can be treated as a new construction process once the cables are installed in the home.

- **The ability of the project stakeholders to influence the outcome of the project is highest at the start of the life cycle.** As you might imagine, not all project stakeholders are well versed in the area of residential integration. They may know what

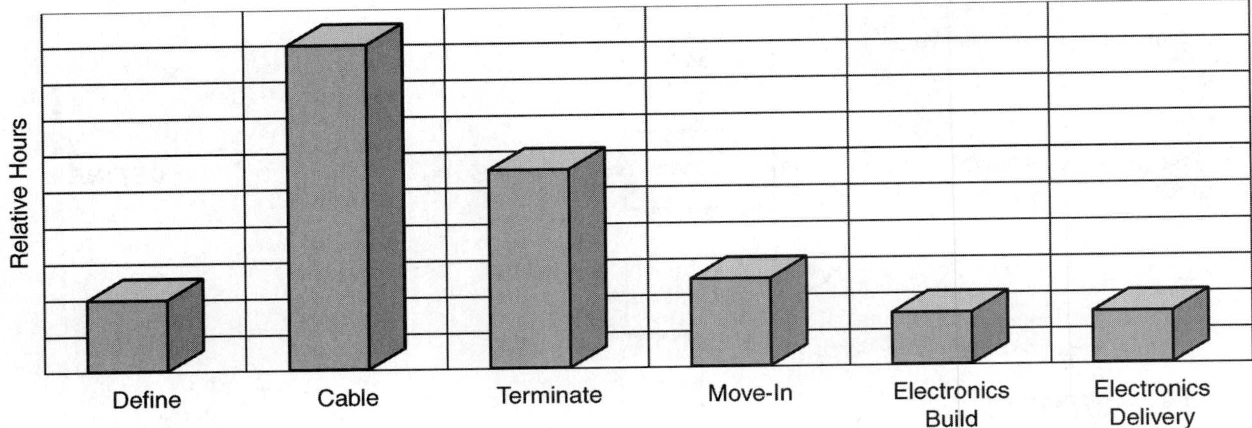

FIGURE 2–3 Relative Project Hours for Each Phase for ABC Integrations

they want, they may think they know what they want, or they may be quite uninformed about the technology. So decisions that are made during the initiating phase of the project may impact the project during the implementation phase. As a case in point, speaker locations and types of speakers are chosen during the initiating phase of the project, and may be changed slightly or drastically during the cabling phase. However, at the end of the project when the speakers are installed, the placement and type of speakers is fixed. The savvy project manager works from the onset of the project to mitigate changes that can influence the timeline and costs.

Critical Thinking Question

- As a project manager, why would it be important to be in on conversations with the sales representative from the onset of discussions with a potential customer?

Study/Career Tip

As you build your career in the residential integration industry, consider staying current on industry trends and equipment by subscribing to industry publications, such as *Electronic House, CE Pro,* and others. The cost for a yearly subscription to these types of journals is minimal, and in some cases publications are free. Timely industry articles can be helpful to you not only in your studies, but in helping to build your career as well.

The Project

The process of initiating the project encompasses several key functions. These functions occur before the physical project begins, throughout the length of the project, and after the project is completed. These include:

- Lead generation, qualification, and contract
- Project scope
- Cost estimates

To begin to define a project, the sales team generates leads, qualifies the leads, generates the scope documents, and closes the deal that results in a signed contract. All this takes place before the physical project begins. During this part of the initiating phase, the project manager assists the sales team in initiating the project scope. In addition, engineering creates the cost estimate, permits are pulled through the local building office, schedules are set, and project folders and documentation are created for each of the project teams.

Throughout the length of project, the sales representative works with the sales engineer and the project manager to ensure that company standards are being met, and at the same time, balances the needs of the customer. For example, a customer may have a particular requirement to ensure that the satellite dish is not visible from the front of the home. This information is passed through project management and on to the installer actually performing the work. Once installed, the sales representative might review the work to ensure that indeed the dish is not visible from the front of the home. If he is dissatisfied with the work completed, a request is made to the project manager to rectify the situation.

The sales representative also gathers the customer's specific preferences during this phase, such as the color of wallplates, temperature setpoints, and keypad locations. The project team uses the final scope document, which defines the project requirements, and the customer information sheets, which list the specific number and location of keypads, touchscreens, thermostats, and other system details, to further define the project and develop the company's pricing strategy for the project. In Chapter 6, The Project Scope Statement, we develop the scope statement for our case study, ABC Integrations.

When the project is closed, the project manager passes responsibility for the project back to the sales person, who continues to maintain a relationship with the customer for possible future sales and service.

Planning the Project

A residential integration project planning phase starts after the customer has signed the contract and has submitted the initial payment. The project manager must make sure that he allows adequate time to direct all of the functions performed in this phase. In other words, the residential integration company cannot take a check on Tuesday and promise the customer that installers will be there on Wednesday to start the cabling.

Why can't the company commit to starting the project the following day? The reasons are simple. To illustrate, in order to complete the cabling phase, resources must be aligned prior to the start of the work. If these resources are not in place then the cabling phase will

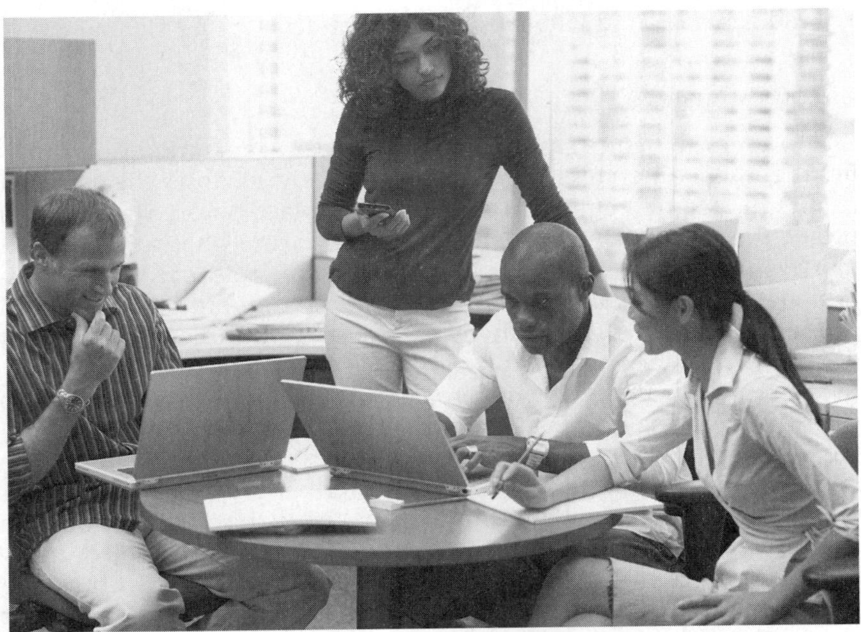

FIGURE 2–4 The Project Team Is an Integral Part of the Planning Phase

not finish on time, the project will suffer from budget overruns, and ultimately the project will lack quality. These resources include cables, mounting brackets, permits, installers, and the installation truck. In addition, planning issues such as field locating the wallplates and the speakers is required to avoid costly mistakes.

Planning the project includes ordering the integrated system, consumer electronics, system components, and a variety of electronic parts; notifying related trades; placing stickers in the home that designate the locations for wallplates, keypads, touchscreens, and thermostats; and creating extensive project folders for each team, i. e., sales, project management, installation, and engineering teams (see Figure 2–4). The planning phase may also include collecting programming preference information from the customer, as well as other preferences such as wallplate colors.

Executing the Project

Clearly, the biggest and most complex phase of a residential technology integration project is the executing phase. Here is where the majority of the work is completed and where the project can and will most likely get off schedule, and over budget, if not closely monitored.

Because of the complexity of many residential integration projects, it is common for an executing phase to begin before the completion of a prior phase. The practice of overlapping phases is known as **fast tracking,** a process used to compress the project schedule to meet project and customer demands. In order to meet the life cycle timeline our customer expects, shown in Figure 2–1 (page 27), the project must be compressed. The result is that some of the phases will overlap. To illustrate, ABC Integrations Project Manager Owen Ginizer compresses the overall schedule by sending field technicians to terminate the cables at the wallplates for the project at the same time the in-house technicians build the equipment at the office site, and other technicians prepare the house for a Certificate of Occupancy for the move-in phase.

Project Executing Phases

For the purposes of our discussion, let's look at how ABC Integrations identifies the implementation phases:

- **Cable** This phase also known as the rough-in or prewire phase in the residential integration industry. In this phase, the cables are installed in the home.
- **Terminate** This phase combined with the move-in phase is sometimes called the trim-out phase. During this phase each of the cables are terminated and the distribution panels are installed.
- **Move-In** In this phase the customer is moving into the home, and critical items such as television and telephone are made live. In addition, the home is made ready to receive a Certificate of Occupancy.
- **Programming and Electronics Build** This phase is completed in the shop and includes building the electronic systems, programming, and demonstrating the completed systems to the primary stakeholder.
- **Electronics Installation** In this phase, the electronic systems are delivered and installed in the home.

Figure 2–2 (page 29) illustrates the life cycle of a residential integration project contracted by ABC Integrations. In this figure, the phases are shown in a project flowchart rather than a timeline chart. The chart indicates how the project information flows from one phase to the next one. In addition, each of the installation phases has five subphases. The subphases are *initiate, plan, execute, monitor and control,* and *close.* All of the subphases in each phase are completed prior to progressing to the next installation phase. This means the phase marked "Cable" has instructions detailing the cabling plan, its execution, how it is monitored and controlled, and finally closing the cable phase. Once all three subphases are completed, the project manager can begin the next phase, marked "Terminate." Initiate, plan, execute, monitor and control, and close instructions and procedures are performed in that order in each subphase before moving on to the next phase. The need to complete the installation subphases before moving to the next phase does not apply to full phases, which can be fast-tracked, or compressed, and worked on at the same time to meet customer or project deadlines. A full phase, such as "programming and electronics build," can be accomplished at the same time as the cable phase. However, in our case study, each phase is considered a single unit with three subphases or, in other words, several sequential steps. And each step within the phase must be completed before moving to the next step.

Critical Thinking Question

The advantage of fast-tracking or compressing phases in a project is obvious.

- Why, then, wouldn't a project manager look to fast-track each subphase for the same reason?

Initiating and Scheduling Executing Phases

There are a number of other ways to define and schedule the various implementation phases of a project. For example, some companies group the "terminate" and "move-in" phases into a single phase, which is commonly called the "trim" phase. This works well for smaller undertakings that do not require a great deal of termination time. Another company might decide to have a single planning phase prior to the installation phase, which become subphases of the executing phase. The project would then close with a single close phase. What is important is that an organization determine its own consistent way of initiating the project life cycle. It is important because the project life cycle answers the following questions:

- **What is the technical work to be accomplished in each implementation phase?** To illustrate, in our case study the cables are installed in the home during the cabling phase. The cables are terminated and tested to make sure they are working properly during the termination phase. It's the project manager's responsibility to prepare a list of the specific types of technical work to be completed for each phase.

Critical Thinking Question

- What might the project consequences be if the validation process is not run during each implementation phase?

Study/Career Tip

Do you live in a part of the country that is considered a hot bed for residential integration projects? If so, consider visiting one or more companies in your area, especially if they happen to have a showroom that highlights the equipment they install and the types of jobs they typically complete. Become the customer. Ask how jobs are completed and what is involved.

- **When are deliverables to be generated in each implementation phase and how is each deliverable validated?** In our case study, the programming deliverable is to be generated in the Programming and Electronics Build phase. During that same phase, the programming is put through an internal **validation** process to ensure its accuracy. This involves testing each keypad and touchscreen command to make sure it produces the correct response. Here's a case in point.

- **Who is involved in each executing phase?** In our case study project life cycle, Joe Trade, the lead installer, and his team cable the home during the cabling phase, while Tom Tek, the lead technician, and his team make the final system connections in the electronics installation phase. These job descriptions and others are spelled out in Chapter 3, The Project Team.

A CASE IN POINT
Preferred Customers

Sales Representative Sal Moore gathers programming preferences from the customers, Fred and Wilma Smith, prior to the start of the Programming and Electronics Build phase. That means Sal Moore asks Fred and Wilma how they want the integrated system to respond. Sal Moore records all the preferences, such as specific lighting scenes for the day or the night, presetting AM and FM radio stations, and presetting temperatures for integrated HVAC systems. In the case of the installation of a home theater system, customer preferences may include a list of favorite stations for cable television or satellite. This information is given to Project Manager, Owen Ginizer, who then provides it to Tom Tek and his technical team.

Depending on the organizational structure of the company, the project manager or procurement manager orders the equipment required for the project. At ABC Integrations, Pam Prod, procurement manager, places the orders. When it arrives at the shop, Tom Tek or another in-house technician assembles the equipment and installs it into the equipment rack. Engineer and programmer, Maggie Pi, programs the system per the customer preferences detailed earlier.

Prior to the delivery of the system, sales representative Sal Moore and the customers, Fred and Wilma, closely review and validate the equipment assembly and programming of the system to ensure that it meets the requirements of the scope statement. For example, lead technician Tom Tek presses each of the favorite station buttons on the touchscreen to confirm that the satellite receiver actually changes to the correct station. He does this by connecting every piece of gear in the system and setting it up on a test bench in ABC Integrations's shop.

Following the creation of the scope document, the project execution or implementation ensues. This is when all the fun starts. The project manager uses the scope statement as a map to assign each task to each team member. ABC Integrations has designed its scope statement with an executing subphase in each of the five installation phases:

- Cabling Phase
- Termination Phase
- Move-In Phase
- Electronics Building Phase
- Electronics Installation Phase

As mentioned earlier, each one of these implementation phases has a subphase that includes the plan, implement, and close steps. Let's look at each one of these in ABC Integrations's implementation phases.

FIGURE 2–5 Cable Installation. (Photo courtesy of constructionphotographs.com)

Cabling Phase

The bulk of the cabling phase involves installing the cables throughout the home (see Figure 2–5). This makes sense since it is called the cabling phase! In addition, brackets for the wallplates and speakers are installed. The head-end panel foundations are built, making them ready for the panels to be installed during the termination phase.

Critical Thinking Questions

- Why might it be important for the sales representative to double check the work being completed during the implementation stages at the job site?
- Doesn't that undermine the work of the installers, and more importantly, the project manager?

Try to think of three advantages of doing this step.

Study/Career Tip

Research the concepts related to these topics further. Use the Web, the library, or other resources, including your classmates. This will not only help you move forward in your career in the residential integration field, but will help you learn and retain the material better.

FIGURE 2–6 Terminate Phase. (Photo courtesy of constructionphotographs.com)

Terminate Phase

The terminate phase occurs when the technicians terminate, test, and label all of the cables throughout the home (see Figure 2–6). In addition, the head-end panels are installed and made ready for the connection of incoming services such as telephone and television.

Move-In Phase

During the last few days of construction, the move-in crew attaches wallplates, installs speakers, makes active connections for telephone and television, and prepares the home for a Certificate of Occupancy (CO).

In order to get a CO, the low-voltage cabling must pass an inspection, generally conducted by the electrical inspector. Other work on the home, such as HVAC, plumbing, and electrical wiring also must pass inspection. When all of the categories of work pass the building inspection, a CO is issued and the owner may move in.

To simplify the inspection process and to cut down on the possibility of failing inspection, it is advisable to delay installing the consumer electronics until after the inspection process is completed. The project manager's goal during this phase is to reach the point in the project that allows the customer to move into their new home. To do that it is unnecessary to connect the HVAC system, the lighting system, and/or the music system to the house-wide integration system. Instead, each system is made operational by its respective subcontractors; it is fully functional, yet not tied into the automation system. Those connections are not required to receive a CO. Once the CO is issued, the touchscreens can be installed, the music system can be brought in and installed, and the security and HVAC systems can be connected to the main integration system. Some customers prefer waiting to move into their new home until all these systems are in place, even though they could move in earlier. As a project manager, it's wise to discuss this aspect of the move-in phase with the customer at the outset of the project. In theory, this helps prevent headaches, customer dissatisfaction, and cost overruns.

The final connections to HVAC and security systems are made after each of the respective subcontractors is through the tinkering process and the customer is happy with each system. This may be up to a month after the customer has moved in. Do not integrate any subsystem until that system has been thoroughly tested and used by the customer as a stand-alone system. Each system should be designed to allow a CO without the need for installed electronics. This greatly alleviates the pressure to get the systems online in time for the CO deadline.

Electronics Building Phase

The electronics building phase, for the most part, is conducted off site at the company's office site or facility. In this phase, the engineering group takes all of the customer preferences from the "Customer Information" sheets, which are included as files PM02, PM03, PM04, and uses the information to program the standard modules to these preferences. We create these templates as part of Chapter 5, Information Gathering. The equipment is purchased and installed into equipment racks, and tested with all of the keypads and touchscreens. ***Very Important:*** Do not deliver any system that has not been through this process or has not been thoroughly tested. It is better to deliver a working system late than an incomplete, untested system on time. See the case in point on page 39.

Electronics Installation Phase

In the electronics installation phase, the electronics team delivers the equipment rack, connects cables to the head-end or distribution panels, and installs keypads and touchscreens and connects them to the appropriate cables. If the other phases are completed correctly, this phase then represents only a small fraction of the overall installation time. Table 2–1 shows the relative amount of time typically spent on each implementation phase of a residential integration project.

TABLE 2–1
Estimated Labor per Phase

Cabling Phase	*Termination Phase*	*Move-In Phase*	*Electronics Build & Programming Phase*	*Electronics Installation Phase*
50%	25%	13%	6%	6%

A CASE IN POINT
Field Work

It was going to be the largest project ever for ABC Integrations. The project included 16 zones of music, over 200 zones of lighting, touchscreens throughout the home, integration with the HVAC system, and more than 40 motorized shades. The home itself was 20,000 square feet and was owned by a VP of Engineering for a Fortune 500 company. The customer request was to integrate everything in the home and make it easy to operate.

All of the lighting was run to dedicated panels, and thermostats were replaced with temperature sensors. This design decision made it impossible to use the home without the electronic systems. The difficult part of the project was that the owners would be taking possession of the home one suite at a time. Each suite needed to be completed one at a time, individually programmed, and installed.

Contractors would finish their work on a suite and then call ABC Integrations to come the following day to "get things online." This resulted in ABC Integrations always needing to be at the ready. Even if other projects were in progress, they had to drop everything and rush to the site to get the next suite up and running. Because parts were arriving daily to the suite and because of the ever-changing requirements of the project, there was never enough time allotted for building and testing the system at the office facility. All of the work needed to be completed at the project site.

The time required to get the system operational in the field was greatly increased compared to working in the shop. Having contractors, and even the customers, constantly interrupting to make changes in other parts of the home and to check out the "cool new stuff" adversely affected the project completion schedule for this phase. In addition, part failures caused an immediate time crunch and caused the homeowners to question the capabilities of ABC Integrations. The result was a highly disorganized project that was less profitable than projects that were much smaller in scope.

What could ABC Integrations executives have done to avoid these pitfalls? Let's look at three possible solutions. ABC Integrations could have designed the systems to be operational without the need for programming during the construction process. How could they do this? Rather than remote dimmers and thermostats, inline dimmers could have been used initially and the integrated system could have been added at a later time. Inline dimmers work like regular dimmers and allow the customer to use the lighting without the need of keypads and touchscreens, which could have been activated later. ABC Integrations could have been more direct and determined in its position with both the customer and contractors in terms of the company's response and the length of time required at each step. Unfortunately, the company had made the customer's problems its problems. Invariably this results in increases in project cost. The company's decision makers could have even considered not taking on such a project. It may be better to do 10 jobs at 10K each than one job at 100K. If the 10 jobs at 10K can be accomplished more quickly than the 100K job, then the profitability would be higher in the total of the smaller jobs. Plus, there's the reputation of the company on the line. As mentioned in Chapter 1, it doesn't take much for dissatisfied customers to send out smoke signals to their friends that the work was unsatisfactory.

Monitoring and Controlling the Project

During the monitoring and controlling phase, a unique set of controls is employed to ensure the quality of the project deliverables. To illustrate, at the end of the cabling phase the project manager walks the site to count the number of wallplates and compares it to the scope statement document. In addition, the job site is checked to ensure it is clean and the work is professional. Finally, each phase contains a close segment, which applies the appropriate approval for each phase, thus allowing the life cycle to continue to the subsequent phase. For example, at the close of the cabling phase, the project manager performs an audit of the system in the home, ensuring that it is cabled to meet the specifications of the scope statement. In order to perform the audit, the project manager visits the home and checks that the appropriate type of cable is used, that each cable is installed properly in the appropriate location, and that the cables are in good condition without nicks or frays. He also checks, or validates, that the terminations are secure. This process continues through each product deliverable in the scope document.

Closing the Project

If all has gone well and the homeowner is happy with the installation job and the project is nearing completion, then it's time for the project manager to finish the project. As you might imagine, several concrete subphases must occur in order for the job to be considered "complete."

The closing of the project phase includes:

- System testing
- Customer training
- Final documentation
- Project audit
- Turnover from project management to the service department.

System Testing

The system testing process evaluates the entire system and tests every feature. This step is always performed in the shop, away from customers and other stakeholders. We are all human, and the workers at the facilities that build the equipment are too. We all make mistakes, and the system testing phase is designed to catch all of those mistakes. By testing every electronic device and all of the programmed features, ABC Integrations ensures there will be very few problems for the customer on delivery.

Customer Training

When the system is designed and installed properly, very little customer training is needed. Do not allow the customer to use the system until the system installation is completed. It is best to unplug it at the end of every day during the programming and build equipment phase. Set up a mutual time for training and demonstrating the system to the customer and notify them that the final payment is due at that time. It is important to gather the family together for training, because most service calls are based upon a lack of understanding of the system. With the entire family present, each member has an opportunity to critique the system and voice concerns before those concerns become a service call. Make the event special and bring a gift that is suitable for the level of system purchased. For a system that costs $10,000, a gift costing about $25 would be appropriate. Explain the **service policy** at that time.

Even though training may seem simple to the project manager, it may take the customer some time to adjust to a new way of doing things. Be prepared to be patient and answer customer questions. The new integrated system provides convenience, comfort, and an enhanced lifestyle to the customer, but it probably will mean a brief period of adjustment for the family.

Study/Career Tip

When reading this textbook, try paraphrasing each chunk of information in your own words. Then, go back and review each section. Make a note of where you are having difficulties with a concept and try paraphrasing again.

Final Documentation

The final documentation is determined by the size of the project as well as the degree to which other contractors are documenting their systems. In smaller projects, such a home theater installed in a retrofit situation, the scope statement can serve as the final documentation. If there have been changes during the project, the scope statement is updated to reflect these changes, resulting in an accurate picture of the final project deliverables.

In large homes most of the systems are documented on CAD on floor plans. The HVAC, electrical, and plumbing systems may all be initially designed using this technique. In addition, changes are updated on the plans. In this case, the low-voltage systems should also be updated in a similar fashion. These homes, which are typically 10,000 square feet or larger, need the documentation since the service of the home can be complex, if not impossible, without such documentation.

Project Audit

Now that the project has been documented with the actual deliverables, the process of auditing the costs of the project ensures good financial practice. Just because the project is finished and documented well does not indicate that it was profitable. During this phase, each product and service is accounted for and compared with what was charged to the customer. This ensures that the customer is charged accurately for the goods and services provided. If the contract is fixed and items are provided at zero cost, due to poor cost estimating, this process will highlight those parts and services. This information is then used to feed back into the cost-estimating process to ensure future projects are accurately cost estimated.

Turnover from Project Management to the Service Department

The final step in the project is to hand over all final documentation to the service department. Intermediate documents may be archived, however, only the final documentation is handed to the service department. This is done in order to provide an accurate picture of the systems installed and not to confuse the service department with intermediate steps.

The service department now is fully in charge of the project, dealing with service calls and customer issues during the service period. The service department may also be in charge of selling system upgrades and extending the service contract.

Closing the Project before the Home Is Completed

Even if the customer's home is not finished, the project manager can and should finish (close) the project. Here's an example:

As part of the house-wide systems, ABC Integrations is supposed to deliver a functional system in the finished basement. However, it is learned that the work on the basement will not be completed for three more months. At this point, the company already has completed most of the preliminary work and is ready to deliver and install the system. It has purchased the consumer electronics equipment, has tested it thoroughly, and has had the customer come to its place of business to sign off on the items. The company is ready to direct the installers to install the items in order to be able to close the project. Let's look at three possible solutions that ABC Integrations Project Manager Owen Ginizer should consider. Here are three cases in point.

A CASE IN POINT
Leave It

Project Manager Owen Ginizer can deliver the materials and equipment to the homeowner's site and leave the items that are ready to be installed in a location that is mutually agreeable to the company and the customer. Then he can write a change order removing the labor required to install the equipment and close the project. Project Manager Owen also provides the customer with an estimate of the cost to install the equipment later, when the customer's home and basement are completely finished.

Critical Thinking Questions

- What are the advantages of using this solution?
- What are the disadvantages of this solution?

A CASE IN POINT
Store It

ABC Integrations Project Manager Owen Ginizer can close the project and place the equipment in storage, either at ABC Integrations' shop or an off-site storage facility. A monthly storage fee will be charged to the customer. A change order to the contract should be written to reflect this agreement as well. Remember, always get it in writing!

A CASE IN POINT
Install It

ABC Integrations Project Manager Owen Ginizer can notify the customer that the warranty and service periods will not apply to equipment installed in an inappropriate location. This can be done first verbally and then followed up with a written change order, removing any charges for the service period.

Customers can always be led down a path of success for both the customer and for ABC Integrations. Here ABC Integrations Project Manager Owen is working with the customer and never saying no, but he is setting limits on what can realistically be provided. If the equipment were to be installed in a location that would suffer from construction dirt, then the company would no longer be able to stand by the long-term effectiveness of the equipment. A written statement to that effect should be put into the change order so the customer is aware that the equipment could be compromised by being installed during a period of construction.

Critical Thinking Questions

- What are the advantages of using this solution?
- Disadvantages?

Critical Thinking Questions

- What are the advantages of using this solution?
- Disadvantages?
- Of the three possible solutions, which one would you recommend to the homeowner?

Summary

In this chapter, we discussed a typical residential integration project life cycle, and defined the project and the roles of the project manager and the project team. We also discussed the important issues of each phase in a residential integration project, and common characteristics of a home integration life cycle. We looked at how certain key functions define and plan a project. We discussed typical implementation phases of a home integration project. We also looked at how to close a project.

Important points in this chapter are:

- In general, home integration projects are divided into sequential phases, collectively known as the project life cycle. This life cycle connects the beginning of the project to the end.
- It is important that organizations standardize how they approach the project life cycle. Industry standards can be used to formulate each company's own set of phases and the segments of each phase. Standardizing the project life cycle is important because it leads to the operation of an efficient and cost-effective company.
- The project life cycle defines important issues of each phase. It defines the technical work to be performed, when deliverables are to be generated, how each deliverable is validated,

the people involved, and how each phase is controlled and approved.

- Every project life cycle shares common characteristics, including sequential phases, low cost and staffing levels at the start and end of the project, the highest level of uncertainty at the start of the project, and the stakeholders' strongest influence at the start of the life cycle.
- Initiating the project encompasses key functions that occur before the physical project begins, throughout the length of the project, and after the project is completed.
- The sales team generates and qualifies leads, generates the scope documents, and closes the deal that results in a signed contract. The project manager assists the sales team in initiating the project scope. The project team uses the final scope document, which defines the project requirements.
- Planning the project begins once the contract is signed and the initial payment is made.
- The executing phases of a home integration project are the cabling phase, the termination phase, the move-in phase, the programming and building equipment phase, and the electronics installation phase.
- The close phase includes system testing, customer training, final documentation, project audit, and turnover from project management to the service department.

Key Terms

Fast tracking A specific project schedule compression technique that changes network logic to overlap phases that would normally be done in sequence, such as the design phase and construction phase, or to perform schedule activities in parallel.

Project life cycle A collection of generally sequential project phases whose name and number are determined by the control needs of the organization or organizations involved in the project. A life cycle can be documented with a methodology.

Service policy An organization policy or procedure for servicing projects after completion. It includes what services and products are included in the warrantee period as well as charges outside the limits of the product warranty.

Validation The technique of evaluating a component or product during or at the end of a phase or project to ensure it complies with the specified requirements.

Review Questions

1. Why is it important to develop a standardized project life cycle for a residential integration company?
2. What are the advantages of fast tracking or compressing a project?
3. What are the five phases of a project life cycle?
4. Why is it important to complete each subphase before moving on to the next phase?
5. In general terms, what are the five executing phases of any residential integration project?
6. Why is it important for the sales representative to capture the customer's preferences at the beginning of the project?
7. Which executing phase represents roughly 50 percent of all work that needs to be completed?

The Project Team

OBJECTIVES

After studying this chapter, you should be able to:

- Describe and discuss the role and responsibilities of the project manager.
- Explain how the project manager's authority varies in companies with different organizational structures.
- Define functional, projectized, and matrix-based organizations.
- Describe the roles and responsibilities of the project team and project team members.
- Define the role of the customer in the residential integration industry.
- Explain the importance of customer service and customer satisfaction.
- Define the customer.
- List the six principles of managing customer service.

OUTLINE

Introduction

If you were stopped on the street by a reporter and asked to describe a successful residential integration project manager, what would your portrayal include? Someone who knows the industry and who can wear many hats? Certainly. A person who has years of experience and the ability to supervise? Most assuredly. Would your description end there or is there more to it than these few characteristics? Let's take a look at how you might see yourself in the role of a residential integration project manager.

You as the Project Manager

The residential integration **project manager** is a strong leader who plans and directs the project's flow from start to finish; provides guidance to the project team, outsourced installers, and subcontractors; initiates customer contact at the start of project; conducts site reviews; consults with the sales representative; and prepares documentation.

The person in this role has a solid understanding of residential integration, design and installation of home theaters, music, lighting, telephone, HVAC, security systems, and software required for user interfaces. He also has a thorough knowledge of millwork design, construction standards, low-voltage cabling, safety codes, construction documentation, and the communication skills required for stakeholder interaction. He may also carry one or more certifications from the various organizations associated with the residential integration industry. (See Appendix C for a complete list of these organizations.)

Critical Thinking Questions

- What skills, knowledge, and personal attributes do you bring to the role of a residential integration project manager?
- What do you still need to learn and how will you learn it?

In addition to maintaining a recognized level of expertise in the custom installation and residential integration industry, the project manager should have total familiarity with many of the project management areas of knowledge. This is because the concept of project management is comprehensive and complex. The successful project manager, then, should have an appreciation for these content areas. See Appendix A for the nine areas of project management knowledge as derived from *A Guide to the Project Management Body of Knowledge*.

Study/Career Tip

Visit the Project Management Institute (PMI) at **www.pmi.org** for more in-depth information on the areas of project management knowledge.

Study/Career Tip

Take the time to write down all the knowledge, skills, and personal attributes you feel you bring to the role of a residential integration project manager. Revisit this list and add to it as you acquire new competencies. Also, keep an ongoing list of the things you feel you must still master and come up with a plan for acquiring those abilities.

A strong project manager has also garnered an established set of soft skills, education, and abilities that are vital to good project management. Some of these skills and abilities are:

- **Comfortable with change.** The very nature of a project lies in its unique characteristics. Every project will present an ever-changing set of challenges. For example, imagine that upon completing a home theater installation project, the team is proud of the completed work, but the customer is unhappy with the placement of the equipment within the cabinetry. Not only must the project manager negotiate with the customer and possibly agree to the requested change, she must then motivate the entire team to move the equipment to the customer's satisfaction. In another instance the customer may change her mind on electronic equipment, placement of wallplates, whether to include security integration or other segments of the system, and many other minor details.

  Each project has many variables that could generate numerous changes throughout the project, so a project manager must be capable of handling each and every one of them. At the same time, it is paramount that the project manager maintain a positive relationship with the customer. Through experience, trial and error, and good humor, the successful project manager will attain this vital workplace skill.
- **A solid understanding of organizational structures.** An **organization** can be function-based, project-based, matrix-based, or a composite of two or more styles. A key to good project management is having a solid working knowledge of each organizational type. A project manager needs to know the organizational framework of his own company to understand his role within that organization. He also needs to know the other types of organizational structures to better understand his competitor's organization, if it has a different structure. Through that understanding the project manager positions himself to help the senior management team develop strategies to surpass their competition in the marketplace. We will cover organizational structures in detail later in this chapter.
- **Understanding the company's unique product strategy.** Project managers need to be well versed in their company's products, their product pricing, and how and to whom the products will be marketed and sold. Questions they need to consider in developing a pricing strategy are varied. What products will the organization provide for its customers? How will the delivery and price affect these products? How does the company vision and mission support the **product strategy?** Project managers should know the answers to these questions, and they should be involved in formulating and championing company policies that provide the answers to these and other questions central to an organization's success.

Critical Thinking Questions

- What do you think would be a good and lasting way to understand a company's unique product strategy?
- How do you think a company's mission statement can help support their product strategy?

- **Develop soft skills.** Strong management, communications, leadership, and corporate political skills are instrumental in achieving customer satisfaction, the satisfaction of every project stakeholder, and employee satisfaction. Project managers can develop and improve their **soft skills** by attending appropriate seminars, by reading a range of management and self-improvement books, by attending trade shows and conferences, and by simply learning from the daily challenges of being a project manager. Soft skills are not innate; they can be learned and improved upon, but they also take time and effort to develop.
- **Education and training.** The residential integration industry generally draws its project managers from two sources—career managers with four-year, management, or technical degrees, and upwardly mobile installers and technicians with years of experience and/or associate degrees in management or other fields. Those who have developed the skills to be a leader and who can smoothly manage teams of people and multiple projects are prime candidates for a career in project management. A project manager can find that the challenges of her career can be as great as the rewards.

Understanding the Company Mission Statement

What Is a Company Mission Statement?

The company mission statement is generally created by the company directors. It is used to measure day-to-day business decisions, and it provides a clear view of what the company does for its customers. It also serves as an overarching mandate to its employees of how they should conduct themselves in the workplace as representatives of the company.

Without a mission statement, the company's success is based on haphazard luck. Although many theories exist regarding the purpose of mission statements, it simply comes down to the concept of being a unified declaration of what you do for whom, and why. *Everyone* in the company should be able to quote it verbatim and embrace its message.

Some companies go beyond the mission statement and add additional concepts that are known as "core beliefs" that further spell out the company's operating philosophy. Typically, these are included in bullet form beneath the mission statement. Some organizations go so far as including a vision statement as well, which is considered even more global in scope.

Let's take a look at the ABC Integrations Mission Statement.

THE ABC INTEGRATIONS MISSION STATEMENT

ABC Integrations provides high quality turnkey integration systems only for entertainment, comfort, and convenience in the modern digital home.

Core Beliefs

- ABC Integrations provides standardized, cost-effective, and high-performance systems that exceed our customers' expectations.
- We agree to requirements and deliverables with the customer before work has begun, establish milestones, and monitor progress throughout the project.
- We strive to minimize on-site installation time.

Critical Thinking Questions

- What does the ABC Integrations mission statement say to you?
- What pops out at you that explains what the company does and its operating philosophy?

What does the ABC Integrations mission statement tell us? Based upon the wording, ABC Integrations is dedicated to working the middle market of the residential integrations industry, meaning that it will install and support systems that require a level of expertise beyond the capabilities of most homeowners.

The company will provide repeatable systems, which are systems that can be installed again and again for many customers with slight modifications from customer to customer. ABC Integrations has systems that can be serviced by anyone in the company with minimal documentation of the specific system. Customers will have a reliable and satisfying experience with ABC Integrations. Problems will rarely arise, and when they do, they will be resolved quickly and efficiently. Through inference, the ABC Integrations mission statement also tells us that the company will refrain from contracting large custom projects, because such projects require custom solutions that the company is not geared to handle.

Why Should the Project Manager Care about the Corporate Mission Statement?

With a corporate mission statement in hand, the project manager is able to make day-to-day decisions in accordance with established company policies and procedures and generally without fear of managerial reprisal. The mission statement also provides guidance as to what business choices he can safely make that help the company's bottom line. Without a mission statement, the project manager could unwittingly undermine the financial health of the organization by accepting projects that the company is not prepared to handle. He is also hampered by the inability to make sound business decisions because he doesn't know what those business decisions should be. Here's a case in point.

A CASE IN POINT
The Problem

ABC Integrations Sales Representative Sal Moore is on a commission-based salary, so the more he sells, the more he earns. Sal approaches ABC Integrations Project Manager Owen Ginizer with a great new project. The project involves all of the standard systems provided by the company, except for a telephone system that has never been installed by ABC Integrations to date. This poses a dilemma for Owen because, while the overall project looks attractive and within the company's current business structure, the telephone system is outside of that realm.

A CASE IN POINT
The Solution

How will the ABC Integrations mission statement aid Owen with his decision on whether or not to accept the project?

The company mission statement clearly states that the company will provide turnkey systems to its customers. That means a new type of system will become part of ABC's product line only if it can be provided to many customers, not just one. Owen considers the issue and brings company engineer Maggie Pi into the picture to review the proposed system. Engineer Maggie considers whether or not this system meets the needs of the entire customer demographic for ABC Integrations. She must determine if the company that provides the telephone system has a reputation for providing a solid product that is reliable and serviceable. Based upon her research, Engineer Maggie offers two possible business solutions to Owen. The new system could be standardized to ABC Integrations turnkey specifications. Or the system could be outsourced for this particular customer because the company providing the telephone system has a solid reputation. In either case, the customer would receive a telephone system that meets his needs, and ABC Integrations stays within the boundaries of its mission statement.

Critical Thinking Questions

- If you were Project Manager Owen Ginizer, which of the two possible solutions would you have chosen?
- Why?
- What factors did you consider when making your decision?

Organizational Structure

Every organization must have an identified structure built around it in order to succeed. Anything short of that is simply organized chaos. Businesses fail every day because of this shortcoming. So how does a

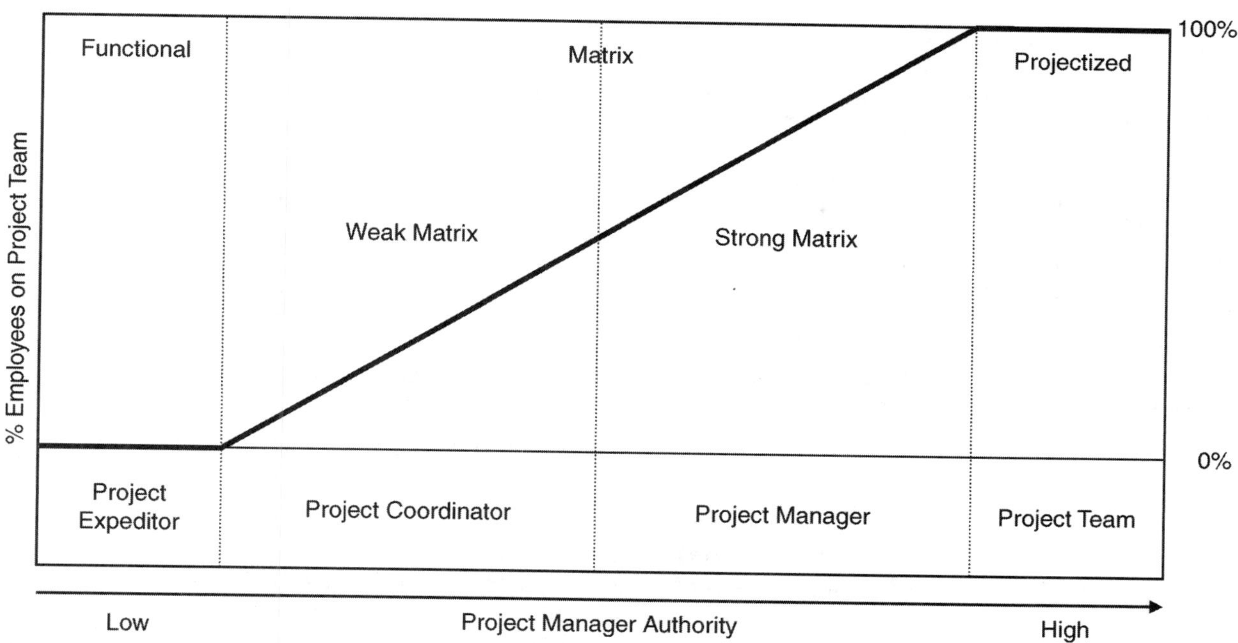

FIGURE 3–1 Organizational Authority

company choose what type of organizational structure to use as its model?

One method for determining the structure of an organization is first to establish what **authority** will be given to the project manager. Top management delegates varying levels of authority to the project manager for time, costs, project resources, scheduling, and employees, as well as decision-making duties and the ability to give approvals. Authority can range from a small amount in the functional organization to a high level of control in a project-based organization. Figure 3–1 indicates the continuum of authority that can be delegated to the project manager. As the graphic illustrates, the project expeditor and the project coordinator possess little formal authority in a functional and weak-matrix organization. A moderate level of control is exercised by the project manager in a strong matrix, while the **project team** possesses complete control. We will discuss each organizational structure in order of authority.

Study/Career Tips

- Do you work at an outside job?
- If so, what type of organizational structure is in place at your company?

If you can't determine that through observation, ask management for its organizational chart to decide your company's organizational structure.

- Where do you fit in?

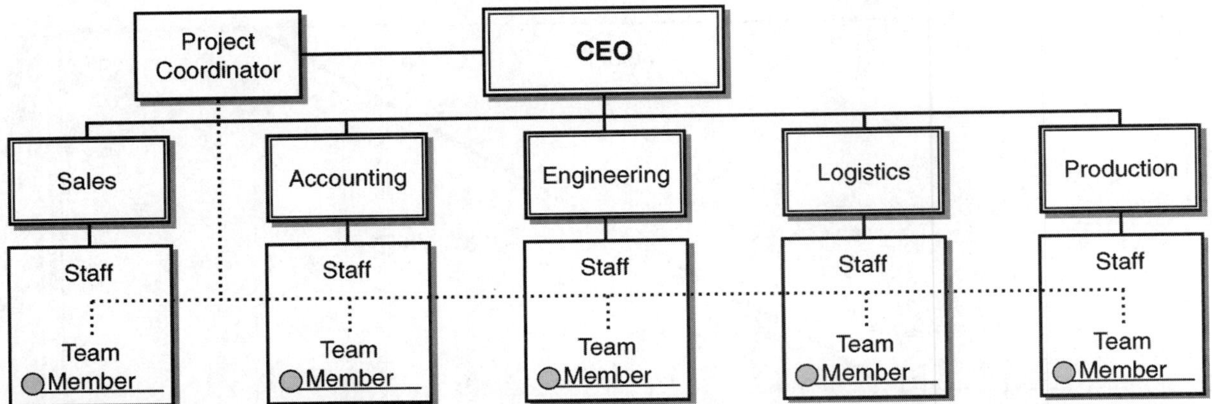

FIGURE 3–2 Functional Organization

Functional Organizations

A **functional organization** is prevalent in companies that do not take on project-based work as part of their regular business. A company with a functionally-oriented structure is based on the theory that staff divisions of responsibility have approximately equal delegations of authority. For example, this structure is typically found in a chain of retail stores. The marketing department's responsibilities are completely separated from the sales and engineering departments' tasks. As such, there is no need to create a team of specialists to work with the customer. In this organization, the marketing team creates advertising and marketing programs. The engineering team researches new products for the organization, and the sales team works the sales floor. All this can happen without interaction between the departments, and in many situations, they exist in different buildings across the globe.

Difficulties develop with this structure when multiple projects are introduced in the business. Conflicts arise over the relative priority of each project. In addition, project team members place little importance on the project, as their career path lies in their functional specialty, not in the project itself. In the residential integration industry, employees in each of these areas are required to interact with the customer; therefore, the functional organization is seldom the preferred organizational structure. Figure 3–2 depicts a standard functional organization chart.

Projectized Organizations

At the other end of the spectrum is the **projectized organization.** In this organization, the project manager has complete authority over the project team (see Figure 3–3). The team is created and assigned to the project manager on a full-time basis, while the parent organization remains functionally organized. The parent organization assigns its personnel to provide their services specifically for the project. It takes a special individual to stop working on a temporary basis for his functional supervisor, go to work for the project manager, and return to the functional supervisor at the completion of the project.

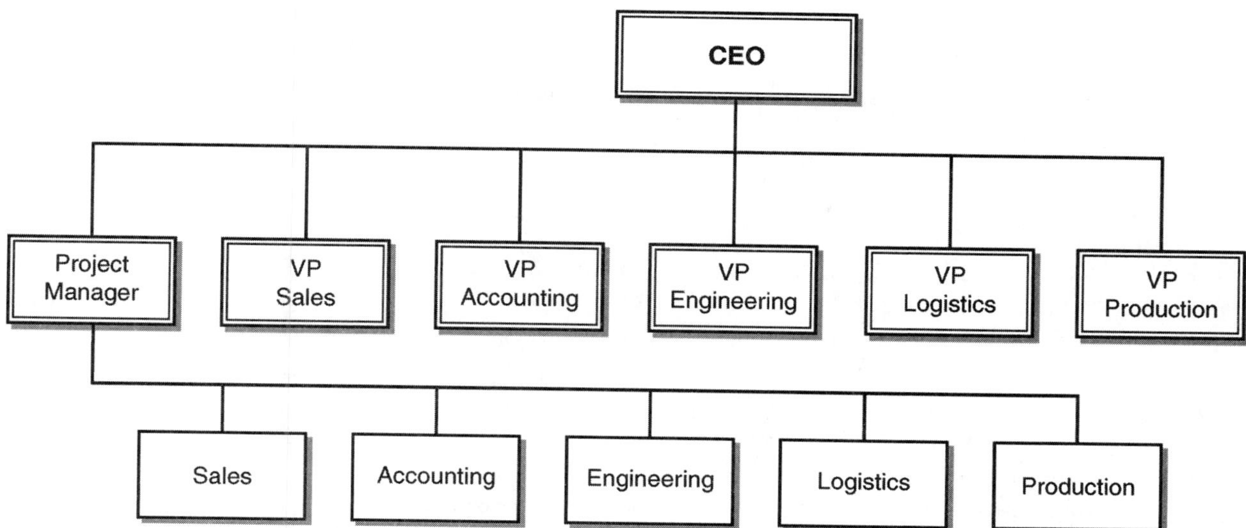

FIGURE 3–3 Projectized Organization

Creating this new "sub-organization" creates specialized problems. There may be a duplication of facilities and an inefficient use of resources. To illustrate, if a documentation engineer is assigned to the project, it may be necessary to allocate a second documentation engineer to fulfill ongoing functional requirements during the time the first engineer is away from his normal position. Another issue is that project team members work themselves out of a job by completing the project. Upon project completion, they return to the functional organization and look for another project to join.

This type of organizational structure is common in very large government contracting enterprises. In these companies, projects can last years, if not decades, and require the full-time efforts of hundreds, and perhaps even thousands of employees. In the residential integration industry, projects are much smaller so the use of full-time project team members is rarely required. As a result, this type of organization is not preferred in most situations.

Critical Thinking Question

- What do you consider the strengths and weaknesses of both a functional organization and a projectized organization to be?

Matrix-Based Organizations

We have looked at project management authority at both ends of the scale. They both have strengths, and they both have inherent weaknesses from being either too far to one end or the other. The compromised organization contains a matrix-based structure. It is in this **matrix-based organization** that we find the most appropriate structure for the residential integration industry. Where a particular

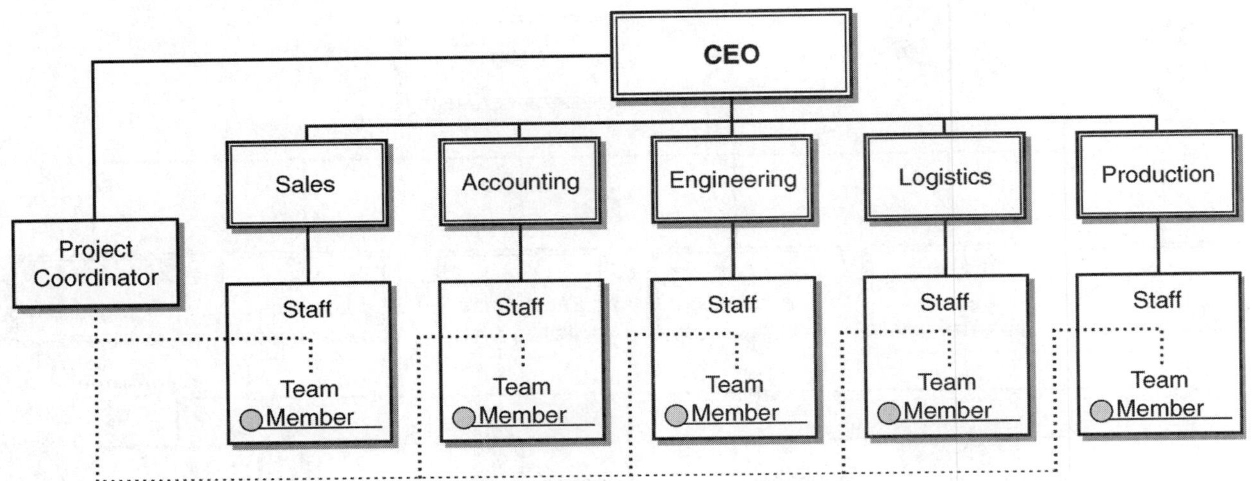

FIGURE 3–4 Weak Matrix Organization

organization lies within the matrix depends on the products and services that the company offers to its customers.

The matrix organization attempts to maximize the strengths and minimize the weaknesses of both the projectized and functional organizations. It maintains the functional (vertical) lines of authority, while establishing a relatively permanent horizontal structure. The term *matrix* refers to the intersecting horizontal (project) lines of authority, and vertical (functional) lines of authority.

The Weak Matrix Organization

In this organization, a project coordinator (rather than a full-time project manager) is assigned to accomplish project goals (see Figure 3–4). The project coordinator has the authority to assign work to individuals in the functional organization. However, the functional manager still has authority to conduct performance reviews. The functional manager shares authority and resources with the project coordinator. In addition, team members work to satisfy both the functional manager's and the project coordinator's requirements.

This type of organizational structure works well for residential integration companies that specialize in projects that are small in scope and time. For example, a retail-based chain that adds installation services to augment product sales may choose this structure. In this case, most sales are still made in the same functional manner; however, a limited number of sales require creating an installation team. A customer visits the showroom, selects her home theater system, and decides she would prefer that the store deliver and install the system. In this case, the project teams need only be assembled for a day or a few days. There is no need for long-term controls, reporting, or financial analysis.

The Strong Matrix Organization

As projects become larger and require weeks and months to complete, and the scope becomes sufficiently complex, the duties of the project

TABLE 3–1
Organizational Structure and Project Characteristics

	Functional	*Weak Matrix*	*Strong Matrix*	*Projectized*
PM Authority	Little to None	Limited	Moderate to High	High to Complete
Resource Availability	Little to None	Limited	Moderate to High	High to Complete
Product Budget Control	Functional Manager	Functional Manager	Project Manager	Project Manager
PM Role	Part-time	Part-time	Full-time	Full-time
PM Admin. Staff	Part-time	Part-time	Full-time	Full-time

coordinator begin to break down. When this happens, the company may decide to adopt a strong matrix organizational structure. In this organization, the project manager replaces the project coordinator who has more authority over the project team members. This authority extends into the following areas:

- **Time spent on the project.** In the strong matrix organization, team members spend the bulk of their time working on projects. In these scenarios, the sales staff may not spend hours in a public showroom, but instead may work with customers on a one-on-one basis in their homes.
- **The physical relationship.** In a strong matrix organization, the team members are located in the same area close to one another, rather than being separated by department. To illustrate, rather than every engineer in the company occupying his own space in an engineering department, each engineer is placed with other project team members in a team space.
- **The administrative relationship.** In a strong matrix organization, the project manager's authority is raised. It should be noted, however, that the project manager and the functional managers share in both the project authority and reporting status. This means the project manager and the functional manager both participate in team member performance reviews. In addition, the functional manager and project manager work together on project status and financial reports. The two roles are seen as equals within the structure of the organization. Table 3–1 identifies the level of authority and control for the project manager in each of the organizational structures.

Study/Career Tip

Conduct an Internet search on "organizational structures." Read as much as you can and take notes on points that are not covered in this chapter. If you have not yet started a folder or journal about project management, now is the time to formalize your information gathering on this subject.

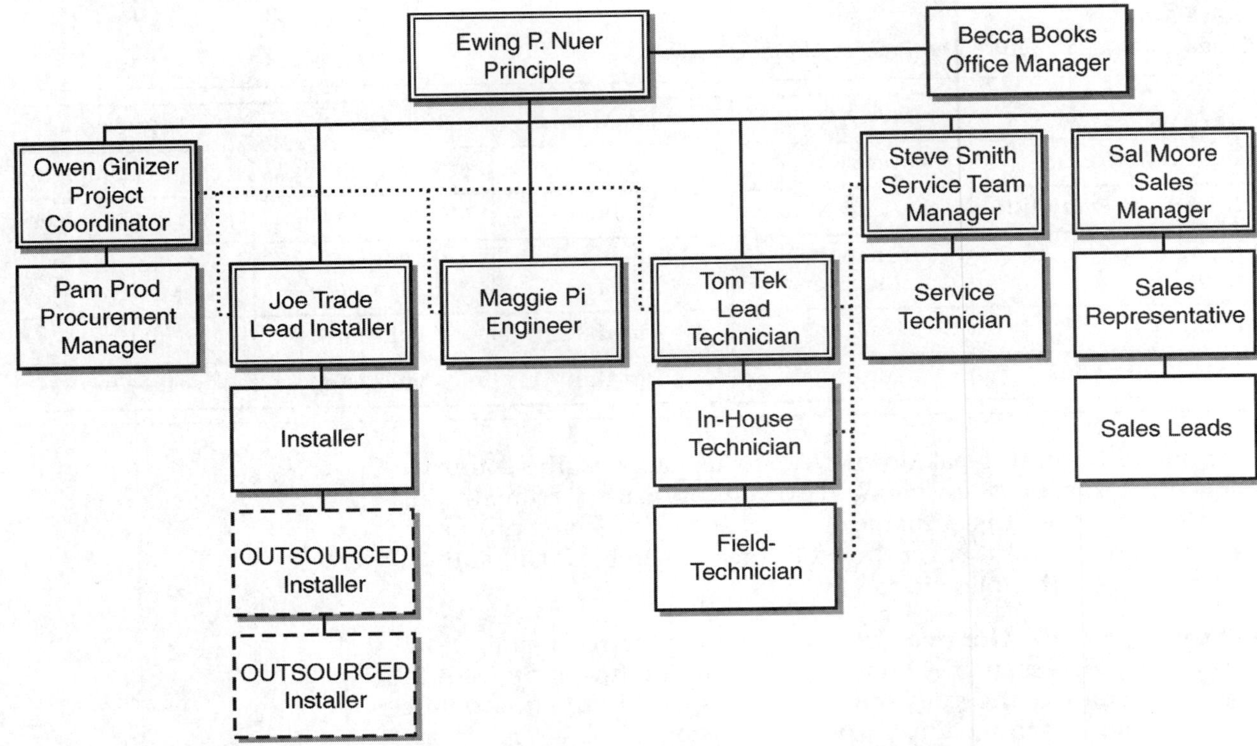

FIGURE 3–5 ABC Integrations Organizational Chart

ABC Integrations Organizational Chart

Figure 3–5 represents the **organizational chart** for our case study company, ABC Integrations. The ABC Integrations organizational structure is based upon a strong matrix organization and it places much of the day-to-day responsibilities on the project manager. Here's a case in point (page 57).

The Project Team

As we discovered in our review of the ABC Integrations organizational chart, the residential integration project manager interacts with a variety of people during the course of most custom installation projects. His primary focus, of course, is to work closely with members of the project team. However, there are two additional interested groups with whom the project manager must collaborate: the authorizing party, and any project stakeholders. Let's explore who all of these individuals are and the roles they play in the residential integration industry.

A CASE IN POINT

Get Organized

Project Manager Owen Ginizer has a procurement manager who reports directly to him. The installation, engineering, and technician groups indirectly, or "dotted line," report to Owen. That means he has authority over them for day-to-day operations and project work. However, the company's principal, Ewing P. Neur, maintains functional authority over these groups. It remains his responsibility to administer performance reviews, hire new employees, and handle any employee terminations. Principal Ewing makes those decisions with input from Project Manager Owen, who has a much closer daily working relationship with each group.

Joe Trade is the leader of the installation team. He is responsible for the Cabling, Terminate, and Move-In phases. Reporting to him is a single full-time installer, and two other outsourced positions. These positions are filled on an "as needed" basis. When the project load runs heavy, Project Manager Owen makes the decision to expand the team for a specified time until the project load reaches a manageable level by the two-man, full-time team.

Tom Tek, ABC Integrations' lead technician, is responsible for the Electronics Build phase, as well as the Electronics Installation phase. He and the two employees who report to him, report to the service manager, Steve Smith, on an "as needed" basis to perform service and maintenance work on existing systems. In addition, they report to the project manager for project-based work.

Sal Moore, the ABC Integrations sales manager, is responsible for every aspect of the sales channel. The sales representative who reports to him handles the smaller, well-defined projects. The lead generation employee is responsible for developing leads through architects, builders, and homeowners, and is a direct report.

Critical Thinking Questions

- Are you currently, or have you been a member of a residential integration project team?
- If so, to whom did you report?
- Were the lines of communication sufficient for you to get your work done in a timely manner?
- What would you have changed if given the opportunity?

Authorizing Party

The **authorizing party** initiates the project. In residential integration, the authorizing party typically is the homeowner or homebuyer—in effect—the customer. On some projects there may be an owner's representative instead, who acts on behalf of the owner, issues payments, and makes final decisions on the project. When the company markets to builders and developers, they also may be the authorizing party. The authorizing party has authority over all aspects of the project scope. That means that the homeowner, his representative, or the builder has the authority to approve design plans for the integrated system and any cabinetry involved, equipment purchases, programming plans, change orders, and the final outcome of the project. The authorizing party also writes the checks to pay for the work.

Stakeholder

A stakeholder is a person with an investment in the outcome of the project—the customer, company owner, general contractor, other subcontractors, or other involved parties. There may be many stakeholders on a project. Although each stakeholder has input on scope decisions, the authorizing party has final authority.

An architect is also considered a stakeholder and he can help in deciding placement of speakers and other devices throughout the home, but ultimately the customer (the primary stakeholder and authorizing party) has the final say in the matter. In fact, everyone who comes in contact with the project can be viewed as a stakeholder: the electrician, the HVAC contractor, the cabinetmakers, the builder, the interior designer, and so on.

Project Scheduler

The project scheduler is responsible for the **project schedule,** which includes planning activities within projects and servicing requests. The project scheduler communicates with the general contractor and other trades to coordinate the schedules. For example, in new construction, the scheduler must be aware of the builder's schedule for applying the drywall and schedule the installation of the cables before the drywall is installed. Just imagine what it would be like to have installers arrive on the job to find the drywall already in place. What a headache it would be to fish all of the cables behind the drywall and have to repair all the holes that would be created to accomplish the cable installation. Needless to say, in addition to the work, the increased cost for the added time it would take to finish the cabling would also add to that headache. The company couldn't recover the cost from the builder or the customer. The installers would bear the burden of the work and the company would bear the cost.

Administrative Manager

The administrative manager tends to the staff by assuring that standard activities, such as employee training, vacations, and other planned activities are listed on the daily work schedules. The administrative manager is also responsible for various office functions, such as preparing project folders, certain company correspondence, and filing various company documents. She could also keep track of receivables, recording the delivery dates and assuring they are delivered to the correct party within the company.

Project Team Member

A project team member is a staff member who performs the work to be managed. A list of typical team members and their job descriptions is presented below.

Sales Representative

The sales representative initiates sales leads, develops long–term customer relationships, and sells company products and services. He has a thorough understanding of the residential building industry and networks with architects, builders, interior designers, subcontractors, and homeowners. In addition, the sales representative acts at the customer advocate during the project life cycle.

Software Engineer

The software engineer plans, documents, and maintains system programming including user interface (keypads and touchscreens) designs, infrared files, and programming files. He develops and maintains archives of customer programming requirements and documentation, ensures programming is complete and tested in advance of installation, supervises in-house technicians, prepares customer user manuals, and provides technician training.

Hardware Design Engineer

The hardware design engineer plans, designs, and develops hardware solutions, and creates bid and estimation documents. She works closely with the project manager and other engineers, applies principles of engineering to resolve hardware design problems, and researches third-party products to meet customer needs.

Mechanical Engineer

The mechanical engineer plans, designs, and creates detailed architectural solutions using 2D CAD software. He provides technical support, installation instructions and equipment specifications; and determines potential hardware conflicts, sizing, ventilation, and structural support.

Lighting Engineer

The lighting engineer designs, documents, programs, manages, and services lighting control systems. She applies engineering principles in solving design issues and works with electricians and lighting subcontractors. She keeps the project manager informed of any system changes, assists in customer training, coordinates "final" lighting adjustments, and visits the site after the customer has moved into the home to confirm that the job was done correctly.

Lead Technician

The lead technician understands in-house and field technician roles. He provides guidance and directs activities of the team, oversees the team schedule to meet the priorities of the project manager, estimates labor hour requirements for project completion, and interprets company standards and enforces safety regulations. He monitors team performance and suggests changes for improved efficiency, and analyzes and resolves technical problems when they arise.

In-House Technician

The in-house technician uses drawings and documentation to assemble, verify, and test systems and components prior to delivery to the customer. She prepares keypads and touchscreens for programming, coordinates with the software engineer to schedule time for programming the system, and works closely with purchasing to order rack parts/components.

Field Technician

The field technician delivers the equipment rack to the project site and connects cables at distribution panels to equipment rack; installs keypads, touchscreens, and system telephones; performs field tests; and assists in customer education/training. He understands distribution panels and connections to the integrated system.

Lead Installer

The lead installer understands the installer role, provides guidance and directs activities of the team, and oversees the team schedule to meet the priorities of the project manager. She estimates labor hour requirements for project completion, interprets company standards and enforces safety regulations, monitors team performance, suggests changes for improved efficiency, and analyzes and resolves technical problems.

Installer

The installer pulls low-voltage cable in residential structures. He installs conduit; uses cable-testing equipment; handles basic cable dressing; and installs modular connectors and wallplates for data, telephone, audio, television, and other low-voltage applications. He terminates cable using proper cable connector tools, approved connector installation techniques, and correct connector pin-out standards. He also installs and mounts electronic devices.

Study/Career Tip

Unify and simply your knowledge: A textbook presents the subjects in a particular form, as does an instructor. By their very nature, however, textbooks and lectures tend to present subjects sequentially. Take the extra step of understanding the material in your terms, which may involve recognizing relationships that could not be conveniently expressed in the order presented in the text and lectures.

Who Is the Customer?

So, ABC Integrations has a competent and experienced project manager in Owen Ginizer, a solid mission statement that all employees understand and adhere to, an organizational structure that is especially conducive to the residential integration industry, and a talented set of project team members at the ready. But who are all these people serving?

Everyone in the organization has a responsibility to serve and support the customer. How do we define the customer? Is he just the person who writes the check, or is it his family, or anybody who comes in contact with the installed system? The short answer is that everyone is viewed as the customer. This includes people both internal and external to the organization. A list below describes the various types of customers.

Internal Customers

- The *project team* acts both as a supplier to the external customer and as an internal customer on each project. When the software engineer calls for information to complete the project, he is acting on behalf of the external customer and, as an internal customer to the project, must receive that information in a timely manner.
- The *company managers* are very important customers who can affect the outcome of the project in an indirect manner, such as ensuring that payroll is implemented on time.

External Customers

- The *homeowner* and his family are certainly viewed as the customer on the project. However, in some cases the residential integration company has little interaction with the owners and final users of the system. In those cases, the homeowner has a representative speak on her behalf during the installation process. That representative then becomes the authority on the project, as discussed earlier.
- The *builder, architect, interior designer, electrician*, and other related *trade professionals* are all acting on behalf of the homeowner/buyer and should be served in the same manner as the customer by project team members of ABC Integrations.
- *Vendors* and *suppliers* are serving the company on behalf of the customer, so the organization must serve their needs as well in order to form solid partnerships. The company does this by providing guidance, feedback, and critical information about their products and services. Helping them succeed in the market is another way of helping the customer.
- The *community at large* can also be viewed as the customer. Responsible organizations initiate a variety of environmental, volunteer, and charity-based projects for the local community. This goes to the core of providing good customer service. By viewing everyone as a customer and treating them as a customer, the entire company's standards are reinforced and raised.

The way in which an organization relates to its internal customers is a direct indicator of how it relates to its external customers. To illustrate, if Project Manager Owen consistently ignores e-mails from the company principal, then he can expect his team to do the same when information is requested from, say, the builder or architect. If Project Manager Owen is constantly rushing from task to task without completing any one task, then that poor working habit will likely prevail throughout the entire project team because that is how Owen has modeled his behavior in front of them. Instead, if Owen allots an appropriate amount of time to complete each and every task and delegates his authority appropriately, the team will see how day-to-day business is conducted and act accordingly.

Summary

In this chapter, we discussed the roles and responsibilities of the project manager and the project team. We also defined the customer and explored the six principles of customer service.

Important points in this chapter include:

- The project manager plans, leads, and directs the project from the start to the finish. He provides guidance, maintains customer contact, conducts site reviews, consults with the sales representatives, and prepares documentation.
- The project manager should be comfortable with change and have a solid understanding of organizational structures. She should work to acquire soft skills, such as strong management, communications, leadership, and corporate political skills.
- Functional organizations generally do not take on projects as products. A project manager has little authority in this organizational structure.
- In a projectized organization, a project manager has total authority of the project and the project team. The parent organization remains functionally organized.
- A matrix-based organization, a compromise between functional and projectized organizations, is the most appropriate structure for a residential integration company.
- A matrix-based structure maintains functional lines of authority while establishing a relatively permanent horizontal (project) structure.
- As a residential integration company grows, it will likely move from a weak matrix-based structure to a strong matrix-based structure.
- The authority of the project manager is determined by the type of organizational structure the company adopted.
- The overall project team consists of the authorizing party, stakeholders, the project scheduler, and the administrative manager. It also includes project team members—sales representative, project manager, software engineer, hardware design engineer, mechanical engineer, lighting engineer, lead technician, in-house technician, field technician, lead installer, and installer.
- Everyone is viewed as the customer, both internal and external to the organization. Customers include the project team; company managers; the homeowner and his family; the builder, architect, interior designer, electrician, and other trade professionals; vendors and suppliers; and the community at large.

Key Terms

Authority The right to apply project resources, expend funds, make decisions, or give approvals.

Authorizing party In residential integration, the authorizing party typically is the homeowner or homebuyer, in effect, the customer, or the person who initiates the project.

Functional organization A functional organization is prevalent in companies that do not take on project-based work as part of their regular business. A company with a functional-oriented structure is based on the theory that staff divisions of responsibility approximately have equal delegations of authority.

Matrix-based organization The most appropriate structure for the residential integration company, it is a compromise between the functional and projectized organization. Where a particular organization lies along the spectrum depends on the products and services that the company offers to its customers.

Organization A group of persons organized for some purpose or to perform some type of work within an enterprise.

Organizational chart A method for depicting interrelationships among a group of persons working together toward a common objective.

Product strategy A product roadmap used to determine what direction to take, how to get there, and why that direction is expected to be successful.

Project manager The person assigned by the performing organization to achieve the project objectives.

Project schedule The planned dates for performing schedule activities and the planned dates for meeting schedule milestones.

Project team All the project team members, including the project management team, the project manager, and for some projects, the project sponsor.

Projectized organization The project manager has complete authority over the project team. The team is created and assigned on a full-time basis to the project manager, and the parent organization remains functionally organized.

Soft skills A set of nontechnical skills that enhance the ability of the project manager to successfully accomplish his work.

Review Questions

1. In general terms, what type of organization structure is best suited for the residential integration industry? Why?
2. Name two types of internal customers.
3. Name three types of external customers.
4. What is the formula for satisfaction?
5. What are some of the soft skills that a project manager should acquire for use on the job? Why?
6. What can happen if a company does not develop its own mission statement?
7. What are "core beliefs"?
8. Why should the project manager care about the corporate mission statement?
9. What are the responsibilities of the authorizing party?

Defining the Project

Introduction

During the defining portion of the project, the project charter formally authorizes a new project or a project phase to begin. In this section of the textbook, we develop the processes that are shown in Figure S2–1. They are: project charter, preliminary scope statement, Work Breakdown Structure (WBS) template, information gathering, scope definition, WBS, and cost estimating.

Project Charter

The *project charter* is the document that is created and used to formally authorize the project. It is the first step in defining the project. At ABC Integrations, the executive team creates the charter.

Preliminary Scope Statement

The *preliminary scope statement* is a document created and used to provide an in-depth and complex explanation of the project scope. At ABC Integrations, it is prepared by the executive team.

WBS Template

The *WBS template* is a process created and used to assist in preparing the WBS, which subdivides the major project deliverables and project work into smaller, more manageable components. At ABC Integrations, the WBS template is prepared by the executive team.

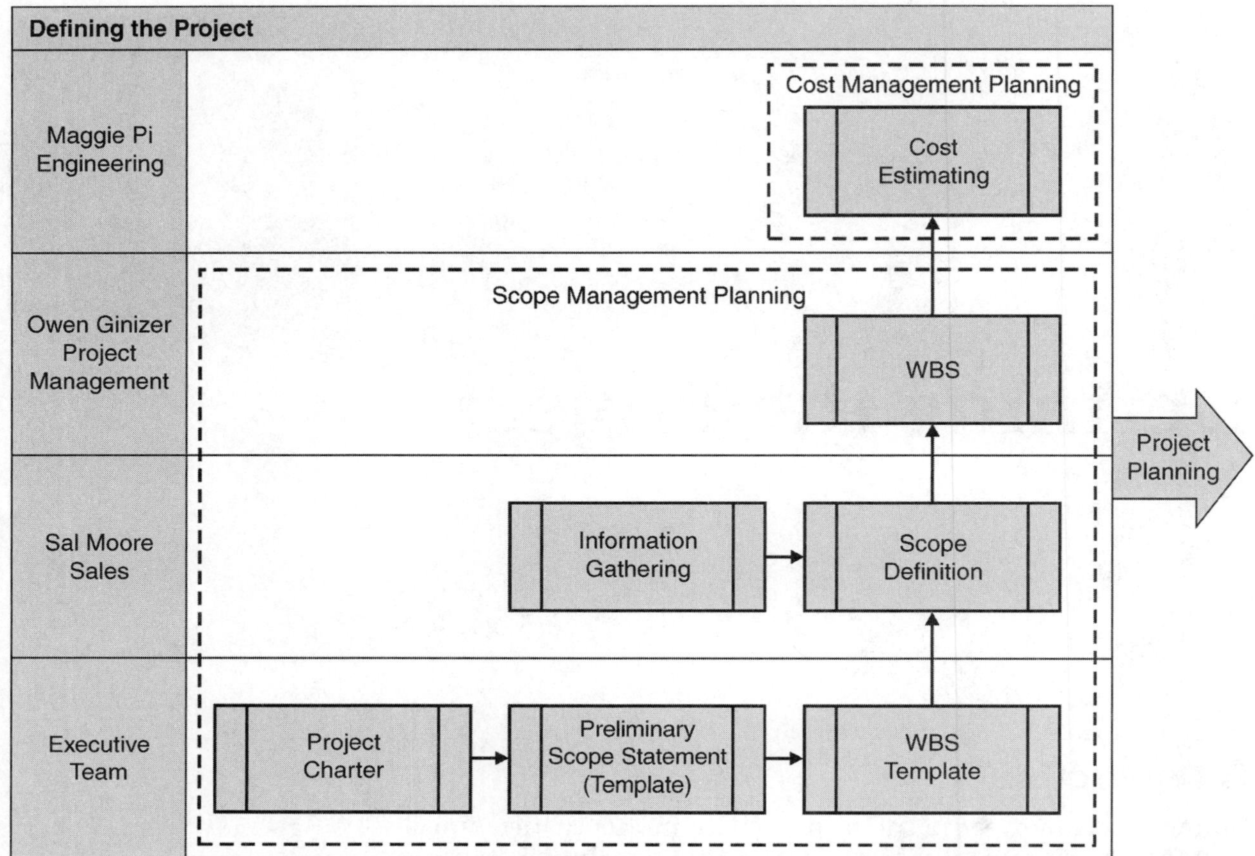

FIGURE S2–1 Defining the Project

Information Gathering

Information gathering is the process created and used to obtain facts about the site, the customer, and the system or systems to be installed. The sales representative and the project manager use a variety of forms to collect the information. The information is then used to prepare a scope statement that explains how to execute the project. At ABC Integrations, information is gathered by the sales representative.

Scope Definition

The *scope definition* is the process created and used to describe the characteristics of the product and services that are offered in the project scope statement. The scope definition provides both the company and its employees and the customer a common document to which they can refer to answer disputed items. At ABC Integrations, the scope definition is prepared by the sales representative.

WBS

The *Work Breakdown Structure,* called the WBS, is a document that is created and used to provide the basis for planning, estimating, and managing the project. The WBS defines the total scope of the project and breaks it down into smaller, more manageable pieces of work. A project manager can more easily schedule, cost estimate, monitor, and control these smaller pieces of work, referred to as *work packages*. At ABC Integrations, the WBS is prepared by the project manager.

Cost Estimation

Cost estimation is a process created and used to develop an approximation or estimate of the cost of the resources required to complete a project. The scope statement provides the input for the process of cost estimating. At ABC Integrations, the cost estimate is prepared by the engineer.

Understanding Cost Accounting

OBJECTIVES

After studying this chapter, you should be able to:

- Define cost accounting.
- Discuss why it's important for a project manager to understanding accounting.
- Explain standard accounting practices, activity-based accounting practices, and project accounting practices.
- Explain the basic terminology and function of accounting.
- Define revenue, costs, profit, and cash flow.
- Discuss the difference between revenue and contract value.
- Explain the difference between a cash-basis company and an accrual-basis company.
- Explain progress payments and state when they generally are made.
- Define tangible and intangible costs, direct and indirect costs, and sunk costs.

OUTLINE

Introduction

In the residential integration industry, as in many industries, two different positions are responsible for the company's accounting procedures. The first and more prominent position is that of a CPA (Certified Public Accountant) who sets up the type of accounting process that will be used by the firm. The second position is that of the bookkeeper, who is in charge of the day-to-day entry of data and the firm's financial reporting mechanics. Although not responsible for cost accounting, the residential integration project manager must be familiar with basic accounting principles to maintain the profitability of each project and the overall profits of the company. This chapter focuses on the basics of cost accounting, basic accounting principles, and why a residential integration project manager needs to understand their underlying structures.

What Is Cost Accounting?

Cost accounting is a process used to record and analyze costs—both labor hours and resources—incurred by the company in order to produce a product (a home theater installed in a customer's home, for example). These costs are those taken by the company, the cost for its purchases and labor requirements for the project. This differs from the cost to the customer that is known as revenue for the company. Cost accounting is also known as job costing. The costs are shown as dollar amounts. In accounting, a cost is a price paid, or otherwise linked with an economic transaction (the home theater costs $100,000). In general terms, all accounting methods record, classify, summarize, interpret, and communicate financial information about a company.

Study/Career Tip

For more complete information on cost accounting and accounting principles, conduct an Internet search using key terms from this chapter. Make notes of what you've discovered. Remember, these accounting principles do not just apply to the residential integration industry; these are foundational concepts found throughout the business world.

There are a number of established ways to approach accounting. Some examples are standard costing, activity-based costing, and project accounting, which is popular in residential integration companies, along with some lesser-known accounting methods. Two established methods of bookkeeping within project accounting are cash-basis and accrual-basis bookkeeping systems. We will give a closer review to these bookkeeping methods later in this chapter.

Critical Thinking Question

- Why do you think the method of project accounting is popular in the residential integration industry?

How Accounting Was Developed

The term "accounting" is derived from a Latin word meaning "constant," because originally individual costs were considered constant or fixed. As organizations developed and grew, it quickly became apparent that some costs were constant, but other costs rose and fell depending on the amount of work that was done or the products that were produced. The result was that costs were divided into two categories—fixed or variable. Fixed costs were associated with business administration and did not change during slow and busy business cycles. Examples of fixed costs are salaries, insurance, rent, office supplies, advertising, and vehicle leases.

Critical Thinking Questions

- What are some fixed costs that you incur in your daily life?
- What about variable costs?
- Do you track these costs on a monthly basis?
- Why do you think you should?

Variable costs are associated with work that produced company products, and naturally rose and fell with the success of the company. Variable costs are expenses that change in direct proportion to the products produced by a company. Examples of variable costs are project supplies, audio/video equipment, commissions, and temporary help.

As increasingly complex organizations developed in the early part of the twentieth century, managers needed an easier way to make decisions about products and prices. As a rough guide for these decisions, managers added up the variable costs and ignored the fixed costs. They could determine the profitability of a product without knowing the fixed costs of the company. To illustrate, let's say a company spends a total of $1,000 a month on fixed costs (for rent and other overhead items); therefore, it is the variable amount (project supplies, equipment, commissions, and other items directly related to the project) that provides the opportunity for profit. Decisions on whether to produce a particular product can be made by just knowing how much profit can be expected. Adding the fixed costs to the business mix does not change the amount of profit available to the organization. Let's say a telephone system can be installed for cost of $3,000, and the system is sold for $5,000 installed. Table 4–1 and Figure 4–1 show two examples. Example A shows the impact of selling two systems per month

TABLE 4–1
Fixed versus Variable Costs

	Example A	*Example B*
Profit	$ 3,000.00	$11,000.00
Fixed	$ 1,000.00	$ 1,000.00
Variable	$ 6,000.00	$18,000.00
Systems	2	6
Standard Cost	$ 3,500.00	$ 3,166.66

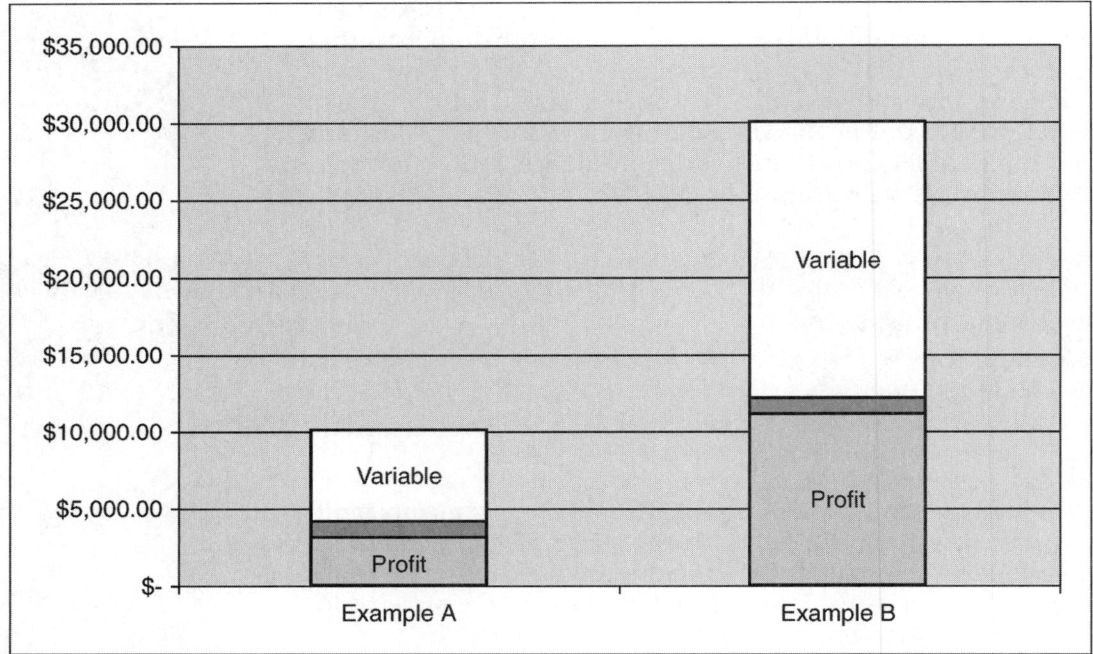

FIGURE 4–1 Variable and Fixed Expenses

while Example B shows six systems per month. While the variable costs increase from Example A to Example B, the opportunity for profit is also increased, as the fixed costs stay the same. The more systems that are sold and installed, the greater the profit opportunity for the organization. The same fixed cost would be added to both sides to determine the total cost, but adding the fixed cost is not necessary to determine which group of telephone systems would have the potentially higher profit.

Standard Costing

Taking the concept one step farther, the need for standard costing came to include an average amount for fixed costs and treating it like one more variable cost, referred to as **overhead.** Standard costing divides the fixed costs by the number of items produced. This added step enabled managers to include a share of the fixed cost in the price of each product unit, making sure the fixed costs are considered in the final pricing structure.

Study/Career Tip

Go to the grocery store and make a simple purchase. Consider the price you are paying for that particular item. Before any taxes that may be added at the checkout line, how much of the total cost of the product do you think is attributable to overhead? Write down what you think some overhead costs are as they pertain to this item.

Let's say our residential integration company installed six telephone systems per month, and the fixed costs were still $1,000 per month. That would mean each telephone system could be said to incur an overhead of $167 (fixed costs = $1000 divided by the number of telephone systems (6) produced, or $167 in overhead). Adding this overhead amount to the variable cost of $3,000 per telephone system produces a unit cost of $3,167 per system.

There is a slight distortion of the unit cost when the standard costing method is used. However, in companies that mass-produce a single product, and enjoy relatively low fixed costs, the distortion is very minor.

Now let's assume ABC Integrations installed two telephone systems one particular month. The standard cost would then become $3,500 per installation (variable cost of $3,000 per installation, plus the fixed cost ($1,000) divided by the number of installations produced (2), or $500 in overhead costs. If ABC Integrations sets reasonable production quotas for the systems they provide, then the standard cost will vary only slightly from month to month. If the quota for telephone systems is six per month, the standard cost will vary only slightly if five or seven systems are delivered; however, if the company does not stay close to its targeted quota then the standard costs will vary greatly, as shown in Table 4–1.

Development of Standard Costing

As time went on, the practice of paying workers by the "piece" or "piece-rate" changed to paying employees on a per hour basis, or an hourly rate. Companies with a wide range of products or services had many tasks common to several finished items, thus making payment for piece-work impractical and too expensive.

Even though electronic equipment for a home theater remains a relatively high variable cost in the residential integration industry, most companies today have relatively low variable costs (primarily supplies and raw material, commissions, or temporary workers) and very high fixed costs (interest payments on business loans or mortgages, salaries, insurance). As a result, today the terms "direct costs" and "indirect costs" are more commonly used instead of variable and fixed costs. This newer industry technology more closely reflects how allocation of overhead is actually calculated. Indirect costs (often the larger of the two types of costs) are usually allocated in proportion to either direct costs or some physical resource that is used. That means allocating fixed costs causes much more distortion in the unit cost figures.

Critical Thinking Question

- Why do you think that "overhead" is such a huge variable when it comes to pricing goods and services?

Activity-Based Costing

Activity-based costing (ABC), (no, not our case study company), is costing by activities. In this approach to accounting, activities are regular actions that are performed inside a company. Talking with a customer about payment questions is an activity. A cost is assigned to this activity in order to allocate costs to products and services. Let's take a closer look.

Company accountants assign 100 percent of the time each employee spends on all the different activities he performs inside a company. In many companies, an accountant will ask employees to fill out detailed surveys to collect data on how workers would assign their own time to the different activities. The accountant can then determine the total cost spent on each activity by adding up the percentage of each worker's salary spent on that activity.

The activities performed within the company determine each product or service that is produced and delivered. The accountant can then assign the different activities to the appropriate products using an appropriate allocation method.

A company can take it one step farther and use activity-based costing to manage the company business. This type of management is called activity-based management.

Managers use the activity cost data to figure out where and how to improve their operations. For example, a project manager may find that a high percentage of installers are spending their time loading their vehicle every morning with project supplies and tools. Using activity-based costing, the accountant provides a dollar amount for the activity called Vehicle Load Time. Management can use this data to decide how to resolve the issue of time spent loading the vehicle and whether the cost to do so is worth it.

Project Accounting

Project accounting, a popular approach in residential integration companies, generates financial reports designed to track the financial progress of projects. Project managers use the financial data to aid them in their decision making on a project-by-project basis. Financial statements record business financial flows and levels. As we discussed in Chapter 1, Introduction to Project Management, a project is temporary and it creates a unique product or service. Project managers are responsible for achieving financial goals while optimizing the use of resources needed to complete the project (time, money, and people).

Standard accounting procedures monitor the financial progress of organizational elements of an organization over a set period of time—weeks, months, quarters, and years. Project accounting, on the other hand, differs in that projects frequently cross various departments, and may last for a few days or weeks or even years. Employees generally fill out a timesheet in order to generate the data to allocate and track project costs. Budgets may also be revised many times. Scheduling changes and revised budgets are especially common in the residential integration industry. The installation of a house-wide sound system can be completed in a few days. Yet the installation of an integrated system in a 20,000 square foot house may take 18 months. Invariably, a customer changes his mind about something and the budget is revised.

Budget generally refers to a list of all planned expenses and revenues. Costs and revenues are allocated to projects, which may be subdivided into a Work Breakdown Structure (WBS—see document PM05). Project accounting allows reporting at any level in the WBS, and allows the project manager to compare historical and current budgets. A WBS defines a comprehensive tree structure, organized from general to specific tasks and deliverables that complete a project. (To read more on the WBS, see Chapter 8, Work Breakdown Structure).

Other Costing Methods

In recent years, more costing methods have developed to better fit different business needs. Approaches include the inventory costing method, the process costing method, the average costing method, and the target costing method. It is beyond the scope of this textbook to discuss these varying accounting methods. It should be noted, however, that most accountants still consider the time honored, traditional costing methods to be generally the most widely accepted in industry.

Study/Career Tip

To find out more information about these newer accounting methods, check your local or on-line library, or conduct an Internet search. You may be surprised that one or more of these methods may be of use to you when you are a project manager at a residential integration company.

Basic Accounting Principles

To help understand cost accounting in the residential integration industry, let's look at some basic accounting principles that are used in project management. Even though the project manager typically is not responsible for the company accounting role, she and the sales engineer need to have a solid understanding of the basic terminology and function of accounting. Let's say the company comptroller sets a policy that all projects must have a gross profit margin of 40 percent. To meet that goal, a project manager must understand what that means and why it is important. Let's review four important elements of cost accounting: revenue, costs, profit, and cash flow.

Critical Thinking Questions

- How much do you think you know about revenue, costs, profit, and cash flow?
- Where do you feel you need more emphasis?
- How do you define these terms in your own mind?

Understanding Revenue

Revenue is money received for goods and services that have been delivered. It differs from **contract value** in that contract value is the agreed upon amount for goods and services that are expected to be delivered. It is easy to confuse revenue with the contract value or progress payments, and the funds paid at specific milestones (initiating the contract, completing various installation phases, final payment). Here are some questions to confuse you even more.

- When a customer authorizes a contract, is the contract value considered revenue?
- When a customer hands over a check to put toward a project, are the deposited funds revenue?
- When a system leaves the shop for the customer's installation, is the amount of equipment leaving considered revenue?
- How is revenue determined for labor? Is it at the completion of each project, by daily time sheets, or at the end of the entire project?

Study/Career Tip

Try to answer the bulleted questions. Write down your answers and review this chapter to see if you are correct. Think of similar questions as you read through the text. Create your own "End-of-Chapter Review Questions" to help you study.

The answers to these questions depend upon the accounting system used by your organization. In a cash-basis company, the revenue is booked (recorded) when the money is received, and in an accrual-basis company, the revenue is booked when the equipment is delivered and as the work is completed. We will be reviewing cash-basis and accrual-basis accounting in greater detail later in this chapter.

Regardless of the accounting method chosen by the organization, a project manager measures three key financial indicators on each project—*sales revenue*, *contract value*, and *progress payments*. Remember, sales revenue is a measurement of the products and services already delivered to the customer. Contract value is the measurement of the products and services agreed upon to be delivered in the future to the customer. And progress payments are periodic payments for products and services that have been or will be delivered (milestones). Let's take a closer look at each and how they impact projects being completed by our case study, ABC Integrations.

Contract Value

The contract value is the money payable by the homeowner under the **contract** documents. This is the total price charged to the customer for the contract. Table 4–2 shows contracts awarded for a three-month period. By monitoring the contract value on a monthly basis, a project manager can determine if the future workload will increase, stay the same, or decrease. If the average life cycle of a project is six months, then a drop in contract value will result in a drop in revenue in approximately six months. Table 4–2 is a simple example of the sales generated by ABC Integrations over a three-month period. It shows

TABLE 4–2
Contract Value (Numbers represent thousands of dollars)

	January	*February*	*March*	*April*	*May*	*Totals*
Contract A	10					10
Contract B		15				15
Contract C			10			10
Totals	10	15	10			35

the sales team has closed Contract A in January, which is $10,000, Contract B in February, which is $15,000, and Contract C in March, which is $10,000.

Sales Revenue

As previously stated, sales revenue is the money received for goods and services provided. As we have discussed, there are two methods for recording revenue: cash-basis and accrual-basis. In a cash-basis organization, the revenue is recorded when money is received for good and services. In an accrual-based system revenue is recorded when goods and services are delivered to the customer. Most residential integrations companies utilize an accrual accounting system. That's because progress payments often are received prior to delivering and even ordering equipment. Our case study company ABC Integrations is running an accrual-based accounting system.

ABC Integrations sales revenue is received for the delivered product and completed work on the project. Table 4–3 shows the contracts as they are signed, and it demonstrates how the value of the products and services is delivered during the project life cycle. Although the total amount of sales revenue is equal to the contract value, the revenue is realized at different times. This is caused by the nature of recording those sales when the product and services are delivered versus when the contract is awarded.

TABLE 4–3
Sales Revenue (Numbers represent thousands of dollars)

	January	*February*	*March*	*April*	*May*	*Totals*
Contract A	2	9				11
Contract B		3	13			16
Contract C			2	7	1	10
Totals	2	12	15	7	1	37

Critical Thinking Question

- Why do you think it's important to track the difference between contract value and sales revenue in tables (or other means) such as the ones shown here?

Progress Payments

Typically, project payments are made at project **milestones,** such as completion of cabling, prior to installation of electronics, and upon completion. These payments are known as progress payments. Table 4–4 tracks the progress payments for our three projects. Again, the total in progress payments is equal to the total in sales revenue and contract values; however, notice that the payments are received at different times than when the product is delivered and the contracts are awarded. See the case in point on page 79.

Cash-Basis Bookkeeping

Cash-basis accounting is a method of **bookkeeping.** In this method, payments receive from customers and bills paid are accounted for immediately. Let's say you have a lemonade stand. On Tuesday you spend $10 on your supplies: lemons and sugar. That's considered a $10 expense. You record the expense immediately on your books. On Wednesday you sell your lemonade and at the end of the day you have collected $30. That is your revenue. You record that amount immediately and find you have made a profit of $20. You are recording your expenses and revenues as they happen.

Revenue is recorded when cash is received and expense is recorded when cash is paid. In cash-basis accounting, revenues and expenses are also referred to as cash receipts and cash payments. Credit transactions are not recorded in cash-basis accounting. A cash-basis balance sheet, which shows the assets (money and items owned) and liabilities (money owed) of the company, does not have **payables, receivables,** or prepaid **expenses.** Cash-based accounting is easier to manage because it lacks credit-based transactions. Cash-basis accounting is a poor choice for cost accounting in residential integration operations because it fails to meet two important principles of **GAAP** (Generally Accepted Accounting Principles). Those principles are:

TABLE 4–4
Progress Payments (Numbers represent thousands of dollars)

	January	*February*	*March*	*April*	*May*	*Totals*
Contract A	2	6	2			10
Contract B		3	9	3		15
Contract C			2	6	2	10
Totals	2	9	13	9	2	35

A CASE IN POINT
Revenue

Local suburbanites Fred and Wilma Smith have recently signed a contract with ABC Integrations for the installation of a structured cabling system throughout their new home that is being built. The Smiths have agreed to pay a total of $6,000 for the cabling job. This total amount is the contract value. In order to get the job started, the Smiths make a down payment of $2,400. This initial sum is considered revenue by ABC Integrations. As the work continues on the cabling system, ABC bills the Smiths an additional $2,400 as part of the agreed upon progress payment. Again, these funds go on to ABC Integrations's books as revenue. When the job is completed to the customer's satisfaction, the final payment of $1,200 is sent to ABC and is again booked as revenue.

- Revenue should be recognized (recorded) when it is realized. Remember our lemonade stand? Expenses were recorded when the money was spent; revenues were recorded when the money was received. In the typical residential integration project, progress payments are made at different times, and not necessarily when the work is accomplished. Therefore, in a cash-basis system, the revenue tracks project payments but does not accurately track the output of goods and service provided by the company.
- Revenue should be matched to the expense if possible. (e.g., Sales to **COGS**—Cost of Goods Sold)

Let's look at some sample figures. Let's say you pay your rent ($1,000). Your landlord records the funds—called income—when you make the payment. The landlord records an expense when he pays the rental agent the fee for your apartment. This type of accounting method generally is used by companies that have limited payables or receivables or whose revenue and expense cash flows are closely associated with each other in time.

Table 4–5 shows a simplified Income Statement and Table 4–6 shows a Balance Sheet for cash-basis accounting.

There are two types of cash-basis accounting—strict cash-basis and modified cash-basis. Strict cash-basis follows an exact record of cash flows. Our lemonade stand uses strict cash flow. Modified cash-basis includes some elements from accrual-basis accounting. That means a

TABLE 4–5
Income Statement—Cash-Basis Accounting

Revenue	$1,000
Expense	$ 800
Net Income	$ 200

TABLE 4–6
Balance Statement—Cash-Basis Accounting

Cash	$5,500
Total Assets	$5,500
Liabilities and Stockholders' Equity Common Stock	$5,500
Total Liabilities and Equity	$5,500

bookkeeper in a small residential integration company could record information like inventory data or purchases of major assets like trucks.

Most households, probably your own, can use cash-basis accounting. Small businesses also can use this method because the financial statements are used by very few people.

Accrual-Basis Bookkeeping

Accrual-basis accounting records financial data that change your net worth (the amount owed to you minus the amount you owe to others). Standard practice is to record revenues and expenses within the period of time (frequently the month) in which they are due. Even though no cash or credit is received, the revenues and expenses are recorded because they are future income and cash flows of the company. Accrual-basis accounting meets GAAP principles.

Let's look at an example of accrual-based accounting. Your rent is due the first of the month. That means your landlord will record the amount on the day your rent is due (you owe it to him). He records the fee owed to the rental agent that month when it is due for your apartment (he owes it to the agent). In addition, in a residential integration company, the bookkeeper records the actual cash flows and their timing.

Table 4–7 shows a simplified Income Statement and Table 4–8 shows a Balance Sheet for accrual-basis accounting. (Note that receivables and payables are recorded.)

TABLE 4–7
Income Statement—Cash-Basis Accounting

Revenue	$1,200
Expense	$ 800
Net Income	$ 400

Comparison of Cash-Basis and Accrual-Basis Accounting

In cash-basis accounting, revenue and expenses are recorded only when cash is received or paid to others. In accrual-basis accounting, receivables and payables are recorded (and, for tax purposes, profit and loss data is maintained), even though no cash has been received or paid to others. Cash-basis accounting delays recording all credit transactions until the cash is received. That's why it is considered a more conservative approach to accounting. A small business—our lemonade stand, which buys its lemons and sugar daily for cash at a wholesale market, sells the lemonade for cash, and throws away what

TABLE 4–8
Balance Sheet—Cash-Basis Accounting

Cash	$5,500
Accounts Receivable	$ 200
Total Assets	$5,700
Accounts Payable	$ 200
Liabilities and Stockholders' Equity Common Stock	$5,500
Total Liabilities and Equity	$5,700

didn't sell—can get an accurate picture of its profits or losses using cash-basis accounting. However, a residential integration company that allows progress payments, and that buys materials on account from manufacturers and dealers, must use the accrual method to have an accurate record of its financial picture.

In accrual-based accounting, a Statement of Cash Flow is created to show company cash flows. In the past, accrual-basis accounting was more expensive than cash-based accounting because a bookkeeper needed to record substantially more transactions. Today, accounting software packages have made the cost differences minimal.

The Residential Integration Selection

A small residential integration company, especially a start-up company, could decide to use cash-basis accounting. It would be easier to institute, result in the capability of deferring tax payments, and eliminate the need to extend or receive extensive company credit. Small companies often favor cash-basis accounting because taxes are deferred until the cash is received. Companies that extend or use credit regularly will use the accrual-basis method of accounting. In the United States, the Internal Revenue Service may require a company to use it. The Securities and Exchange Commission requires all publicly traded companies to follow GAAP. That means all publicly traded companies must publish their financial statements in accrual-basis accounting.

In deciding which accounting method to be used, three external groups should be considered: creditors, stockholders, and the IRS. Creditors and stockholders generally consider cash-basis accounting inadequate because it does not allow them to forecast the future income and cash flow of the company. The company also may look less appealing because, by deferring credit transactions, the revenues and assets often seem less than they are. Many small residential integration companies start out using the cash-basis method of accounting. However, as corporate growth dictates the need for creditors and potential stockholders, larger size residential integration companies will change their books to an accrual-basis accounting method.

Tracking the Company's Financial Health

If the company produces the same amount for all three key indicators every month—contract values, sales revenue, and progress payments, it is in good financial health, keeps projects flowing into the company as fast as they are completed, and cash from payments is maintained to pay expenses. In a growing company, the contract values are slightly higher than the sales revenue. In the event the opposite happens and sales revenues outstrip contract value over a period of several months, then the company is sure to be in the midst of a financial crisis. If your project team is able to install more systems than the sales force is able to sell, it won't be long before a layoff is necessary for company health and survival. Monitoring these key revenue figures ensures the long life and steady growth of the organization. If the progress payments are lagging behind sales revenue and contract values, this is a good indicator that final payments are not being made. That generally happens when projects are not adequately finished.

Understanding Costs

Another major accounting principle is costs. Tangible and intangible costs are categories for determining how definable the estimated cost is for a project. **Tangible costs** can be quantitatively measured or valued, such as the cost of the products associated with the project. **Intangible costs** are difficult to measure in monetary terms. Installation productivity is an example of an intangible cost. Because intangible costs are hard to measure, they are difficult to justify in a cost estimate.

As we explained above, direct and indirect costs (formerly referred to as fixed and variable costs) are categories used to determine whether a cost is attributed to a given project or over many projects. **Direct costs** for a project includes the cost of installed equipment, and material and labor directly involved in the physical construction of the project. All costs that do not become a final part of the installation, but that are required for the orderly completion of the installation, are considered **indirect costs**. They may include, but are not limited to, field administration, direct supervision, capital tools, startup costs, contractors' fees, insurance, taxes, and similar items.

Sunk cost is money that has been spent in the past and is nonrecoverable. This money is gone, like a sunken ship, and can never be recovered. Never include sunk costs in determining whether to accept or continue a project.

Critical Thinking Questions

- What are some "sunk costs" that you have incurred over time? For example, did you purchase a concert ticket in advance and then not attend the concert?
- If you purchased the ticket in advance and did not attend, but sold the tickets, is that considered a sunk cost?

Understanding Profit

Profit, sometimes referred to as income, is revenue minus costs. It can be measured on a single project both prior to and after taxes and depreciation. **Gross profit** for a project is the contract value less the associated direct costs. **Operating profit** is the earnings or income after all expenses (selling, administration, and depreciation) have been deducted from gross profit. **Net profit** is the earnings or income after subtracting miscellaneous income and expenses (patent royalties, interest, capital gains) and federal income tax from operating profit. **Profit margin** is the ratio between revenue and profits. If revenues of $100 generate $10 in profits, then there is a 10 percent margin.

Review Table 4–9. Note that the owners of the company set a desired net profit for the organization (row 18). In addition, they set a sales goal for sales revenue (row 2), progress payments, and contract values.

TABLE 4–9
Profit from Ongoing Projects (Numbers represent thousands of dollars)

	Project A	*Project B*	*Project C*	*Total*	*Row*
Sales Revenue	10.0	14.0	10.0	34.0	2
Direct Costs					3
Cost of Goods Sold	4.0	6.0	4.0	14.0	4
Cost of Services	1.4	2.3	1.4	4.3	5
Gross Profit	4.4	6.8	4.4	14.8	6
Gross Profit Margin	44%	44%	44%	44%	7
Expenses (Indirect Costs)					8
Rent	1.4	1.4	1.4	4.4	9
Advertising	0.4	0.4	0.4	1.4	10
Administrative Payroll	1.2	1.2	1.2	3.6	11
Insurance	0.2	0.2	0.2	0.6	12
Training	0.4	0.4	0.4	1.4	13
Operating Profit	0.6	2.9	0.6	4.1	14
Taxes and Depreciation					15
Depreciation	0.1	0.1	0.1	0.1	16
Taxes	0.1	0.4	0.1	0.7	17
Net Profit	0.4	2.3	0.4	3.3	18

From these numbers, and through understanding expenses, a solid financial plan of the company emerges. When the organization sticks to its financial plan, net profit is achieved as predicted. The financial plan can include setting a desired gross profit margin on each project (row 7), since the project manager has direct control over this factor.

If the company drastically surpasses estimated sales goals, then the company suffers from too much growth. The result is a shortage of trained employees capable of carrying out the project. Too much work can be worse than not enough work if a growth plan is not in place. If the company drops below its targeted sales goals, then there is insufficient work to keep everyone employed. Worse yet, the overhead, which was projected for a certain size company, leaves the company top heavy in management and overhead expense. That again leads to a company's demise. As already explained, financial goals must be delineated with accuracy.

Understanding the gross margin on a project helps a project manager in selecting projects. By taking the gross profit and dividing by the number of resource hours, a gross profit per hour of labor can be

TABLE 4–10
Gross Profit per Man Hour Analysis

Project	*Gross Profit*	*Labor Hours*	*Gross Profit/ Labor Hour*	*Accept Project?*
Plasma TV	$3600	6	$600	Yes
DLP TV	$1800	3	$600	Yes
CRT TV	$ 60	2	$ 30	No

calculated. Let's look at three projects: the installation of a plasma television, the installation of a rear projection DLP television, and the installation of a 20" CRT television. Let's also assume that the company comptroller has determined that the gross margin per labor hour needs to be a minimum of $40. Looking at the numbers in Table 4–10, we see that even though the DLP television produces less profit, it results in the same amount of gross profit per labor hour. However, with the CRT television, the profit per labor hour is lower than what is required to maintain overhead expenses in the company. That means this work should not be accepted. Keep in mind that this is a simple example; and when a complete system is installed, it is comprised of both low margin and high margin products. The overall system gross margin per labor hour is the critical element in the decision-making process of whether to accept a particular project or not.

Understanding Cash Flow

Cash flow, the last major accounting principle, is the movement of money in and out of the organization. Payments are received from customers in accordance with the payment terms of their projects, and that money is used for rent, advertising, payroll, products used on the project, and other day-to-day business operations.

Inflows are the movement of money into your organization. Inflows are most likely from progress payments, which occur when the customer makes a payment toward her contract value. Other inflows come from the proceeds from a bank loan, or investments from the owners.

Outflows are the movement of money out of your organization. Outflows are generally the result of paying expenses. The largest outflow is most likely to be for the purchase of products used on projects (projectors, screens, stereo equipment and similar electronic equipment). Other outflows include purchasing fixed assets (cable trucks, office building), paying back loans, payroll, and accounts payable (various supplies).

Cash flow management is vital to the health of the organization. A successful organization has more money flowing in than flowing out each month. It should be noted that profit is not the same as cash flow. It is possible to show a healthy profit at the end of the year, and yet face a significant cash shortage at various points during the year. Unfortunately, cash outflows and inflows occur at different times, and rarely occur together. More often than not, cash inflows lag behind your cash outflows, leaving shortages of money. This money shortage is considered the **cash flow gap**. The cash flow gap represents an excessive outflow of cash that may not be covered by a cash inflow for weeks, months, or even years. Managing cash flow well allows the organization to narrow or completely close the cash flow gap. By examining the different items that affect the cash flow of the business, a project manager can control the cash flow.

Cash flow alone can make or break a business. Let's look at our case study company as an example. ABC Integrations receives deposits from customers and with it purchases equipment for several projects. The company is currently holding $400,000 in customer deposits, money that is paid toward a project but not invoiced (no bill has been sent); that will happen when the product and services are delivered. If ABC Integrations spends $400,000 in inventory, the company is left with no cash. Even if those projects are profitable, there is no money left to make payroll. The employees are not going to accept inventory in lieu of a paycheck. Without employees, the projects would not be finished, and further progress payments would not be made. ABC Integrations would find itself in a Catch-22 situation, meaning it has placed itself in a self-defeating, no win position. No cash to get the projects done, no company! The only options are for the owner to invest more funds in the company or borrow additional funds, because without funds the company would have to close its doors. Again and again in business, you will hear: cash is king!

Another option for ABC Integrations is to create a payment schedule for each project that will keep the cash inflows slightly ahead of the cash outflows during the project life cycle. Payments made by the customer can trigger events such as purchasing products for a given phase and authorizing the project team to start work on the given phase.

Summary

In this chapter, we have defined cost accounting and explained why it's important to the residential integration project manager. We reviewed standard accounting, activity-based accounting, and project accounting practices. We discussed the basic terminology and function of accounting principles and explored revenue, costs, profit, and cash flow.

Important points in this chapter are:

- Project accounting and accrual-basis accounting are favored choices for residential integration companies.
- The project manager needs a solid understanding of the basic terminology and function of accounting. He must have a solid understanding of revenue, costs, profit, and cash flow.
- In a cash-basis company, the revenue is booked (recorded) when the money is received; and in an accrual-basis company, the revenue is booked when the equipment is delivered and as the work is completed.
- The total amount of sales revenue is equal to the contract value; however, the revenue is reached at different times. This is caused by

the nature of recording those sales when the product and services are delivered versus when the contracted is awarded.

- Project payments are made at project milestones, such as completion of cabling, prior to installation of electronics, and upon completion.
- Profit is revenue minus costs. It can be measured on a single project both prior to and after taxes and depreciation.
- Cash flow is the movement of money in and out of the organization.
- Cash inflows are the movement of money into your organization, which includes progress payments, bank loans, or investments from the owners.
- Cash outflows are the movement of money out of your organization, which includes purchase of products used on projects, fixed assets and paying loans, payroll, and accounts payable.

Key Terms

Accrual-basis accounting A method of accounting in which each item is entered as it is earned or incurred regardless of when actual payments are received or made.

Bookkeeping The process of maintaining, auditing, and processing financial information for business purposes.

Cash-basis accounting A method of accounting in which each item is entered as payments are received or made.

Cash flow The movement of money in and out of an organization.

Cash flow gap An excessive outflow of cash that may not be covered by a cash inflow for weeks, months, or even years.

Cash inflows The movement of money into an organization.

Cash outflows The movement of money out of an organization.

COGS (Cost of Goods Sold) In accounting, the cost of goods sold describes the direct expenses incurred in producing a particular good for sale, including the actual cost of materials that comprise the good, and direct labor expense in putting the good in salable condition. Cost of goods sold does not include indirect expenses such as office expenses, accounting, shipping, advertising, and other expenses that can not be attributed to a particular item for sale.

Contract A contract is a mutual binding agreement that obligates the seller to provide the specified product or service or result and obligates the buyer to pay for it.

Contract value The agreed upon amount for goods and services that are expected to be delivered.

Costs Price paid, or otherwise linked with economic transaction.

Direct costs Costs that are directly related to the project life cycle.

Expense Expense is a general term for an outgoing payment made by a business or individual.

GAAP Generally Accepted Accounting Principles (GAAP) are the accounting rules used to prepare financial statements for publicly traded companies and many private companies in the United States.

Gross profit The earnings from an ongoing business after direct costs of goods sold have been deducted from sales revenue for a given period.

Indirect costs Costs that are shared over many projects.

Inflows *See* Cash inflows.

Intangible costs Costs that cannot be quantitatively measured or valued.

Milestone A significant point or event in the project.

Net profit The earnings or income after subtracting miscellaneous income and expenses (patent royalties, interest, capital gains) and federal income tax from operating profit.

Operating profit The earnings or income after all expenses (selling, administrative, depreciation) have been deducted from gross profit.

Outflows *See* Cash outflows.

Overhead An average amount used for indirect costs that is treated like a direct cost.

Payable Accounts payable is one of a series of accounting transactions covering payments to suppliers owed money for goods and services.

Profit margin The ratio between revenue and profits.

Receivable Accounts receivable is one of a series of accounting transactions dealing with the billing of customers who owe money to a person, company or organization for goods and services that have been provided to the customer.

Revenue Money received for goods and services that have been delivered.

Sunk costs Money that has been spent in the past and is nonrecoverable.

Tangible costs Costs that can be quantitatively measured or valued.

Review Questions

1. Define sales revenue.
2. If you worked for a residential integration company, which kept its books on an accrual basis, how would the revenues be booked?
3. What are three key financial indicators on a project that a project manager must track?
4. Why is it important for a residential integration project manager to be familiar with basic accounting principles?
5. In general terms, what do all accounting methods do?
6. What are at least three examples of fixed costs?
7. Why was it important that the concept of standard costing was developed?
8. How is the financial data generated by the project accounting method used by residential integration project managers?
9. Define activity-based accounting.
10. When are progress payments typically made during a project?

Information Gathering

OBJECTIVES

After studying this chapter, you should be able to:

- List the three areas of information needed to develop the project scope statement.
- Gather and record site condition information.
- Explain how site conditions affect a project.
- Explain the importance of gathering customer information.
- Discuss how to collect standard and non-standard systems information.

OUTLINE

Introduction

A successful residential integration project, whether done on a new home, as a remodel, or as a retrofit, is only as successful as the information gathered to support the wishes of the stakeholder(s). To that end, it is vitally important that a logical set of procedures be followed, using solid written documentation, in order to meet the needs of the client and of those who help support the project. In this chapter we will explore ways in which the project manager should undertake the information-gathering process, and identify areas that need to be covered and how this information is used in order to develop the project scope statement.

So far we've learned about the basic tenets of project management, the residential integration life cycle, and the project team. The next step in the evolution of a project is to gather information about the site, the customer, and the system or systems to be installed. As a rule, a project manager makes use of a variety of industry-specific forms to collect the information. This information is then used to prepare a scope statement that explains how to execute the project. Writing the scope statement will be discussed in the next chapter.

The first step is to gather and record complete information about the three residential integration areas—site, customer, and system(s)—to be addressed for the project. Let's take a comprehensive look at each area.

Study/Career Tips

Are you planning on taking CompTIA's HTI+ certification exam, or other industry-related certification exam? If so, it's not too early to start prepping for any of these exams. Visit the organizations' Web sites and download the appropriate information, including exam objectives. Consider purchasing a book or two about how to prepare for a specific exam. Employers seek out individuals who have certifications next to their name.

Site Conditions

A residential integration **site condition** review is actually comprised of a number of components. These include:

- Site Location Issues such as distance, map, directions, parking, and tools and supplies.
- Existing Construction
- House Layout
- Equipment Location
- Unusual Circumstances

To get a jump-start on the project at hand, a project manager or another member of the project team will likely conduct a preliminary screening of the job. A Site Conditions Survey Form (see Figure 5–1) is a very efficient way of capturing the initial information that is used by all members of the project team. A site conditions survey contains basic information about the home as it pertains to the installation project. This includes such items as the construction type, size of the home, room names, ceiling type, and any other pertinent information. It should also identify the location of the site, parking availability for the work crews, and hotel accommodations in the area if overnight travel is necessary. Other relevant information includes cellular telephone coverage, current and future broadband Internet access, and the onsite telephone number.

Additional site condition information can be gathered in a number of ways. A project manager can place a phone call to the general

SITE CONDITIONS SURVEY FORM

PROPOSAL NUMBER: ______________________ DRIVING DISTANCE: ______________________

PROPOSAL DUE BY: ______________________ CUSTOMER NAME: ______________________

SITE ADDRESS: ______________________

PROJECT SCOPE: ______________________

____

BASEMENT

CEILING HEIGHT: ______________________ TYPE: ❑ OPEN ❑ DROP ❑ BOARD & PLASTER ❑ __________

SQUARE FEET: ______________________ FINISH: ❑ YES ❑ NO

CONSTRUCTION: ❑ NEW ❑ RETROFIT

ROOM NAMES:

__________ __________ __________ __________ __________

__________ __________ __________ __________ __________

1st FLOOR

CEILING HEIGHT: ______________________ TYPE: ❑ OPEN ❑ DROP ❑ BOARD & PLASTER ❑ __________

SQUARE FEET: ______________________ FINISH: ❑ YES ❑ NO

CONSTRUCTION: ❑ NEW ❑ RETROFIT

ROOM NAMES:

__________ __________ __________ __________ __________

__________ __________ __________ __________ __________

2nd FLOOR

CEILING HEIGHT: ______________________ TYPE: ❑ OPEN ❑ DROP ❑ BOARD & PLASTER ❑ __________

SQUARE FEET: ______________________ FINISH: ❑ YES ❑ NO

CONSTRUCTION: ❑ NEW ❑ RETROFIT

ROOM NAMES:

__________ __________ __________ __________ __________

__________ __________ __________ __________ __________

____

CEILING HEIGHT: ______________________ TYPE: ❑ OPEN ❑ DROP ❑ BOARD & PLASTER ❑ __________

SQUARE FEET: ______________________ FINISH: ❑ YES ❑ NO

CONSTRUCTION: ❑ NEW ❑ RETROFIT

ROOM NAMES:

__________ __________ __________ __________ __________

__________ __________ __________ __________ __________

(Company Name):: 123 Main Street, Anytown, MA 01234:: (999) 555-1212:: www.company.com Page 1 of 1

FIGURE 5–1 Site Conditions Survey Form PM02

contractor to learn when to anticipate future construction milestones, and visit the site to learn the current status of the home construction. It is important to keep in mind that the general contractor may provide dates for completion of significant tasks such as the electrical work, but these dates should not be taken at face value. The general contractor's information may not be accurate and the completion date could slide. The project manager needs to confirm the dates with the subcontractor who is actually going to do the work. Read the case in point below.

A CASE IN POINT
The Bottom Line

Let's say the general contractor says that electrical work will be completed by the 15th of the month. That would mean the low-voltage cabling could be installed starting on the 16th. However, the project manager calls the electrical contractor and finds out that the electrical work is not expected to be completed until the 20th. The project manager must resolve this discrepancy before he assigns installers to go to the worksite. He will also need to get the electrical contractor and general contractor to agree on an actual completion date. And as that date approaches, the project manager needs to follow up with the electrical contractor to ensure the deadline will be met. What would happen if the electrical work was incomplete when the installers arrived on the 16th to begin their work? It's unlikely that the installers could start their cabling, but if they could start, it would be difficult to work around the electrical contractor. The bottom line is time and money are lost and the reputation of the project team is compromised.

Critical Thinking Questions

- What might be some additional ways that a project manager could gather information about the site conditions?
- Aside from the general contractor, whom might the project manager contact to get specific information?

A good way to keep milestone information is to use Microsoft Project, a software program used by many project managers in many industries. It can track not only the residential integration milestones, but also those of related contractors whose work affects the completion of the residential integration systems. In our case study, ABC Integrations has developed a project-tracking template, PM20 (see Figure 5–2) that utilizes Microsoft Project. We cover the use of Microsoft Project in detail in Chapter 11, Risk Management Planning, and Chapter 12, Procurement Management Planning.

Study/Career Tips

If you don't otherwise have access to it, visit **www.microsoft.com** and download a 60-day free trial of Microsoft Project. It's also available on CD. Many residential integration firms use this software package, so it would benefit you to have some familiarity with the program. Consider using it to track your school work and other personal projects as a start.

		Task Name	Predecessors	Work	Duration	Resource Initials	Start	Finish	Resource Names
1		**Utility Rough-Ins**		**0 hrs**	**0 days**		**Fri 5/6/05**	**Fri 5/6/05**	
2		Rough-in HVAC complete		0 hrs	0 days		Fri 5/6/05	Fri 5/6/05	
3		Rough-in plumbing complete		0 hrs	0 days		Fri 5/6/05	Fri 5/6/05	
4		Rough-in electrical complete		0 hrs	0 days		Fri 5/6/05	Fri 5/6/05	
5									
6		**1.0 Cabling Phase**	1	**73 hrs**	**13.13 days**		**Fri 5/6/05**	**Wed 5/25/05**	
7		**1.1 Plan**		**22 hrs**	**4 days**		**Fri 5/6/05**	**Wed 5/11/05**	
8		1.1.1 Sales to Operations Turnover Meeting		2 hrs	0.25 days	PM	Tue 5/10/05	Tue 5/10/05	Project Management
9		1.1.2 Secure low voltage permit		1 hr	4 days	PM	Fri 5/6/05	Wed 5/11/05	Project Management[3%]
10		1.1.3 Scope Verification		4 hrs	1 day	PM	Fri 5/6/05	Fri 5/6/05	Project Management[50%]
11		1.1.4 Risk Assessment		1 hr	1 day	PM	Fri 5/6/05	Fri 5/6/05	Project Management[13%]
12		1.1.5 Preliminary Schedule		2 hrs	1 day	PM	Fri 5/6/05	Fri 5/6/05	Project Management[25%]
13		1.1.6 WBS Review		1 hr	1 day	PM	Mon 5/9/05	Mon 5/9/05	Project Management[13%]
14		1.1.7 Cost Estimate Review		1 hr	1 day	PM	Mon 5/9/05	Mon 5/9/05	Project Management[13%]
15		1.1.8 Resource Commitment		1 hr	1 day	PM	Mon 5/9/05	Mon 5/9/05	Project Management[13%]
16		1.1.9 Critial Path Analysis		1 hr	1 day	PM	Mon 5/9/05	Mon 5/9/05	Project Management[13%]
17		1.1.10 Finalize Project Schedule		2 hrs	1 day	PM	Mon 5/9/05	Mon 5/9/05	Project Management[25%]
18		1.1.11 Project Kickoff Meeting		2 hrs	1 day	PM	Tue 5/10/05	Tue 5/10/05	Project Management[25%]
19		1.1.12 Welcome Letter to Builder		1 hr	1 day	PM	Tue 5/10/05	Tue 5/10/05	Project Management[13%]
20		1.1.13 Welcome Letter to Client		1 hr	1 day	PM	Mon 5/9/05	Mon 5/9/05	Project Management[13%]
21		1.1.14 Install Project Binder Released		2 hrs	1 day	PM	Tue 5/10/05	Tue 5/10/05	Project Management[25%]

FIGURE 5–2 Milestones Shown in Microsoft Project

Site conditions affect many factors involved in defining a residential integration project. Let's look at a specific example of how site conditions may affect the scope and cost of the project. Read the case in point below.

A CASE IN POINT

Cable Vision

Barney and Betty Jones have decided to undertake the daunting task of remodeling their home of 20 years. As such, they felt this would be the ideal time to upgrade their home's wiring and have hired ABC Integrations to install a structured cabling system while the remodeling is in progress.

Project Manager Owen Ginizer has learned through a conversation with the homeowners that all the walls will be open and available for cabling the home while the remodeling continues. Given this information, Owen estimates that the time to install the cabling system will only be slightly more than that of a new home.

When Owen talks to the general contractor he learns that, although the walls will be opened, each room will be remodeled one at time. Thus, it will take many trips to cable each room according to the remodeling schedule. This new information now has an impact on the cabling system installation schedule. That's why it's essential to interview each stakeholder on the project—the architect, the builder, the homeowner, and the subcontractors.

Critical Thinking Question

In the residential integration industry, don't assume that no news is good news.

- What are several ways for you as a project manager to track the various aspects of the project? Be specific, keeping in mind your own organizational skills and habits.

Distance to the Site

A project manager needs to consider many questions before a project begins. How long is the commute to the project? Will overnight accommodations be needed for the installation crew? What is the cost and availability of overnight accommodations? Residential integration companies frequently work on homes in vacation areas. In these locations, hotel accommodations are very reasonable off-season and extremely high during the peak season. Seasonal times need to be annotated in the scope statement with additional charges for project work completed during high season.

The project manager and perhaps senior management also should consider how far away the company should routinely travel for work. Plotting territories on a map and providing a travel rate for each territory is helpful. Table 5–1 shows an example of how ABC Integrations charges for travel. The travel charge is represented as a percentage to be added to normal installation hours. Let's look at a specific example. If a project requires a total of 300 hours and is 98 miles from the shop, the customer will be charged for 300 hours for the labor and an

TABLE 5–1
ABC Integrations' Travel Policy

Distance from shop	*Travel Charge (% of labor charges)*	*Hotel & Per Diem Charges*
<30 miles	None	No
<60 miles	5%	No
<90 miles	10%	No
>90 miles	20%	Yes

additional 60 hours for travel time. That means ABC Integrations will charge the customer a total of 360 hours for the installation. In addition, the company will pass on any hotel and per diem charges incurred by the crew to the customer, as applicable.

ABC Integrations has developed a travel policy to help the project manager determine the various charges to include in the customer's estimate. In this way a project manager can provide the customer with more accurate pricing.

Study/Career Tip

Look through the Yellow Pages or go online using a search engine to find residential integration firms doing business in your area. Don't forget about other trade firms and retail stores who not only sell the system solution but do the installation as well. Consider contacting one or more firms or visiting showrooms to ask questions. Become the customer.

Critical Thinking Questions

- What are some advantages of taking on a residential integration project that is outside a firm's normal travel territory? It's more than just money, remember.
- What are potential disadvantages?
- Write down your answers to see which list is longer.

Site Map and Directions

Whether the project is planned for a home in a vacation area, a big city, or a country estate, a map to the **site location** and directions should be included with the project documentation. That means no exceptions—not even when members of the project team are familiar with the job location. New streets and construction change the best route to follow, and sometimes even street names change. And what if the installation job is at a new construction site or development? Most installers would have virtually no idea of street names, and there probably aren't any street signs up at that point either.

Including an accurate map and directions with the project documentation is just a good habit to follow. Think of the time, and thus money, lost if the installation team spends an extra 20 minutes per day trying to find the job site. Now imagine the cost to the company over a one-year period. Consider losing 20 minutes per day, five days per week, over roughly 40 weeks per year (subtracting holidays, sick days, vacation, training, and other non-installation days.) That is a total of 67 hours per installer per year, which at a billing rate of $60 per hour is $4,000 per installer, per year. Having the right information, such as directions, in the installer's hands saves precious time and helps to keep costs under control.

In addition, contact information at the job site should be included in the installer documentation. The installation team members need to know if there will be someone at the job site who they need to contact before they begin their work, and they need to know the names of the homebuyer, the builder, the architects, the electrical contractor, and other pertinent subcontractors, in case they are at the site at the same time. Installers should notify workers already at the site when they arrive. This is a common courtesy, and it helps to maintain a professional attitude at the job site. Phone numbers of pertinent people should also be part of installer documentation in case an installer needs to call someone to answer a question or to report a problem. It's just good practice to provide the above information to every project team member who might have a need to be at the job site.

Site Parking

Parking at the job site can become a major issue. The project manager cannot assume there is a place for worker vehicles on the property. For example, working in a major city creates a unique set of constraints on the project. Is parking available on the site? If not, does the city allow parking with a construction permit? How far is the closest parking garage? A project manager needs the answers to these and similar questions to help develop an efficient scope statement.

Bringing Tools and Supplies to the Site

A project's location directly affects what tools and supplies can be taken to the project, and if those tools can be stored at the site location. A location that requires a great deal of effort to bring tools to will require a job box or a closet for securing supplies and tools. For example, installers working on a high-rise apartment in the city probably will not want to carry their tools back and forth to the job site every day. On the other hand, if the job site is a single home with ample parking in the driveway, then tools and supplies can be stored in a truck and brought to the site each day. The project manager needs to gather specific information to determine how tools and supplies are to be handled and then communicate this information to the installation team.

> **Study/Career Tip**
>
> As the residential integration industry grows, so do the publications (both in print and online) that support it. Consider subscribing to one or more magazines that target this industry. Many subscriptions are complimentary for those in the field. Student discounts are also widely available. Don't just keep up with the glitzy new equipment, keep up with consumer trends as well.

Existing Construction Projects

In the case of remodeling an **existing construction,** i.e., a home, information about the state of the home is important. A residential integrator will need to know which walls will be left standing. Which walls will be exposed? Are there adequate chases in the home to install cables? The age of the home may also be a factor in determining the difficulty of the project.

Older homes, built before current building code standards were put in place, introduce an entirely separate set of project risks. A project manager must be aware of and consider the affect of asbestos insulation, lathe and plaster wall construction, unknown quality of power, and unknown framing techniques. Many of the materials currently in the home may no longer be available; that means items such as millwork can't be removed and replaced to install cabling. All of these issues should be well documented and passed along to all members of the project team.

House Layout

The **house layout** is essential information needed for the scope statement. Each floor of the home should be noted on the information gathering forms, and include ceiling height, room names, and dimensions (see Figure 5–1, Site Conditions Survey Form). The house layout can be obtained from an accurate set of blueprints or floor plans. If these plans are unavailable from the architect or general contractor, the project manager can visit the job site and draw a simple plan of the layout on graph paper. The house floor plan is essential in laying out locations of wallplates, speakers, and audio/video equipment within the home.

Utility areas, such as unfinished basements, attics, crawl spaces, and utility closets need to be marked on the floor plans as well. It is also important to note whether these areas have finished walls and ceilings. The project manager should consider these questions:

- Could each utility area be connected with conduit to allow for future system upgrades?
- Are there areas of the home that can't be reached from a utility area?
- If so, can access panels be provided to allow for future upgrades and system maintenance?

Let's look at an example where conduit and access panels are critical to the success of the project. See the case in point box on page 98.

Reality dictates that we can never really predict the future of technology, only anticipate what services might be forthcoming. To that

A CASE IN POINT

The XM Factor

Fred and Wilma Smith hire ABC Integrations to install a structured cabling system in their home, as well as a house-wide music system. During the sales process, Fred and Wilma have the impression that by installing the structured cabling system, the home will be ready for any technology that might be developed in the future.

Toward the end of the project, Fred and Wilma request that the music system include XM radio. XM radio requires its own special cable from the antenna to the location of the radio. This type of cable is not part of a standard structured cabling system. Since it was not cabled as part of the original system, additional cabling is required to get functionality for the XM radio. ABC Integrations previously installed both conduit and access panels throughout the home. That allows ABC Integrations to easily install the special cables for XM radio without major cost overruns. That now means the project manager can say yes to Fred and Wilma's request and continue to satisfy customer expectations.

end, the savvy residential integrations firm will recommend and install the proper cabling, conduit, and access panels now to accommodate those services when they become available.

Study/Career Tip

Create review materials to aid you in remembering the concepts and facts you've learned in this chapter. Make study checklists to use for standardizing your study routine, and mark off items as you master them.

Electronics Location

The electronics or **equipment location** is the place in the home where the integration and audio/video electronics are kept. A common practice in many areas of the country is to install a specially built closet for housing the distribution panels for television, telephone, networking, HVAC controls, security panels, music system, and lighting controls. This location needs to be selected to be consistent with the size requirements of the distribution panels. By laying out the distribution panels on paper first, an overall square footage of wall space requirements will be evident. Smaller systems may only need a single 4′ × 8′ sheet of plywood to mount the panels, and more complex home installation projects may require 10′, or even 20′ of wall space.

In California, Florida, North Carolina, and other southern states where most homes are built with no basement, an appropriate-sized closet can be constructed in a centralized area of the home. Sometimes this closet is behind or near a home theater. In some cases, the equipment must be split into separate utility areas. If so, it is important to consider whether the areas can be interconnected with conduit.

Unusual Site Circumstances

Every state and some cities have their own set of requirements for installing residential integration systems. Often these sites have peculiar or **unusual circumstances** that require special attention. Building methods and codes are always subject to change. In hurricane-prone areas, concrete block outer walls are standard; in colder environments, 2 × 6 wood-studded walls are preferred. A residential integration organization becomes accustomed to a set of standard building methods and requirements for their area. The project manager should note any items outside those standard methods and requirements when gathering information. This additional information is used in the scope document and in cost-estimating the project. Here are a few conditions a project manager should note in the scope statement.

- **Working hours.** What are the overtime laws in a location? What are the working hours set by the general contractor?
- **Bonding and insurance requirements.** What does a location and the general contractor require for bonding and insurance, such as worker's compensation?
- **Conduit requirements.** Does a location require all cables to be installed in conduit? How must conduits be terminated and interconnected?
- **Cabling by others.** If cables are already in place, what is the process that the local electrical inspector uses for approving existing versus new work?
- **Special permits.** Are special permits required for a location, and what is the cost of these permits? How difficult is it to secure such permits?
- **Power/no power on site.** Will power be present during the installation of the project?

Knowing the details of the site conditions affects the bottom line of the project. Here's a case in point.

A CASE IN POINT
Cha, Cha, Cha

ABC Integrations Project Manager Owen Ginizer bids on a project in Massachusetts. The project is to upgrade a lighting control panel for a ballroom of a busy hotel. Owen has never worked in a hotel environment before. He never thought to discuss his installation crew's working hours with the hotel managers. After accepting the project, Owen found out that his installation team was severely limited in when it was allowed to work on the project. Installers could only work from 2 AM to 6 AM when the ballroom was closed. The ballroom was booked solid for months and Owen had to install all the lighting upgrades when no one was using it. Owen's team had four hours in the middle of the night to install sophisticated lighting control upgrades in time for the ballroom to open the next morning for dance classes. Cha, Cha, Cha. That unknown site condition certainly complicated the project installation timeline. To finish the work within the period allowed, Owen had to use and pay additional installers. And that meant severe cost overruns and much smaller profits. Had he gathered site information on the availability of working hours, Owen could have avoided his dilemma by simply increasing the labor costs to accommodate the additional installation staff.

Customer Information

The residential integration industry prides itself on providing custom solutions. As we've seen, these systems are customized to the preference of the homeowner. It's the job of the sales representative to gather and record customer information during the interviewing process (see Figure 5–3) to ensure that complete customer satisfaction is realized. To that end, it's paramount that the proper **customer information** is captured and documented in order to build customized systems that exceed the customer's expectations. The following groups should be engaged in the interview process:

- Customer
- Family
- Extended Family and Houseguests
- Household Workers

The Customer

In this instance, we define the customer as the main stakeholder in the project. This person can be the husband, the wife, or a single person. It is the person who writes the check to pay for the project. Here's some important information that needs to be captured for use by the project manager and other members of the project team:

- **Contact information.** Include the job site address, the billing address, telephone numbers (cell, home, and business), fax number, e-mail address, IM address, and any other means of contacting the customer. This might also include the name and phone numbers of a personal assistant or a secretary. If someone other than the customer answers the phone, then knowing that person could help get you through to the customer in a timely manner.
- **Lifestyle information.** Capture the day-to-day life of the customer. This includes questions such as: What is your typical schedule on a workday, a weekend? How will the family be using the home—as a principle residence, as a vacation home, or for another purpose? Is the family active, sedentary? Do family members use the Internet? Does the couple entertain frequently? How much is the family at home? It's also a good idea to request the family's specific preferences for the system (colors and location of wallplates and touchscreens, temperature settings, lighting designs, music choices, and other system particulars). In this way the system(s) being installed will match the preferences and needs of the family.
- **Risk tolerance information.** Assess the customer's comfort level in handling risk. Determine if the customer can adapt to new, untested equipment and design. Does it meet a customer's need for new technology versus a reliable and proven design? If the customer is comfortable with some risk, then that customer may want the latest, greatest electronics and may be willing to trade some level of

CUSTOMER INTERVIEW

PROPOSAL NUMBER: ____________ DRIVING DISTANCE: ____________

PROPOSAL DUE BY: ____________ CUSTOMER NAME: ____________

SALES REPRESENTATIVE: ____________ DATE: ____________

CABLING DATE: ____________ MOVE-IN DATE ____________

PROJECT SCOPE: ____________

BILLING INFORMATION

COMPANY NAME: ____________ CONTACT NAME: ____________

STREET ADDRESS: ____________ ADDRESS 2: ____________

CITY: ____________ STATE: ____________ POSTAL: ____________

PHONE: ____________ FAX: ____________ CELL: ____________ EMAIL: ____________

PROJECT ADDRESS

COMPANY NAME: ____________ CONTACT NAME: ____________

STREET ADDRESS: ____________ ADDRESS 2: ____________

CITY: ____________ STATE: ____________ POSTAL: ____________

PHONE: ____________ FAX: ____________ CELL: ____________ EMAIL: ____________

GENERAL CONTRACTOR INFOMATION

COMPANY NAME: ____________ CONTACT NAME: ____________

STREET ADDRESS: ____________ ADDRESS 2: ____________

CITY: ____________ STATE: ____________ POSTAL: ____________

PHONE: ____________ FAX: ____________ CELL: ____________ EMAIL: ____________

ARCHITECT INFORMATION

COMPANY NAME: ____________ CONTACT NAME: ____________

STREET ADDRESS: ____________ ADDRESS 2: ____________

CITY: ____________ STATE: ____________ POSTAL: ____________

PHONE: ____________ FAX: ____________ CELL: ____________ EMAIL: ____________

(Company Name):: 123 Main Street, Anytown, MA 01234:: (999) 555-1212:: www.company.com Page 1 of 3

FIGURE 5–3 Customer Interview Form

dependability for the proverbial bells and whistles. In that case, the latest on the store shelves may be the right choice for this client. If a couple indicates they are somewhat conservative in their approach to technology, then a tried-and-true system is likely the one for them As such, the risk-tolerant couple might be comfortable with replacing in-wall dimmers and switches with color touchscreens with a graphical display interface to adjust the lighting throughout the home. Our conservative couple would probably be more comfortable with dimmers and switches that are simple to use and less "high tech." Regardless of which type of customer you are dealing with, a careful explanation of the pros and cons of the equipment to be installed is important. Remember, the client is looking to you, your project team, and your company as the experts who will advise them on making the right technology choices for them.

The Family

For the purposes of the interview process, the "family" includes all family members who live at the house and call it home on a daily basis. It should be noted that the family carries as much weight in the design process as the primary customer. The primary customer is defined as the main contact to the family; it is however the family that is purchasing and living with the system. The family is comprised of the spouse, the kids, and even the pets. While the project is likely designed primarily for the customer who pays the bills, the family represents a very essential part of any design decisions. If the customer has a high-risk tolerance, yet the family is very conservative, the system design will likely lean toward the conservative side. So, if we look at the previous example of lighting, the decision might be made to install dimmers and switches throughout the home so that the conservative members of the family can feel comfortable with this level of technology. So as to not forsake the needs of the primary customer, touchscreens also are located throughout the home so he can control the lighting.

Knowing the ages of the children helps determine what systems are best suited for the family. Teenagers want Internet and telephone access in their bedrooms. It's likely that parents will want to be able to monitor an infant's bedroom with audio and perhaps even video. All this information helps the project manager provide the appropriate system for the customer and the whole family.

Critical Thinking Questions

- Why do you think asking a series of open-ended questions to the customer and other family would be so valuable during this interview process?
- Should you ask probing questions?
- If so, what might be a good way to ensure you've asked all the right questions without offending a family member?

The Extended Family and Houseguests

It's helpful to know if there are frequent guests in the home and if they have particular or special needs—perhaps an elderly parent or a frequent guest with poor vision. A project manager should be knowledgeable in these areas and be able to recommend system changes to accommodate these individuals. Here's a case in point.

A CASE IN POINT

Let There Be (Too Much) Light

Project Manager Owen Ginizer installs a lighting control system in the guest suite of a 10,000-square-foot home. Next to the bed, he installs a 15-button keypad capable of controlling the lights in the suite—the bedroom, dressing area, patio, and bathroom. After the installation, Owen called the homeowner to see how everything is working, only to find that there had been a bit of a glitch. Guests who stayed in the suite slept with the lights on all night because they were afraid to touch the keypad. The couple, unfamiliar with digital integration, thought they might set off an alarm or something worse, so they chose to sleep with the lights on instead.

What seems simple to the integrator and even the family after training may not be so simple for visitors unfamiliar with the technology. It's important to explain these types of scenarios to your customer at the outset of the project so that situations such as the one described above don't happen. To accommodate that guest with poor vision, for example, recommending installing standard dimmers that push on and off would be in order. Special frequent visitor information and any special considerations the homeowner wants included in the integration system need to be recorded on the appropriate information gathering form. The sales representative should provide an explanation as to why the information is important to the customer.

Study/Career Tip

Spread out your study time over more days and shorter periods of study. Cramming for quizzes and exams and completing homework too quickly is not as effective as studying small chunks at a time more thoroughly.

Household Workers

The sales representative should gather information on the household staff as applicable. For example, a customer could have a house cleaning service whose workers arrive at the home every Wednesday at 10 AM and leave by 1 PM. The residential integrator can provide a keyless entry at the rear of the home by programming an entry code for the cleaning service that is valid only during those times. The same holds true for lawn/gardening services, pool maintenance people, and other specialized services. This provides both added security for the home and convenience for the workers.

Critical Thinking Questions

- Should the sales representative and the project manager conduct their site and customer visits together or separately?
- What are the advantages and disadvantages to this approach?

System Information

Every system that is offered to a customer requires information to determine the project deliverables. **System information** contains instructions about each system to be installed, a table indicating which rooms in the home get which products installed, a structured cabling installation document, and any other installation data. All this information is an essential part of the project scope statement.

System information can be broken into two sections: those that are standard systems regularly offered by the residential integration company, and non-standard systems offered by the company. We define non-standard systems as those systems that have not been installed, as of yet, by the residential integrator at any job site.

Standard Systems

Let's look more closely at a **standard system.** The Structured Cabling Survey displayed below (see Figure 5–4) shows an excerpt from an ABC Integrations survey form for the structured cabling system. This form indicates a number of products that are needed to build the structured cabling systems defined in the scope statement. Among these products are several wallplates. The wallplates become products that the customer could purchase to create a customized system for her home. Each wallplate offers its own set of features, such as color, Internet, cable television, telephone, and satellite television. Each of these features can be made available in any number of locations in the home. The Structured Cable Survey form uses a table to list the rooms of the home as rows and the wallplates as columns. In this way, the wallplate selections can easily be gathered. The distribution panels that drive the wallplates can be easily determined from the total numbers of each wallplate. Every part of the cable installation job needing to be completed is indicated on this form. This is all part of a standard system.

Another information-gathering form that might be used would be all of the details of the music system. A project manager can create a form that indicates every location that is to receive a speaker, as well as the speaker type and the amplifier that powers the speaker. All of the information-gathering forms are derived from the same process of first defining the system in a project charter, then developing a system design, a scope statement, and finally a form for gathering required information for the system.

STRUCTURED CABLING SELECTIONS

Location	Hanging Telephone	Telephone and Networking	Surround Sound	Music Input Output	Speaker Cabling or Wallplate
Kitchen					
Breakfast Rm					
Study					
Living Rm					
Dining Rm					
Family Rm					
Deck					
Garage					
Master Bdrm					
Master Bath					
Bedroom 2					
Bedroom 3					
Bedroom 4					
Theater Rm					
Game Rm					
Totals					

DISTRIBUTION PANEL SELECTIONS

Distribution Panel	Capacity		
Telephone and Networking	❑ 24	❑ 48	
Networking	❑ 4	❑ 8	❑ 16
Television	❑ 8	❑ 15	❑ 23
High Definition Satellite	❑ 8	❑ 16	❑ 24
Off Air HDTV and FM Reception	❑ Included		

(Company Name):: 123 Main Street, Anytown, MA 01234:: (999) 555-1212:: www.company.com

Page 1 of 3

FIGURE 5–4 Structured Cabling Survey

Non-Standard Systems

A **non-standard system** is one that has not gone through a formal process of introduction into the organization. For example, ABC Integrations has a preferred telephone system that is used in all projects that require a telephone system. However, if a single customer does not want that standard system, but instead is requesting a system that has never been installed by ABC Integrations, that would be considered a non-standard system by the company. After conducting research, the company may opt not to sell this system to other customers as part of the company's common offerings. This decision is made because non-standard systems represent the least amount of profit for a residential integrator as well as the largest amount of grief. It is common for companies to suffer from poor referral business, and even go out of business, when they frequently take on non-standards systems.

Because a different telephone system is considered non-standard, it requires more information than the standard system in order to create the scope statement and cost estimate for this particular job. Additional information that may be required might include the following:

- **System manuals** to include an owner's manual, installation manuals, and design manuals. This information is used to determine the parts required to provide the system and to offer customer guidance for use.
- **Technical support contact information** for the new system. This information will be helpful for the cost estimator in determining the required parts list and estimating the amount of time required for installation. This may also include consumer Help Desk information.
- **System configuration details** information is similar to what is required for the standard systems typically installed in that it includes what components are needed and where in the home they are to be installed. Looking at our non-standard telephone system, we would include the main processor and the telephones to be installed throughout the home.

Summary

In this chapter, we have discussed how to gather information for the project scope statement, the importance of gathering and recording site conditions, and the issues surrounding site location. We also reviewed the importance and relevance of gathering complete customer information to include all family members and other users of the home. Finally, we discussed standard and non-standard system information.

Important points in this chapter are:

- The three key areas in developing the project scope statement are information about the site, information about the customer, and information about the system to be installed.
- Gathering and recording site conditions is important because the condition of the site can affect the project.

- Issues surrounding the site location include distance to the job site, availability of parking, length of commute, and availability and cost of overnight accommodations. A project manager should consider how far away the company would travel for work.
- Gathering and recording the condition of an existing home and the house layout is important to do before beginning a project.
- Gathering and documenting all relevant customer information to include family and extended family members is paramount. Additional users of the home, such as outside workers, should also be considered.
- Each system offered requires information to determine project deliverables.
- Standard systems are those that are offered regularly by the company. Minute details must be collected to customize each system for each individual customer.
- Non-standard systems are those that the organization has never installed.

Key Terms

Customer information Information about the customer such as contact information, family members, and personal habits that help define the project.

Equipment location The location in the home where the integration equipment such as the television and telephone panels are located.

Existing construction A project that is carried out in an existing structure, such as in a home remodel.

House layout A general floor plan of the home, not to scale, designed to show an overall layout of the home.

Non-standard systems Systems that an organization has never installed.

Site conditions The conditions of the worksite.

Site location The physical location of the worksite.

Standard systems Systems that an organization regularly offers to customers.

System information Information about any one of the integrated electronic systems, such as options, placement, and quantities of products desired.

Unusual circumstances Any items that are not included on the information gathering forms, but are required for the purpose of defining the project.

Review Questions

1. What are the three primary areas where information needs to be gathered before a scope statement is written?
2. Why is it important for the project manager to have a house layout?
3. Why is it important to determine the technical know-how of the customer prior to any system installation?
4. Why should unusual circumstances be considered as they relate to site conditions?
5. Name four conditions that might be considered unusual circumstances.
6. Why is it important to ask the homeowner questions to capture the family's lifestyle?
7. What advantage is there to knowing if the residence has any household workers?
8. What's the difference between a standard and a non-standard system?

The Project Scope Statement

OBJECTIVES

After studying this chapter, you should be able to:

- Create a project scope statement using a pre-existing template.
- Explain the importance of a project scope statement.
- Discuss the internal versus external view of a project scope statement.
- Define the major components included in the project scope statement.

OUTLINE

Introduction

Now that we have gathered the important information about the job site, the customer, and the type of system to be installed, it's time to take that information and incorporate it into a standardized project scope statement.

The **project scope statement** explains the project details in easy-to-understand terms and leaves out the technology jargon. The statement does not attempt to impress customers with technical expertise; instead, it is clear and concise. The purpose of the scope statement is to let both the customer and the project team know exactly what is expected of both parties during the installation and when major events are scheduled to occur. In this chapter we will create a template-based project scope statement for our case study company, ABC Integrations, as well as review the major aspects that comprise a scope statement as defined by the book *A Guide to the Project Management Body of Knowledge* (*PMBOK Guide*).

Creating a Project Scope Statement Template

Creating a template-based project scope statement allows the project manager to steer the organization in a single direction because it contains standardized information. The scope statement for one house with a networking panel requirement will contain the same information in a scope statement for the next house with that same requirement. Because the project scope statement template is pre-filled with a considerable amount of information, the project manager can quickly plug in the information that is unique to a particular job. Throughout this chapter we'll review specific sections of this document as they pertain to the project requirements. A complete Project Scope Statement Template (PM05) has been included on the accompanying CD.

Study/Career Tip

Visit **www.projectmagazine.com** and type in "project scope statement" or "scope statement" into the search area. Here you will find a variety of articles about this topic.

Let's look at why a project manager might want to use a template-based scope statement. Consider the automobile industry. What if that industry asked every customer who visited a showroom how they would like the perfect car to be built. Perhaps it would have a Ford engine, a Chevrolet transmission, and seating from a Mercedes Benz. Now tell the customer: "Please come back in six months to pick up your new specially-built car that will cost $230,000." Of course this sounds crazy, but that is exactly what happens in many residential integration organizations. Often, companies insist on reinventing the wheel with every customer. Automotive manufacturers build high quality products that meet the needs of their customers by designing to a demographic (a specific group of the population), rather than a specific customer. For example, sports cars are designed for the young and "young at heart," and minivans cater to families. The manufacturers build several models across a range, and as their manufacturing company grows, the number of models offered generally increases.

Now, let's consider the structured cabling system in a similar fashion; we want to construct a unified system that can be placed in every home, but one that allows for customization, such as the wallplate color, quantity and location of wallplates, and the addition of distribution panels to manage various areas of technology. Every home owner seeking to have Internet access throughout the home will have the same patch panel, router, and hub as every other home with this technology.

Standardizing network panels and other items is one way to ensure steady profits. With a template in hand, a project manager can quickly prepare an entire scope statement in a short period of time.

Critical Thinking Question

- Aside from saving time, what other advantages might there be to using a project scope statement template?

Be specific.

It should be noted that the entire scope statement combined with the cost estimate is the contract with the customer. Its purpose is to create an understanding among all parties as to the overall deliverables of the project and its terms. It is a good practice to have the project scope statement template reviewed by the corporate attorney to ensure it complies with local and state laws.

Internal Versus External View of the Project Scope Statement

As a rule, the overwhelming majority of the content of the scope statement is generated prior to meeting with the customer. However, adding the details learned during the information-gathering stage presents a customized view to the customer. It looks to the customer (external view) that the company is providing a highly customized product to meet their specific project requirements. On the other hand, from an internal (company) view, the systems are very much the same as others it has installed.

Let's take a look at our case study. ABC Integrations provides three quality levels of architectural speakers as part of the music system. In addition, each of these levels is available in ceiling mount, wall mount, bookshelf, and outdoor varieties. Additionally, a name brand and a generic brand are available for each level and type. That means 24 architectural speakers are available to draw on for a custom speaker package for the customer. However, how these speakers are purchased, cabled, installed, and accounted for is identical from installation to installation. Again, the use of a project scope statement template streamlines the process for the project manager and the customer, yet provides for many individual job differences.

Study/Career Tip

Visit a search engine of your choice and type in the word "template." There are a myriad of Web sites that offer a wide variety of templates. Peruse those that interest you, and consider creating a template for your own use. For example, if you do not have a current resumé or standard cover letter in your possession, this is an effective way to create those very important documents.

Aspects of a Project Scope Statement

The scope statement contains many aspects that must be addressed within the statement. This is true for scope statements that are unique for each customer or for template-based scope statements. These aspects provide the key information concerning a project. *A Guide to the Project Management Body of Knowledge* identifies the following aspects as those that are contained within each and every scope statement:

- Project objectives
- Product scope definition
- Project requirements
- Project boundaries
- Project deliverables
- Product acceptance criteria
- Project constraints
- Project assumptions
- Initial project organization
- Initial defined risks
- Schedule milestones
- Fund limitation
- Project configuration management requirements
- Project specifications
- Approval requirements

Let's take a closer look at each of these key components of a project scope statement.

Project Objectives

Project objectives are the measurable success criteria of the project. The objectives must be measurable—just as the chapter objectives are in this textbook. They can be measured in terms of time, cost, schedule, or quality. For example, a project objective could be to stay within the total contract value of $100,000. If the total project cost is $101,000, then the project objective was not met, and the organization could be stuck for the extra $1,000. If the total project cost is $99,900, then the project objective has been met with a few dollars left over. Another project objective could be to complete the project by a certain date. If the project is completed before that date, the objective has been met successfully. If the project is not completed by the objective date, naturally, the objective is not met.

Project objectives also provide a measurement tool to determine the success rate of the completed project. This information can then be used to improve the success of the next project. The quality of workmanship and equipment that is provided can also be presented as

Project Management Plan
Project Scope Statement PM05
Structured Cabling Section

Our Certified Installers will meet every requirement of the EIA/TIA 570A Residential Wiring Specifications.

FIGURE 6–1 Project Management Plan Project Scope Statement PM05 Structured Cabling Section

objectives. They can be measured by the way cabinetry looks, the way a CD sounds, or the way a DVD appears on a screen. Cabinetries with splintered edges, CDs with flat sounds, or DVDs that are washed out on a screen are all objectives that did not meet the quality standards of the company. On the other hand, finely crafted cabinets, robust sound, and crystal clear TV pictures meet quality objectives and customer expectations.

Critical Thinking Questions

- Why do you think objectives, whether for a textbook or a project scope statement, need to be measurable?
- What "action verbs" can you think of that could be used to measure whether or not an objective has been met?

Study/Career Tip

Objectives not only need to be measurable, they need to be observable as well. Observable means that someone can visibly see evidence that a user or the business achieved a specific outcome. Note that terms like "know," "understand," "appreciate," and "feel" are not observable. Steer clear of these terms when writing project objectives.

As an example, the sample project objective in Figure 6–1 specifies what wiring criteria installers will use during the installation process at the customer's home. They could be installing structured cabling, distribution panels, or wallplates. The objective states that *all* installations will meet a known standard. The project scope statement lets the customer know that the project will comply with a standard set by an industry-wide trade association, and that this standard is a key project objective.

Product Scope Definition

The **product scope definition** describes the characteristics of the product and services that are offered in the project scope statement. Defining the company products and services makes it clear from the outset what the customer can and will expect. And it helps eliminate possible confusion and conflicts over such products and services.

Both the company and its employees and the customer have a common document to which they can refer to answer disputed items. These product scope definitions are typically called for by sales, and then thoughtfully created by the designer. For example, the sales representative will call for an XM radio as part of the music system, and the designer will make the final determination as to the specific product and what features will be offered as part of the overall system.

Each product scope definition becomes a small part of the overall project scope statement; it defines a specific product. A product scope definition is important because it makes clear to the customer exactly what is included in each product. That helps the customer to understand the totality of each product and is one way to justify the cost of the entire project. Customers sometimes do not understand why a piece of electronic equipment can be purchased from a discount store for much less than for what a residential integration company sells its products.

Critical Thinking Question

- What might the consequences be if there were no product scope definition? There are several.

List as many as you can, keeping in mind the consequences to both the company and the customer.

The product scope definition in Figure 6–2 describes the telephone and networking wallplate that is to be installed and the parameters for this product. This includes both materials and the labor needed to install the material; however, what is shown in the product scope definition is the functionality of the product, not the pieces (the labor and materials). In the custom install industry, a product is not only the piece of equipment purchased by your company; the product is the total assembly needed, plus the installation that a residential integration company sells. This means the product consists of a number resources, both materials (connectors, cables, terminations, wallplates) and labor (the length of time it takes an installer to complete the installation and connection to the integrated system). As noted above, the finished product for the telephone and networking wallplate includes four telephone lines, a single home networking connection, and the wallplate to which the

Project Management Plan
Project Scope Statement PM05
Structured Cabling Section

Telephone and Networking Wallplate

The telephone and networking wallplate allows connections to four (4) telephone lines and a home network.

FIGURE 6–2 Project Management Plan Project Scope Statement PM05 Structured Cabling Section: Telephone & Networking Wallplate

Project Management Plan
Project Scope Statement PM05
Structured Cabling Selection

STRUCTURED CABLING SELECTIONS

Location	Hanging Telephone	Telephone and Networking	Multimedia	Surround Sound	Music Input/Output	Speaker Cabling or Wallplate
Kitchen		1	1		1	1
Breakfast Room						1
Study		1	1		1	1
Living Room						1
Dining Room						1
Family Room		1		1		
Deck						1
Garage		1				
Master Bedroom		1	1		1	1
Master Bath	1					1
Bedroom 2			1			
Bedroom 3			1			
Bedroom 4			1			
Theater Room		1		1		
Game Room		1	1			1
Totals	1	7	7	7	3	9

FIGURE 6–3 Project Management Plan Project Scope Statement PM05 Structured Cabling Section: Structured Cabling Selections

cables connect. It also includes the installer's work required to make the product functional. When a customer and her family are able to use the four phone lines and the home networking system seamlessly, then the product has met the criteria established for a successful product.

Project Requirements

Project requirements describe the conditions or capabilities that must be met or possessed by the deliverables of the project. In Figure 6–3, each of the products for the structured cabling selection is indicated at the top of the column, and the quantity of each product is indicated for each room in the home. By clearly defining each product (see the Project Scope Statement Template (PM05) in Appendix A) and by providing a location and quantity for each, the project manager can keep track of and easily manage the many project requirements, which usually involve numerous tasks.

In Figure 6–3 a project requirement is shown as a hanging telephone in the master bathroom. The sales representative collects project requirement information from the customer, places it in the appropriate table, and gives that information to the project manager. Using the table, the project manager can quickly determine the project requirements.

Study/Career Tip

Pay careful attention to graphic information. Graphics in the textbook present information in a different way and may also present other information that is not in the text. Draw graphics in your notes. They can often summarize the information effectively.

Project Boundaries

Project boundaries identify what is included and what is not included in the project. Specifying the boundaries helps avoid conflicts between your company and the general contractor or other subcontractors. The boundaries are set by the project manager in consultation with the project team before any of the work begins. Figure 6–4 shows an excerpt of the project scope statement that outlines specifically what

Project Management Plan
Project Scope Statement PM05
Project Boundaries Section

Residential Integrator's Responsibilities:

- We will cable, install, terminate, and finish all telephone, TV, satellite, phone, computer, and music systems, and home theater items as contracted.
- We will maintain Worker's Compensation throughout the complete project.
- We will acquire all necessary permits and observe all national and local building codes.
- We will provide specifications of all contracted equipment installed with built-in cabinetry.
- We will install all finished equipment after the customer has taken possession of the home. This may take up to 60 days depending on the complexity of the system.

What We Need Your Contractor to Do

- Provide various electrical outlets where needed for auxiliary devices throughout the home. These locations are often needed for televisions, computer peripherals, telephone chargers, and other electronic devices. These locations will be laid out during the rough electrical walk through or as specified on the plans.
- Provide two dedicated 20 Amp services near the main panel (usually in the basement or other unobtrusive location) for telephone, television, music, and networking.
- Provide conduit to the street for telephone and CATV services.
- Provide conduit that can be used for exterior speakers or connections to exterior buildings.
- Provide a wood panel to accommodate the low-voltage equipment (usually in the basement or other unobtrusive location) and enough floor area for the equipment.

FIGURE 6–4 Project Management Plan Project Scope Statement PM05 Project Boundaries Section

your company will provide and complete, and what the general contractor is required to provide. This information is contained within the project scope statement, but keep in mind that changes at the worksite, such as a **retrofit** installation, will require changes in the project boundaries from project to project.

In our example, ABC Integrations states their project responsibilities and the responsibilities of the general contractor. For example, ABC Integrations stipulates that television locations will require an AC outlet and that the installation of the AC outlet is the responsibility of the general contractor. It makes sense that a low-voltage contractor is not installing electrical outlets, however, by putting this detail in print it prevents any miscommunication along the project life cycle.

Project Deliverables

Project deliverables are the outputs that comprise the products or services of the project. Project deliverables include all the items and services defined in the product scope definitions. Project deliverables also include ancillary or non-technical results, such as accounting reports or written how-to instructions, project reports, and documentation. These deliverables are typically developed by the executive team and comprise the soft deliverables of the project, which differ from the electronics and other hard deliverables. Figure 6–5 depicts what is known as "Continued Service" that details how long ABC Integrations will commit to delivering service after the completion of the system installation, in this case, 90 days. This statement allows everyone associated with the project to quickly understand the after-service deliverables, again avoiding possible future conflicts.

Project Management Plan
Project Scope Statement PM05
Continued Service Section

Continued Service

The Company Work: The Company warrants the design, integrity, and installation of its work for a period of ninety (90) days from the date of notice of substantial completion of the electronic system.

Electronics: The electronics installed by the Company are subject to independent manufacturers' warranties. The Company warrants service on said finished electronics for a period of ninety (90) days from the date of notice of substantial completion of the electronic system.

All warranty periods commence on the date of the notice of substantial completion, regardless of the existence of punch-list times.

FIGURE 6–5 Project Management Plan Project Scope Statement PM05 Continued Service Section: Continued Service

Project Management Plan
Project Scope Statement PM05
Closing Section

Product Acceptance

When the project is completed, we will accompany you on a walkthrough of your home to complete a review of the system. We will compare the finished installation with the project scope statement and any signed change orders. We will explain each item that has been installed and reference products on the project scope statement. Upon acceptance, you will be requested to provide the final payment and sign a Completion of Work letter.

FIGURE 6–6 Project Management Plan Project Scope Statement PM05 Closing Section: Product Acceptance

Product Acceptance Criteria

One of the most important conversations you will ever have as a project manager with the customer is that of product signoff. Product **acceptance criteria** defines the process that the company uses to gain customer acceptance of the products upon completion of installation. The acceptance process is explained in the scope statement template and should be shared with the customer. Let's take a closer look by reviewing Figure 6–6. The acceptance process is an integral part of the finished product. It lets the customer know the product is ready for use, and it provides an opportunity for the customer to raise any issues he may have regarding his new system.

In the residential integration industry, the lines of completion for a project can sometimes be blurred. Describing a process for project acceptance makes those lines clear. A customer who purchases a stereo from a discount store, for example, walks out of the store with the entire package. That clear. But that's not the case in the residential integration industry. There are many more facets to installing an integrated system,

A CASE IN POINT
Music to Their Ears

Our good customers Fred and Wilma Smith have a contract for a house-wide music system. ABC Integrations installs the system, except for a piece of speaker material (cloth) that they requested near the end of the project. A change order is written and signed. Later the manufacturer of the speaker material notifies ABC Integrations that the material won't be available for three months. That means the project will be substantially completed on time except for that one item that is priced at $289. The final payment for the system is $5,000. That money is needed to meet payroll and other expenses incurred by the company. Project Manager Owen Ginizer needs to secure the final payment from Fred and Wilma. How does Owen meet this challenge? Owen subtracts $289 from the contract price, agrees to send an installer to install the special speaker material when it arrives, and presents the final payment request to Fred and Wilma. The couple makes the payment and signs the Completion of Work document.

Project Management Plan
Project Scope Statement PM05
Project Delays Section

Overtime Charges

In the event of project delays caused by events or actions outside of our control, we may pass on overtime labor fees. These events include limited site access due to actions of the property owner or contractors performing tasks on the project independent of our work. We will notify you of any issues that arise that might contribute to overtime labor fees.

FIGURE 6–7 Project Management Plan Project Scope Statement PM05 Project Delays Section: Overtime Charges

and a customer may think every facet must be completed before the project is substantially complete. Let's consider the case in point on page 118.

This type of situation often occurs throughout a project, and the project manager is responsible for meeting these challenges and resolving them in a manner that continues to meet customer expectations.

Project Constraints

Project constraints describe specific constraints on a project that limit the installation team's options for completing the required work. For example, a homeowner may have inadvertently locked the door to the worksite (his home), forcing the installation team to return on a Saturday in order to finish its work prior to another subcontractor's completion of his work. That means the installation team would accrue several hours of pay at the higher Saturday rate. This cost would be passed onto the customer (see Figure 6–7).

Typically, project constraints are included in the contract provisions and cover additional costs for changes, working hours, union requirements, permits and fees, and any scheduling issues. Delineating the constraints helps to eliminate disputes and misunderstandings of events that might occur, which could result in additional costs to the customer.

By including this information within the scope statement, and explaining the practice to the customer, adding the extra cost is much more readily accepted. Without this information a customer can argue that the company is responsible for the added cost.

Critical Thinking Questions

- What might be some additional project constraints that aren't listed here?
- Do you think the project scope statement should list every possible constraint?
- Why or why not?
- Is that even feasible?

Project Management Plan
Project Scope Statement PM05
Retrofit Cabling Section

Retrofit Cabling

Your project is in an existing home, and therefore it is difficult to estimate the time required to install the cabling in your home. We have provided for 40 hours of cabling labor. Additional hours may be required; however, we will discuss additional charges with you prior to starting the work.

Please note that although we try our best not to puncture holes in the walls in your home, sometimes it is necessary depending on the need for access to selected locations. The patchwork that may be needed to restore the wall surface is *not* included in this scope statement.

FIGURE 6–8 Project Management Plan Project Scope Statement PM05 Retrofit Cabling Section: Retrofit Cabling

Project Assumptions

Project assumptions describe specific assumptions made within the project scope statement and the potential impact of these assumptions, if they prove to be false. If the project at hand is in an existing home, often called a retrofit project, the project manager must add verbiage to the project scope statement that specifies the assumed construction of the home and how the project will change if the cabling cannot be installed as planned (see Figure 6–8). One assumption might be that existing cabling in the home will pass a specific test, and another assumption might be that existing speakers will provide adequate sound. Without stating the assumptions for a particular project at the outset, there is a high risk of misunderstandings and disputes between the homeowner and the company.

Let's see what happens for our case study company. ABC Integrations has allotted 40 hours of time for the cabling of a customer's home. In addition, the project scope statement states that any extra hours are to be billed at a rate of $60 per hour, and that the customer will be informed if additional hours for the cabling are required. Once the cabling of the home is complete, the remainder of the project (electronic equipment installation and programming) will continue with the same hourly requirements as if the home were a new construction project.

Initial Project Organization

If you've ever been around a home construction project or know someone who has gone through that process, you realize that it's important to know with whom you will be interacting during the course of the work. A project scope statement should include an **initial project organization** list that identifies members of the project team as a way of introducing the key players who will be involved with the project. A short paragraph that describes each project team member's role

Project Management Plan
Project Scope Statement PM05
Your Project Team Section

Your Project Team

Sal Moore is your Sales Representative. He will ensure your project requirements are met, and will act as your agent throughout the process. At the appropriate time, he will collect your preferences for color, programming, speaker selection, and handle any custom requests along the way to a successful installation project.
Cell: (999) 555-1212
Email: sal.moore@ABCIntegrations.com

Maggie Pi is your Sales Engineer. She will design the nuts and bolts of your project, ensuring that each product and service has been thoroughly engineered for serviceability, repeatability, and reliability.
Cell: (999) 555-1212
Email: maggie.pi@ABCIntegrations.com

Owen Ginizer is your Project Manager. He is responsible for all actives and deliverables on our project. He will work closely with your architect, general contractor, and other trades to ensure a successful project.
Cell: (999) 555-1212
Email: owen.ginizer@ABCIntegrations.com

Joe Trade is your cabling, termination, and move-in Team Leader. He is responsible for site work during the construction of the project.
Cell: (999) 555-1212
Email: joe.trade@ABCIntegrations.com

Steve Smith is your Service Team Manager. He will work with you at the completion of the project to help you understand all aspects of the system, and provide you with continued service after the project completion.
Cell: (999) 555-1212
Email: steve.smith@ABCIntegrations.com

FIGURE 6–9 Project Management Plan Project Scope Statement PM05 Your Project Team Section: Your Project Team

and what they will be contributing to the project should be included (see Figure 6–9).

Customers are more likely to feel comfortable about the quality of the project when they know the names of the people involved. The initial project organization list can be included in the project scope statement, or the sales representative can present the list to the customer once a contract is signed and the team is ready to move forward. The advantage of introducing the team to the customer is that a more personal contact is provided with each team member. Instead of a group of anonymous people wandering around a customer's home, the team becomes a congenial bridge between the company and the customer. The example in Figure 6–9 describes a project's sales representative and several others on the project team. Other descriptions could include the project manager, lead installer, lead technician, programmer, and any other worker who is expected to be in the customer's home.

Initial Defined Risks

Initial defined risks identify known project risks. In Figure 6–8 we describe an assumption, in this case, the number of hours required for cabling. We also inform the customer of possible cost overages due to the risk of unknown framing in the home. In new construction, a common initial defined risk is the installation of the electrical outlets. If improperly installed, these outlets may cause a humming noise from the speakers. By defining these potential occurrences, the customer is forewarned and thus is less likely to be surprised if a risk becomes reality.

Schedule Milestones

As with any project, it's important to identify the major events that will occur in a residential integration installation. The project scope statement should include an area where **schedule milestones** can be listed. Schedule milestones identify all imposed schedule constraints for each phase of the project. Figure 6–10 comes from our structured cabling project scope statement. It provides estimated start and completion dates for each phase of the project. These dates all have dependants; in the case of the termination phase, all drywall must be installed before termination takes place. If the drywall is not installed until a later date, then the milestone dates will need to be updated and forwarded to all stakeholders. It should be noted that the milestones listed in Figure 6–10 are for our case study company ABC Integrations, and their names may change in other companies within the industry. In many companies the

Project Management Plan
Project Scope Statement PM05
Scheduled Milestones Selection

Schedule Milestones

The following dates are an estimate based upon current project information. These dates could change because of project delays out of our control, such as delays in the schedules of other trades, and limited site access.

Milestone	Start Date	Completion Date	Dependants
Cabling Phase	3/1/05	3/8/05	All HVAC, electrical, and plumbing rough-in work completed
Terminate Phase	4/15/05	4/22/05	Drywall installed
Move-In Phase	5/15/05	5/22/05	Interior painting complete
Electronics Build	5/1/05	6/1/05	System preferency worksheets
Electronics Installation	6/1/05	6/8/05	Certificate of Occupancy

FIGURE 6–10 Project Management Plan—Project Scope Statement PM05 Scheduled Milestones Section: Schedule Milestones

cabling is referred to as rough-in or prewire, the termination and move-in are combined and called trim-out, and the remaining phases are called finish. These are currently the de facto standard terms used; however, in the authors' opinions, they do not accurately describe the actual milestones in a residential integration project.

Project Configuration Management Requirements

Project configuration management requirements describe administrative constraints placed on the project. These constraints include items such as the payment schedule, change order policy, and the termination of the contract. In our example (see Figure 6–11), ABC Integrations provides a payment schedule for the project. An important detail in the payment schedule is the notice of substantial completion, which allows ABC Integrations to collect the final payment at the time the system is, to a large extent, complete (see also Project Constraints). In many cases there are a few punch-list items at the end of a project that may exist after the project is essentially complete. An example of this might be a delay in the delivery of custom colored wallplates or covers for speakers. Although the system is fully functional, the custom color plates will not arrive for several weeks. ABC Integrations needs to be paid upon completion of the work, and will then return to install the wallplates when they arrive from the manufacturer. This allows ABC

Project Management Plan
Project Scope Statement PM05
Payment Section

Payment Schedule

The Original Contract Value shall be $100,000. Payment of the Original Contract value shall be made as follows:*

20%, $20,000.00 upon execution of this agreement for deposit and design fee;

20%, $20,000.00 due 15 days prior to commencement of installation of cable;

30%, $30,000.00 due 15 day prior to planned termination of cable;

20%, $20,000.00 due 30 days prior to planned installation of finished electronics, and

10% due upon notice of substantial completion of system.**

*The addition of any mutually agreed upon Change Orders that affect the Original Contract Value would require a revision of this schedule.

**If punch-list items (minor work that does not materially affect the operation of the electronic system) exist, the Customer may withhold 2% of the Original Contract Value from the final Payment pending completion of the punch-list items to the satisfaction of the Customer.

FIGURE 6–11 Project Management Plan—Project Scope Statement PM05 Payment Section: Payment Schedule

Integrations to keep a reasonable level of cash flow and it also triggers the start of the service period.

A clearly defined payment schedule, again, lets the customer know what to expect and allows the customer to plan his finances accordingly. The project manager should also complete the process of delineating the other administrative constraints in the scope statement. It should be noted that if the customer is unhappy with any of the constraints (company policies), they can be addressed before work begins, rather than in the middle of a project. The more the project manager resolves before the start of the project, the easier it will be to meet the customer's final expectations.

Study/Career Tip

Consider studying the key terms at the end of each chapter before reading the text. Becoming familiar with these terms at the outset will help you to better understand the chapter material and aid you in memorizing these terms as a part of your residential integration vocabulary.

Project Specifications

Project specifications reference any external documents, such as floor plans, and other third-party specifications that affect the project (see Figure 6–12). If a project manager is working with a cabinetmaker to provide a separate cabinet for the equipment installation, then the project scope statement must include a reference to the documents provided by the cabinetmaker for the cabinet. The project scope statement also indicates that the cabinetmaker must receive approval for the equipment designs prior to building the cabinets. The scope statement states who requests the approval and how that approval is forwarded to the cabinetmaker. It also states who is authorized to approve the plans and how that person indicates her approval. It should be noted that the homeowner, the architect, the builder, or other person named to that responsibility can grant approval.

Project Management Plan
Project Scope Statement PM05
Project Documents

Project Documents

ABC Integrations requires the following documents prior to the start of work on the project. In addition, ABC Integrations requires all revised versions within seven days of their release.

- Electrical Plans
- Cabinetry Plans

FIGURE 6–12 Project Management Plan—Project Scope Statement PM05 Project Documents: Project Documents

In the case of our sample structured cabling system, the project scope document requests two external documents, the electrical plans and the cabinetry plans (see Figure 6–12). Also, if the project is small in scope, the need for low-voltage floor plans or elevation plans for a satellite dish will be minimal. However, in managing larger projects, it is likely that the project manager will need to conduct a site visit with the architect to determine the location for a satellite dish, wallplates, and the distribution panels. Any documentation that the architect generates in connection with the project should be referenced in the project scope statement.

Common sense dictates that the project manager needs this information to plan the low-voltage system without interfering with additional systems installed by others, and to know where electrical outlets are planned to efficiently plan a low-voltage system. A project manager will also need the electrical information to request additional outlets in locations where they may be needed for the low-voltage system.

Approval Requirements

Approval requirements are those that can be applied to project objectives, deliverables, documentation, and work. By signing the project scope statement, the customer gives approval for the project (see Figure 6–13). In addition, the project manager can outline what approvals are required prior to any changes in the project scope statement. For example, in our structured cabling system, what happens when a customer requests a change? Upon the completion of the cabling phase, the customer might decide to add another wallplate. Since all the cabling work has been completed, it would require considerably more time to install this one wallplate than it did to install each of the other wallplates. The cabling team will need to return to the home, set up, run the cables, and break down the job. The added cost to complete this request would need to be presented to the customer for approval in the form of a change order prior to this work being started.

An example of a change not requiring approval would be the movement of a wallplate. In that case, when the cabling phase is completed a customer could decide that one of the wallplates needs to be

Project Management Plan
Project Scope Statement PM05
Project Changes Section

Project Changes

Prior to the beginning of each phase, you (the customer) will be given an opportunity to make changes to the project. We will give you a change order for your approval prior to the start of any project changes.

FIGURE 6–13 Project Management Plan—Project Scope Statement PM05 Project Changes Section: Project Changes

moved down the wall by only a couple of feet. Since the cable wouldn't need to be extended, (and let's assume the cabling team is still on site), the extra time required is minimal and the change could happen without the need for extra charges and a formal change order.

Summary

In this chapter, we discussed creating a template-based project scope statement. We also defined the major components included in all scope statements, whether template-based or written totally from the beginning for each customer, noting that a template-based scope statement is the preferred industry method.

Important points in this chapter are:

- The project scope statement explains the project details in language that is easy for the customer to understand and leaves out the technology jargon.
- It is important to distinguish between the internal (company) and external (customer) view of the project scope statement.
- The project scope statement consists of key information concerning the project. These elements are the project objectives, product scope definition, project requirements, boundaries, deliverables, acceptance criteria, constraints, assumptions, initial project organization, initial risks, schedule milestones, fund limitation, project configuration management requirements, specifications, and approval requirements.

Key Terms

Acceptance criteria The conditions that a company sets before beginning a project to gain the customer's acceptance of the products upon completion of installation.

Approval requirements Conditions established in the scope statement that are applied to project objectives, deliverables, documentation, and work that when signed by the customer gives approval for the project. In addition, the conditions required prior to any changes in the project scope.

Assumptions Physical conditions in a project that are accepted or assumed without proof. Within the project scope statement, project assumptions describe the potential impact of such assumptions, if they prove to be false.

Boundaries Conditions that are set before the work begins to identify what is included and what is not included in the project.

Deliverables The products or services defined by the product scope definitions and provided to the customer, also including ancillary or non-technical results, such as accounting reports or written how-to instructions.

Initial project organization The individual members of the teams who work on a project.

Initial defined risks Known project risks.

Product scope definition A definition that explains the details and parameters of a product and is included in the scope statement.

Project assumptions *See* assumptions.

Project boundaries *See* boundaries.

Project configuration management requirements The administrative constraints placed on the project, including but not limited to, payment schedule, change order policy, and termination of the contract.

Project constraints Specific conditions that limit the installation team's options on a project.

Project deliverables *See* deliverables.

Project objectives A project's success criteria that can be measured in terms of time, cost, schedule, or quality.

Project requirements The conditions or capabilities that must be met or possessed by the deliverables of the project.

Project scope statement The narrative description of the project scope, including major deliverables, project objectives, project assumptions, project constraints, and a statement of work, that provides a documented basis for making future project decisions and for confirming or developing a common understanding of the project scope among the stakeholders. The definition of the project scope—what needs to be accomplished.

Project specifications Any external documents, such as floor plans, and other third party specifications that affect the project.

Retrofit A project in which a device or system is installed for use in or on an existing structure.

Schedule milestone A significant event in the project schedule, such as an event restraining future work or marking the completion of a major deliverable. A schedule milestone has zero duration. Sometimes called a milestone activity.

Review Questions

1. What is the purpose of a project scope statement?
2. Why is a project scope statement template so valuable to a project manager?
3. What are some ways project objectives can be measured?
4. Why is the project scope definition important?
5. Why is it important to identify project boundaries at the outset?
6. What constitutes "project deliverables"?
7. Name some typical project constraints.
8. What is the advantage of creating an initial project organization list?
9. What are approval requirements?

The Work Breakdown Structure

OBJECTIVES

After studying this chapter, you should be able to:

- Define the Work Breakdown Structure (WBS).
- Explain formatting approaches to the Work Breakdown Structure.
- Discuss the similarities and differences between the bottom-up and the top-down methods of creating a Work Breakdown Structure.
- Define the Work Breakdown Structure dictionary.
- Discuss criteria for creating a good quality Work Breakdown Structure.

OUTLINE

Introduction

As we learned in Chapter 6, a project scope statement is an overarching document that guides the project manager, project team members, and the customer in the framework of the entire project, and outlines significant aspects of what may impact a successful custom installation.

With that in mind, it's time now to turn our attention to another critical document that is used by the organization to further define specific features of the project so that it can be properly planned and executed. In this chapter, we examine this key informational piece that is the cornerstone of successful project management.

Defining the Work Breakdown Structure

The **Work Breakdown Structure,** generally referred to as the WBS, is a foundation document for the project (see document PM08 in Appendix A). It provides the basis for planning, estimating, and managing the project. Like the project scope statement, the WBS defines the total scope of the project, but it also breaks it down into smaller, more manageable pieces of work through the use of techniques called **decomposition, verification,** and **work packages.** Each descending piece of the WBS represents an increasingly detailed definition of the project work. With this structure in place, a project manager can more easily schedule, cost estimate, monitor, and control these smaller pieces of work. We will be reviewing these techniques in greater detail later in this chapter.

The project manager's ability to devise a complete and detailed WBS through the use of templates not only makes his job easier, it also benefits the company by helping to manage costs and keep the project on schedule. This ultimately benefits the customer, who ends up with a smoothly run project that is delivered on time with few, if any, issues. And a good WBS also allows the project manager to manage many projects at the same time—a boon to any company.

Study/Career Tip

Think of a WBS as a very sophisticated "To Do" List. Although much more granular and detailed in its approach than a working list you might create at home, the principle is the same in that it identifies everything that needs to be completed.

Critical Thinking Question

- Why do you think it would be important to use templates when creating a WBS? List at least three reasons.

Format of the Work Breakdown Structure

The WBS can be represented in one of two ways. The first way is a flowchart that is similar to an organizational chart with the WBS organized around products, phases, or a process. This type of format is popular with many project managers because it presents a visual image of the project and shows how the work packages roll into the next level of assembly. Figure 7–1 shows a flowchart of a residential integration WBS for the Smith residence. In this flowchart, the top level represents the entire project scope statement, the second level details each of the systems to be installed in the home, and the third level is the work package with each of its associated products and services.

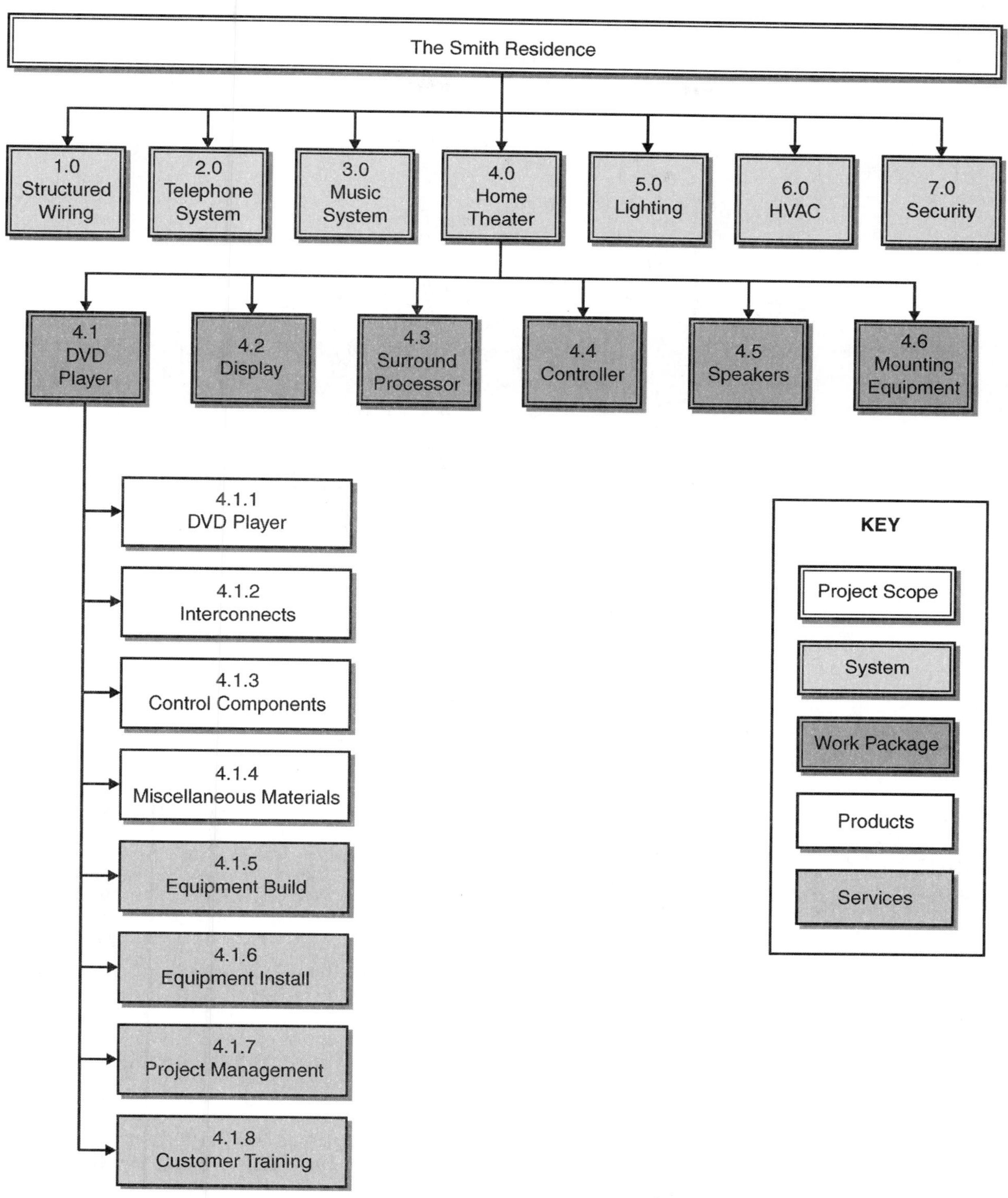

FIGURE 7–1 DVD Player WBS (Flowchart format)

The Smith Residence
1.0 Structured Wiring
...
2.0 Telephone System
...
3.0 Music System
...
4.0 Home Theater
    4.1 DVD Player
        4.1.1 DVD Player
        4.1.2 Interconnects
        4.1.3 Control Components
        4.1.4 Miscellaneous Materials
        4.1.5 Equipment Build and Programming
        4.1.6 Equipment Installation
        4.1.7 Project Management
        4.1.8 Customer Training
    4.2 System Display
    ...
    4.3 Surround Processor
    ...
    4.4 Controller
    ...
    4.5 Speakers
    ...
    4.6 Mounting Equipment
    ...
5.0 Lighting
6.0 HVAC
7.0 Security

FIGURE 7–2 DVD Player WBS (Tabular format)

(Note that we have displayed only the DVD installation details, 4.1, under the 4.0 Home Theater installation. A full WBS would include the information for each system.)

The second method of representing a WBS is in tabular form. It is shown as an intended list of tasks, yet has the same grouping of work as the flowchart method. While the items are the same, the tabular format demonstrates the same material in a different manner. Figure 7–2 shows a tabular layout for the same home theater system installation shown in Figure 7–1. The tabular format is a good choice as a first step in developing a cost estimate in a worksheet format. In our case study example, ABC Integrations uses a tabular form as part of its Microsoft Project file that tracks project status. The "flowchart" format is often preferred due to its visual aspects, allowing the project team to process a complex WBS. The "tabular" format may work well for auditory-type people, but the "flowchart" will work better for visual-type people.

Study/Career Tip

Visit **www.criticaltools.com.** Here you will find project planning software that includes a product called WBS Chart Pro, used to create work breakdown structures. Read through the information and review the screen shots. Then, consider downloading a free 30-day trial. Design your own residential integration project and use this free software to create a WBS.

Critical Thinking Questions

- If you had to create a WBS, would you design it as a flowchart or in tabular form?
- Why?
- What advantages does one design have over the other?

Techniques for Developing a Work Breakdown Structure

As we mentioned earlier, in order to create a good quality WBS, certain project management standards must be applied. These techniques are as follows:

- *Decompose* the higher levels of the WBS into lower level, more detailed, deliverables and subdeliverables.
- *Verify* the degree of decomposition. Since the lowest level is the work package, we need to ensure excessive decomposition does not lead to nonproductive management of the work. The project manager seeks a balance between too much detail and too little detail.
- Structure the work into packages and identify all deliverables related to that discrete *work package.*

Let's take a more complete look at each of the three techniques.

Decomposition

The first order of business in creating a WBS is to decompose each deliverable and subdeliverable into smaller, more manageable components until the deliverables are defined in sufficient detail to support development of specific project tasks, known as work packages. This is accomplished by identifying the resources required to complete those specific project tasks. Resources are divided into two major categories, products and services. The following is a list of the types of resources typically used in a residential integrated system:

Common Products

- The main product to be installed
- Interconnecting cables and connections including audio, video, power, and control
- Miscellaneous materials, which are items less than a small dollar value that has been preset by the company, say $5. These parts may include screws, tie wraps, electrical tape, labels, mounting brackets, and Velcro.

Common Services

- Installation, which can be further broken out by phase, such as cabling, termination, move-in, equipment build, and equipment installation
- Project management, which includes the management of the project, resource scheduling, and materials handling
- Customer training
- Equipment repairs during the service period
- Preventative maintenance, if included, during the service period
- Back-up and recovery of electronic data, if included, during the service period
- Documentation, which includes laying out the system in **riser diagrams** or floor plans, documenting the programming, and providing instruction at the end of the project
- Programming the software for remote controls, lighting controls, telephone systems, and other programmable items.

Let's review the case in point on page 135.

It should be noted that each project has its unique characteristics that influence the final work packages and the overall WBS. However, there's a lot to be said for standardizing the processes and procedures to the extent possible. It provides a solid blueprint for future projects, minimizes project risk, helps to keep the project within budget and time constraints, and ensures a successful project now and in the future.

Verification

Verification of the decomposition process is imperative. Without verifying the validity of the deliverables and subdeliverables identified in the decomposition sequence, the WBS will be inaccurate, incomplete, or mired in minutia, any and all of which will have a negative impact on the project.

Questions to consider when verifying the decomposition process include:

- Are the identified deliverables and subdeliverables both necessary and adequate for the completion of the decomposed item? If not, then the components comprising the decomposed item need to be modified in some fashion (added to, deleted from, or redefined).

A CASE IN POINT
Details, Details, Details

In preparing a work package for the installation of a DVD, ABC Integrations Project Manager Owen Ginizer initiates the decomposition process. First, he determines an appropriate amount of time for the in-house technician to install the product into the equipment rack and test it. Remember that each piece of electronic equipment must be tested before it is delivered to the home. Next, Owen calculates the amount of time the field technician will need to install the rack into the cabinetry in the home. Other areas that Owen must consider are project management time, customer training, programming the user interface, and time to service the unit during the service (warranty) period.

Project Manager Owen then considers what parts will be required to connect the DVD component to the system, including what cables are needed. As such, he asks the engineering team to evaluate and recommend interconnects that will provide a consistent video path through all sources. The engineer selects component video and coaxial digital and this information is added to the DVD work package, along with the equipment rack shelf and the required miscellaneous parts such as cable ties and rack screws.

Owen's final determination is to identify the parts required to control the DVD player. He concludes that infrared control is appropriate for this piece of equipment, based upon what is currently available in the consumer market.

The finished work package consists of many resources, including both common products and common services. By taking the time to thoroughly research and document the installation of a standard DVD player the decomposition process, Project Manager Owen now has a DVD work package that is written in a manner that allows the entire ABC Integrations project team to understand exactly what is being sold and how it is to be installed.

- Is each deliverable clearly defined? If not, descriptions should be refined or expanded.
- Has adequate management control been assigned to a specific organization unit such as a department, team, or person who will assume responsibility for the completion of the item? Have budget and scheduling considerations been taken into account? If not, revisions are required.

Careful management of this aspect of building a WBS is paramount to the overall success of the project.

The Work Package

We define a *work package* as a detailed description of a project deliverable or subdeliverable. It explains all of the components involved in completing the deliverable so that no step is missed in the process. Its purpose can be two-fold: a work package allows the project manager to prepare information on a project deliverable (such as the installation of a DVD player) one time, and then to use that information "packet" repeatedly on a variety of residential installation projects. A work package may also be used by outside contractors who have a greater need for the deliverable detail than the project manager because they are actually performing the work.

As stated earlier, a work package is a task at the lowest level on the WBS. In a bottom-up approach, discussed later in this chapter, the work package represents a single item to be installed as part of the total installation project For example, in Figure 7–1, we show a WBS for the Smith residence. As part of that project, ABC Integrations is installing a home theater system, and as part of that system, a DVD player will be installed. The DVD work package identifies the products and services required to make the DVD player part of the home theater system.

It is typical that a work package is written for generic equipment and may list several brands and models that are acceptable, allowing the sales representative to determine the appropriate model based upon price point, brand awareness, and customer needs, while ensuring the player will integrate with the overall system.

Study/Career Tip

Project management studies have shown that when a WBS is not decomposed the following things happen:

- The project takes about 50% longer than it should.
- The work list expands weekly during the project's entire life cycle.
- The project manager makes vague, unclear assignments to the project team.
- The team spends countless hours in meetings discussing what to do next.
- The project misses milestones, finishes late, and requires substantial rework.

Critical Thinking Question

- How is a "work package" like a recipe?

Types of Work Breakdown Structures

There are two main approaches to developing an effective WBS. These approaches are:

1. **Bottom-up.** The bottom-up approach is more like a brainstorming session than an organized approach to building the WBS. This approach considers the smallest portions of the project first, such as the amount of time that is required to terminate a single connector for a single wallplate
2. **Top-down.** This approach considers the largest portions of the project first, such as the total amount of termination time required on the entire project. The top-down approach begins at the goal

level and successively partitions work down to lower levels of definition until the participants are satisfied that the work has been sufficiently defined.

Let's review each in greater detail.

The Bottom-Up Work Breakdown Structure

In the bottom-up approach, project team members first identify as many specific tasks related to each work package as possible. These tasks are then combined to create the next higher level of assembly. A bottom-up approach works well in production situations where the same work package will be performed many times. It involves estimating individual work items and adding them to the project total. The DVD work package is an example of bottom-up work breakdown structure.

Figure 7–3 contains a WBS for a home theater system. Each item in the home theater has its own set of resources, so we see project management included in each work package. A bottom-up WBS must be detailed, and data is gathered at the lowest level of the WBS. These packages are combined to create the system, and the systems can then be combined to create a complete cost estimate. We will discuss cost estimates in greater detail in the next chapter.

The Top-Down Work Breakdown Structure

To approach the WBS from the top-down, a project manager considers the largest items of the project and divides them into subordinate items. This method organizes the project to provide the basis for the project schedule. Let's look at the same example as that shown in the bottom up WBS and create a top-down WBS. Figure 7–4 shows an example of a top-down WBS for the same home theater system shown in Figure 7–3. In this example, the project has been divided into phases and each type of service is listed only once. The project management for the project is estimated as an overall number, rather than counted inside of each product delivered. This makes sense for a small project such as a home theater, where the project as a whole can easily be visualized. Both the time for installation and management of the project can be determined more readily be seen from the project as a whole.

Larger residential integration projects may have additional systems such as lighting, music, security interface, HVAC interface, motorized shades, several home theaters, and touchscreens; and the list can go on and on. In this type of project, it is more difficult to determine exactly how much time the overall project will require for various tasks, and as such, the bottom-up approach may be the more viable choice for a WBS.

Figure 7–5 shows a top-down approach to creating a WBS for a residential integration project for ABC Integrations. The following steps are involved in creating the WBS:

- **Level One.** Organize the project by phases. The phases are the top level of the WBS.

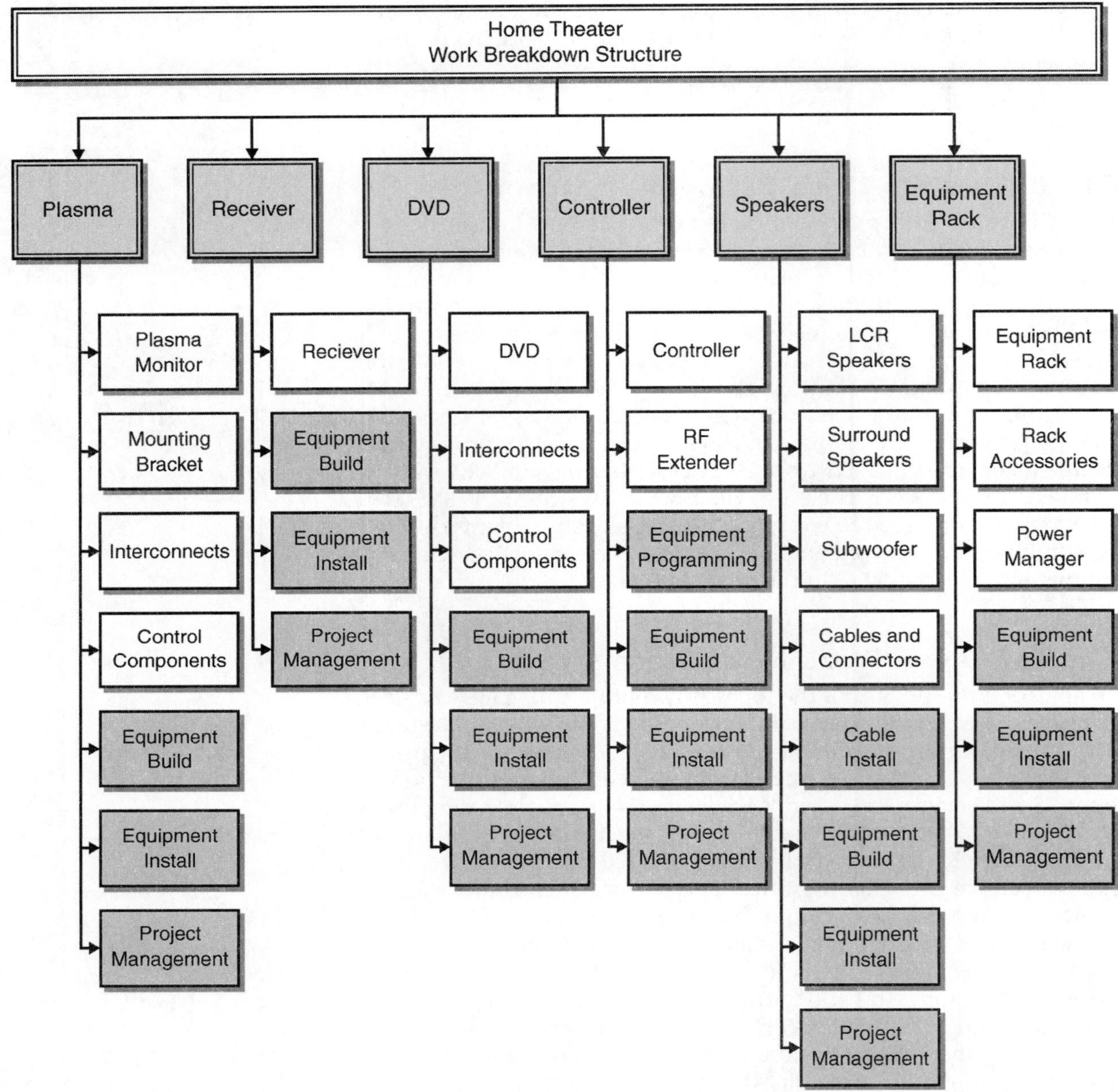

FIGURE 7–3 Home Theater Bottom-Up WBS

- **Level Two.** Organize each phase into smaller segments. Each of the phases is divided into the three project management segments of planning, implementing, and closing. Each phase is treated as a separate project.
- **Level Three.** Individual work packages are created. This level details each task to be accomplished in the project.

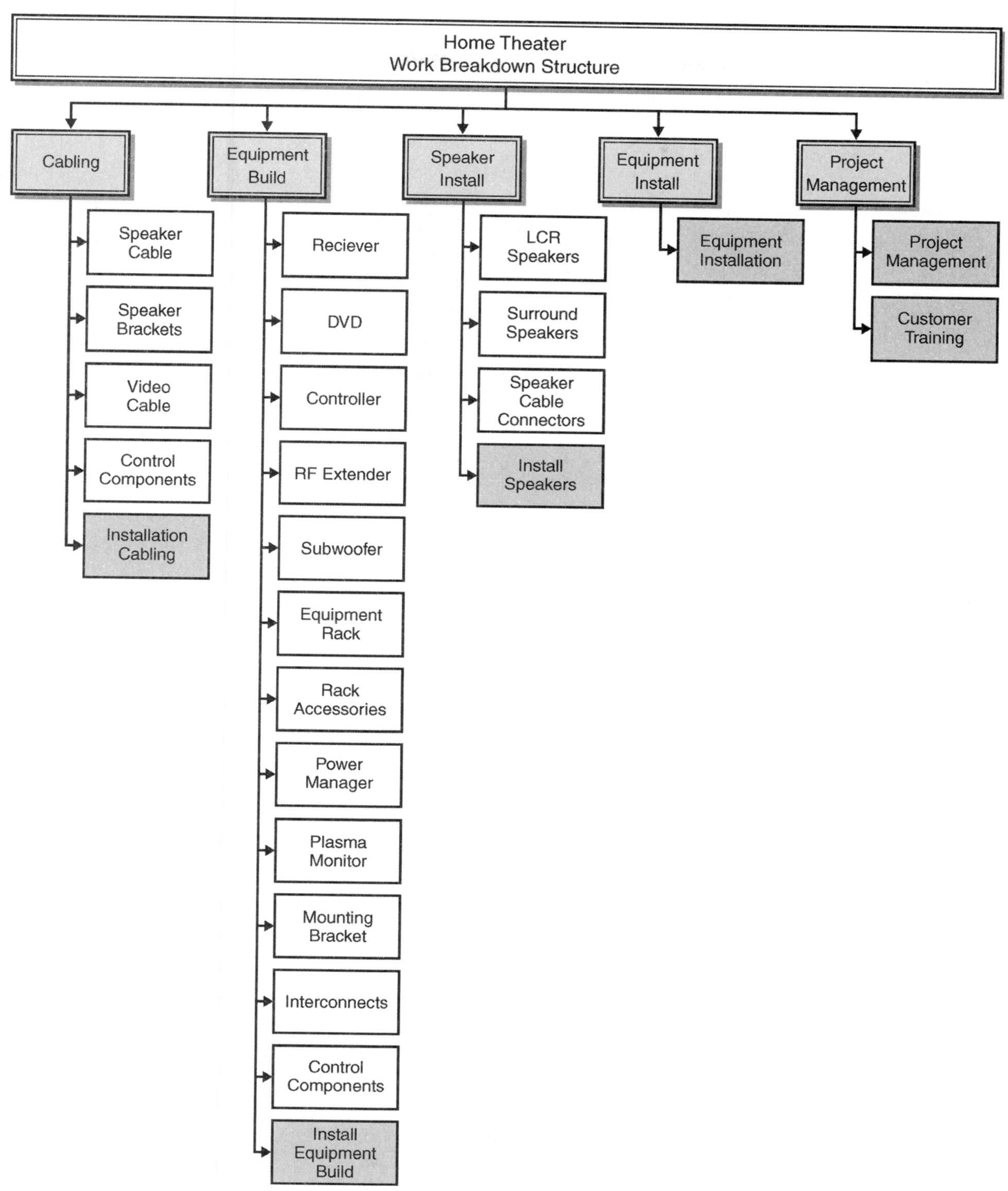

FIGURE 7–4 Home Theater Top-Down WBS

1.0 Cabling Phase **(Level One)**
- 1.1 Plan **(Level Two)**
  - 1.1.1 Sales to Operations Turnover Meeting **(Level Three)**
  - 1.1.2 Secure Low Voltage Permit
  - 1.1.3 Scope Verification
  - 1.1.4 Risk Assessment
  - 1.1.5 Preliminary Schedule
  - 1.1.6 WBS Review
  - 1.1.7 Cost Estimate Review
  - 1.1.8 Resource Commitment
  - 1.1.9 Critical Path Analysis
  - 1.1.10 Finalize Project Schedule
  - 1.1.11 Project Kickoff Meeting
  - 1.1.12 Welcome Letter to Builder
  - 1.1.13 Welcome Letter to Client
  - 1.1.14 Install Project Binder Released
- 1.2 Implement
  - 1.2.1 Procure Supplies
  - 1.2.2 Deliver Project Supplies
  - 1.2.3 Cable 1st Floor
  - 1.2.4 Cable 2nd Floor
  - 1.2.5 Cable Incoming Services
  - 1.2.6 Inspect Work
  - 1.2.7 Create Punch List
  - 1.2.8 Complete Punch List
- 1.3 Close
  - 1.3.1 Schedule Rough-In Inspection
  - 1.3.2 Rough-In Inspection Pass
  - 1.3.3 Cabling As Builds
  - 1.3.4 Change Order Developed, as needed
  - 1.3.5 Change Order Submitted
  - 1.3.6 Change Order Approved
  - 1.3.7 Job Cost: Bid vs. Actual
  - 1.3.8 What went right / What went wrong?
  - 1.3.9 Cabling Complete Letter to Client
  - 1.3.10 Cabling Complete Letter to Builder

2.0 Termination Phase

3.0 Move-In Phase

4.0 Equipment Program and Build Phase

5.0 Equipment Delivery Phase

FIGURE 7–5 Top Down WBS for ABC Integrations

Study/Career Tip

Did you know that the WBS was originally developed by the U.S. Department of Defense? It was defined in the following terms: "A work breakdown structure is a product-oriented family tree composed of hardware, software, services, data and facilities . . . [it] displays and defines the product(s) to be developed and/or produced and relates the elements of work to be accomplished to each other and to the end product(s)."

Critical Thinking Question

- Why would brainstorming with project team members be important when creating a bottom-up WBS?

The WBS Dictionary

Once the WBS is complete, the packages with all of their details are compiled into the **WBS dictionary.** Each of the work packages could include control accounts, lists of schedule milestones, contract information, quality requirements, technical references, and any other information that is relevant to the completion of the work package. Going back to our DVD player as a work package, the project manager will include schematics for connections, a bill of materials, and time requirements for each of the project teams involved with the DVD player. This is, in essence, an installation manual, and a standard that guides the entire project team through all of its projects. The best way to handle the WBS dictionary is to create company-wide standards for each package, and to note only deviations from the company standard within the project's WBS.

Study/Career Tip

Here's an example of a WBS dictionary description:

"WBS Element 1.5.4.5—Systems Integration Test Equipment Planning—This element includes the effort to identify requirements and specify types and quantities of test equipment needed to support the System Integration and Test process. It does not include the design or procurement of such equipment which is covered in Element 1.5.4.6."

Critical Thinking Question

- What do you think would be most important to find in a WBS dictionary description: what is included or what is not included?

Creating a Good Quality Work Breakdown Structure

As we've seen, building a good quality WBS is at the heart of creating an accurate cost estimate and lays the foundation for a successful project. Leaving out small details or underestimating installation labor hours has a dramatic impact on the project.

Here's a case in point.

A CASE IN POINT
A Lack of Support

ABC Integrations Engineer Maggie Pi is responsible for maintaining the company work package dictionary. However, she leaves out the required brackets for touchscreens that are being installed as part of an integration project. Her error is not found until the final day of cable installation. Unfortunately, there are no brackets at the company facility.

Weighing his options, Project Manager Owen decides the best-case scenario would be to ask the vendor to "overnight" the required brackets to ABC Integrations. Then the installer could pick up the brackets and return to the project site for installation. Regrettably, this would occur at the same time another contractor plans the home insulation installation. In addition to the inconvenience of installing brackets around another contractor's work, the bottom line would result in extra shipping and labor hour charges on the project. And that eats away at the profitability of the project.

The worst-case scenario for Project Manager Owen would be that the brackets are on back order and won't arrive until the Sheetrock™ has been installed. Now even more time is required to install the brackets very carefully in the already existing walls.

Poor Owen. The failure to properly plan has cost ABC Integrations time, money, and consumer confidence.

As we can see, WBS errors can create serious financial consequences. In order to avoid project pitfalls such as befell ABC Integrations, be sure to include the following in the WBS, regardless of its format:

- Confirm that the WBS accounts for all project deliverables.
- Ensure that testing and training is scheduled and completed.
- Verify that documentation and review activities and product/service launch and implementation activities are planned.
- Include project management deliverables on the project as well.
- Include any deliverables that must be met or delivered by the customer.
- Obtain manufacturer training. Evaluating the products ahead of time is imperative to knowing what parts are required to install those products.
- Base all WBS documents on templates. Templates are a great way to store historical information. Take the example of the home

theater discussed throughout this chapter. As more and more theaters are installed, the template can be continually tweaked to show a more accurate list of parts, perhaps even small parts such as electrical boxes, and other supplies. In addition, it provides an accurate measure of the time required to install each item in the project.

- Learn from your mistakes. If a part or service is underestimated or forgotten, thorough records will allow you to update the WBS templates to reflect the missing pieces.

Test for completeness of the WBS by:

- Verifying the status/completion of a task is measurable. If someone asks about the status of a task, and the task is defined properly, the question should be easily answered.
- Verify the technical aspects of each work package through a project team member such as Maggie Pi, our engineer. A good project manager must know how to delegate and share responsibility throughout the entire team.
- Clearly define start/end events. Once the start event has occurred, work can begin on the task, and the deliverable is most likely the end event.
- Ensure that each activity has a deliverable. The deliverable is a visible sign that the task is complete.
- Confirm that time/cost is easily estimated. This allows you to roll individual costs together and estimate the total project cost and the completion date.
- Verify that activity duration is within acceptable limits. There is no fixed rule for the duration of an activity.
- Confirm that work assignments are independent. Once work has begun on the task, it can continue without interruption and without the need of additional input.

Study/Career Tip

The Project Management Institute (PMI; **www.pmi.org**) has a document called the *Project Management Institute Practice Standard for Work Breakdown Structures* that provides examples of WBS formats commonly used in several different project areas such as construction and defense. From the PMI Web site, upper right corner, click on Publications & Information Resources, then on bookstore. A keyword search for Work Breakdown Structure will bring up this standard and other WBS references.

Critical Thinking Question

- What other criteria can you think of that could be included in creating a good quality WBS? Be specific.

Summary

In this chapter, we examined the Work Breakdown Structure (WBS), how to format the WBS, the similarities and differences between a bottom-up and top-down WBS, the WBS dictionary, and the criteria for a creating a good quality WBS.

Important points in this chapter are:

- The Work Breakdown Structure (WBS), is a foundation document that provides the basis for planning, estimating, and managing the project. The WBS defines the total scope of the project and breaks it down into smaller detailed, more manageable pieces of work.
- The Bottom-up WBS identifies the specific tasks related to the project and then combines these tasks to create the next higher level of assembly. The bottom-up approach is a great tool for determining project costs.
- The Top-down WBS considers the largest items of the project and divides them into subordinate items. This method provides the basis for the project schedule. The top-down approach is a great tool for project management, as it closely resembles the project schedule.
- Both approaches identify all deliverables in a work package, decompose the higher levels of the WBS into lower levels, and verify the degree of decomposition. Verification of the decomposition is imperative in order to create work packages.
- The resources defined in the WBS can be divided into two major categories: products and services.
- The WBS dictionary is, in essence, an installation manual and a standard that guides the project team in all projects. It is advisable to create company-wide standards for each work package, and to note only deviations from the company standard within the project's WBS.
- There are several proven ways of creating a WBS that will help ensure its quality.

Key Terms

Bottom-up WBS Once project team members identify specific tasks related to a project, the tasks are combined to create the next higher level of assembly. Smaller and simpler tasks are combined to create larger and more complex tasks.

Decomposition The subdividing of the major project deliverables or subdeliverables into smaller, more manageable components until the deliverables are defined in sufficient detail to support the development of project tasks.

Riser diagrams A set of diagrams that show the interconnections between equipment installed in a residential integration project.

Top-down WBS The largest items of the project are divided into subordinate items. This method organizes the project to provide the basis for the project schedule.

Verification Confirmation that the decomposition of a project into deliverables and subdeliverables is correct, leading to the formation of work packages.

WBS dictionary A document that describes each component in the work breakdown structure (WBS). For each WBS component, the WBS dictionary includes a brief definition of the scope or statement of work, defined deliverables(s), a list of associated activities, and a list of milestones. The information may include responsible organization, start and end dates, resources required, an estimate of cost, charge number, contract information, quality requirements, and technical reference to facilitate performance of the work.

Work Breakdown Structure A deliverable-oriented hierarchical decomposition of the work to be executed by the project team to accomplish the project objectives and create the required deliverables. It organizes and defines the total scope of the project. Each descending level represents an increasingly detailed definition of the project work. The WBS is decomposed into work packages. The deliverable orientation of the hierarchy includes both internal and external deliverables.

Work package A deliverable or project work component at the lowest level of each branch of the work breakdown structure. The work package includes the schedule activities and the schedule milestones required to complete the work package deliverable or project work component.

Review Questions

1. Why is the tabular format of a WBS a good choice for assigning tasks to the project team?
2. What are commonalities between the bottom-up and the top-down WBS?
3. When would you consider using a bottom-up WBS over a top-down WBS?
4. When would you consider using a top-down WBS over a bottom-up WBS?
5. What is the main purpose of a WBS dictionary?
6. What are several ways to test that a WBS is complete?
7. Why is it important to base all WBS documents on templates?

Cost Estimating

OBJECTIVES

After studying this chapter, you should be able to:

- List the items included in a bill-of-materials.
- Explain who typically performs the cost estimate in a small, medium, and large residential integration company.
- Discuss the industry, organizational, and scope factors that affect the cost estimate.
- Discuss the main reasons for a poor cost estimate.
- Explain the uses for and the difference between bottom-up and top-down cost estimates.

OUTLINE

Introduction

Cost Estimating

Industry Factors Affect Cost Estimating

Organizational Factors Affect Cost Estimating

Project Scope Factors Affect Cost Estimating

Causes of Poor Cost Estimating

Approaches to Cost Estimating

Introduction

A common question asked of residential integrator project managers from across the country is, "What do you find to be the single greatest problem in managing your projects?" The answers they give are amazingly similar from company to company. "In order to sell the projects, we must bid for nearly nothing. And once the project has been signed, we spend all our time scrambling for resources."

How can a residential integration project manager avoid the issue of committing to a project with insufficient resources? The answer lies in how estimates of the cost of the projects are prepared. In this chapter we will discover the various aspects of cost estimating and how to avoid the pitfalls of poor monetary management.

Cost Estimating

Cost estimating (see document PM08 in Appendix A) involves developing an approximation or estimate of the cost of the resources required to complete a project. The scope statement, as we have seen, provides the input for the process of cost estimating. The output of this process is a **bill-of-materials (BOM),** which details the quantity and costs for each resource required during the installation and completion of the project. As we will uncover in this chapter, there are many views of the bill-of-materials. The customer view is known as the cost estimate, and in some integration companies it is also called the proposal. In the residential integration industry, a bill-of-materials includes a list of equipment required for the project and any miscellaneous materials such as brackets, interconnecting cables, and connectors. It also includes installation services, such as cabling, equipment installation, project management, and training time. Many aspects of the bill-of-materials are kept internal to the residential integration company. This prevents the customer from obtaining the secret recipe for the installation, and minimizes the potential of the BOM being distributed by the customer to obtain a lower price on specific pieces.

Study/Career Tip

Pay careful attention to graphic information such as diagrams, illustrations, and tables in this textbook. These graphics present information in a different way and often present additional information that is not in the text itself. These graphics can often summarize the information effectively, as well as show relationships that may be difficult to explain in words.

Critical Thinking Question

- What are three reasons that the function of cost estimating is important in a residential integration project?

Who Performs the Cost Estimate?

An employee with a complete set of skills and experience in the custom installation industry makes a good choice for the cost estimator position. However, while this may be the optimal scenario, oftentimes there is no clear line that delineates who from the organization will complete the cost estimate. Factors such as job size and complexity, background and knowledge of the person doing the estimating, and organizational infrastructure may influence who is assigned the task of completing a cost estimate.

Many companies assign the responsibility to a sales person. That may or may not be the best choice. ABC Integrations sales representative Sal is an excellent sales person, but is he the best person to prepare the cost estimate? For small jobs—possibly. But for large projects, the cost estimate is probably more wisely prepared by a sales engineer who has a solid background in the installation of integrated systems. That type of person will have a better understanding of the project resource requirements.

Let's review the two job titles that typically fill the role of the cost estimator.

The Sales Representative

As noted above, in many small companies and in some larger companies, if the project is small, a sales representative alone can prepare the cost estimate. In this case it's a matter of the salesman wearing two hats. This may be cost effective for the organization, but it can have its drawbacks. For example, organizations that employ a sales force to both sell and estimate projects encounter a common issue—the sales representative who promises features and services that the project team cannot deliver! This happens primarily because most sales representatives derive their income from a percentage of the gross sale of the project rather than the net profit. This also is due to the close relationship that develops between the sale representative and the customer during the negotiation process. Without the ability to stand back and see the big picture of the project requirements, it is difficult to include every piece of equipment required to accomplish the project objectives. A different set of personality traits generally exists between someone who excels at closing a deal and someone who can detail every aspect of the project requirements.

What some organizations do to solve this issue is to create a separate position for estimating the project that is different from defining the project. A sales engineer, who is responsible for understanding the art and science of cost estimating, typically fills this position.

Study/Career Tip

For more information on cost estimating, visit on-line booksellers such as amazon.com, booksamillion.com, allbookstores.com (and the list goes on and on) or visit a brick and mortar store to peruse books on cost estimating. This could be a valuable addition to your personal library.

Critical Thinking Question

- What do you think the single best reason is to have a sales representative complete a cost estimate?

The Sales Engineer

A sales engineer is the preferred person to be the cost estimator if a residential integration organization can justify the position based upon company size and revenues. Why? Normally sales engineers have both the education and technical background that make them well suited for this job title. And they likely bring a wealth of experience to the position by having been in the field in some capacity for several years. As we will see, the job of cost estimating is much more rigorous than merely getting the big picture—every detail of every project must be not only accounted for, but priced out as well.

Size Matters

Each residential integration organization must determine how it defines its large, mid-sized, and small projects, depending on the size and growth of the company. One company might consider a $70,000 project a small job, while another company would consider it a mid-sized project. Still another company would say it's a large project. Why is this important? It all relates to how a company performs its cost estimating and why it would or would not consider taking on a particular project. In addition, it could directly influence who in the organization will devise the cost estimate. Let's look at an example.

In Table 8–1, ABC Integrations has set a policy regarding who performs the cost estimation for each type of project. The type of work can

TABLE 8–1
Project Categories

Scope	*Type and Value*	*Responsible Party*
Developer Project	Fixed Only	Sales Representative
Integrated Home	Semi-Custom Less than 150K	Sales Representative to Sales Engineer
Integrated Home	Custom any value	Sales Representative and Sales Engineer work as a team. Requires PM signoff.
Media System	Fixed Less than 15K	Sales Representative
Media System	Semi-Custom Less than 50K	Sales Representative to Sales Engineer
Media System	Custom any value	Sales Representative and Sales Engineer work as a team. Requires PM signoff.
Dedicated Theater	Semi-Custom Less than 150K	Sales Representative to Sales Engineer
Dedicated Theater	Custom any value	Sales Representative and Sales Engineer work as a team. Requires PM signoff.

be categorized as requiring no changes to the design (fixed), requiring minimal changes to the overall design (semi-custom), and a newly created design (custom). ABC Integrations has defined four variations of a scope statement. They are:

- **Developer Project.** In this project, the customer is a builder or developer who is requesting that the home be cabled to allow future technologies for the homebuyer.
- **Integrated Home.** In this project, systems such as structured cabling, music, lighting, HVAC integration, and security integration are included.
- **Media System.** In this project, a surround sound system is provided in a multipurpose room such as the family room.
- **Dedicated Theater.** In this project, a surround sound system and home theater are provided in a dedicated room.

As we can see, ABC Integrations has spent considerable time establishing which individuals from the organization should complete a cost estimate based upon the type of job and the dollar value associated with it. Careful upfront planning such as this helps to ensure that each cost estimate is completed accurately and that the gross margin is met for each and every project.

Industry Factors Affect Cost Estimating

Although the scope statement and the Work Breakdown Structure (WBS) are the main and most obvious contributors to the cost-estimating process, there are many other factors that must be considered. These factors can be divided into two major categories, **market conditions** and **commercial databases.** The *A Guide to the Project Management Body of Knowledge* categorizes these factors as **enterprise environmental factors,** which are any or all external environmental factors and internal organizational environmental factors that surround or influence the project's success.

Market conditions account for what products and services are available during the project life cycle. They also account for the availability of these resources and the terms and conditions for procurement. Customers Fred and Wilma want the latest digital projection system from a manufacturer that received a rave review in a recent consumer electronics magazine; however, that model is unavailable until after their desired completion date. What options are open to a project manager? Can a temporary projector be used until the desired unit becomes available? Will Fred and Wilma be happy with a second choice projector that has immediate availability? Can a comparable projector be offered at a lesser price? The answers to these questions and others influence the cost estimate.

Commercial databases house information about resource cost statistics and track skills and human resource costs for standard services such as cabling. Standard parts and materials generally used for a given project activity also are available in such databases.

The availability of database information also affects the cost estimating process. Here's how. The time it takes a residential integrator to install the speakers and the brackets for the first time is an unknown quantity. Also unknown is the cost of any miscellaneous job supplies such as wire nuts, tape, and screws. After installing the same pair of speakers many times, these unknowns become very clear. Companies that generate commercial databases collect that information—installation times and parts—from many companies in the custom installation industry. The result is a reliable cost-estimating resource for both young start-up companies and older, well-established companies.

Industry information generally costs two or three hundred dollars per month and is based upon specific proposal software. Most software packages on the market have a maintenance and data service or have a list of outside consultants who can maintain the data. When you take into account it takes 40 to 50 hours per month for the average company to maintain the data themselves, the price is well worth outsourcing.

Study/Career Tip

If possible, talk to professionals in the workplace about the information you are learning to get their perspective on project management in general and cost estimating in particular. Write down the tips and tricks they offer.

Critical Thinking Question

List as many external environmental factors that you can think of that could influence a residential integration project's success. Do the same for any internal organizational environmental factors. Review your lists.

- What are some ways to eliminate or lessen the impact of these factors?

Organizational Factors Affect Cost Estimating

As we have seen, every organization has its own internal processes and policies that support its mission statement. Among other things, the mission statement determines what approach that individual organization uses to prepare a cost estimate for its products. For example, the ABC Integrations mission statement says . . .

> *"ABC Integrations provides high quality turnkey systems only for entertainment, comfort, and convenience in the modern digital home."*

From this statement, we can determine that ABC Integrations bases its cost estimating process on turnkey systems. That means the same system is installed repeatedly. Therefore, the company uses a set of well-defined templates (see documents PM01, PM05, PM19, and PM10 in Appendix A) to create its cost estimates.

Although the cost-estimating process is similar from organization to organization, the individual steps and responsibilities can vary. At ABC Integrations, the cost estimate is prepared either by a sales representative or a sales engineer and is based solely on templates that the sales engineer creates and manages. It should be noted that while the project manager, along with the executive team, maintains the templates for the project scope statement and WBS, the cost estimate templates require engineering expertise to determine the appropriate products.

Another organization has a mission statement that says . . .

"XYZ Custom Installers provides leading edge technology, using a 'never the same system twice' approach . . . "

Just as with ABC Integrations, XYZ Custom Installers also might employ a sales engineer whose responsibilities may also include cost estimating. However, given the company's mission statement, the process varies in the way the cost estimate is prepared because XYZ's sales engineer does not use cost-estimating templates. That sales engineer prepares each cost estimate from "scratch" and has to rely heavily on involvement with the customer. Two different organizations, two different mission statements, two different approaches to cost estimating.

Now, let's look at some of the factors that can affect a project. Can you determine why they would vary depending on the organization carrying out the project?

Study/Career Tip

How are your professional skills for the workplace? Make an honest assessment of the skills you have and those you need to develop further. Consider both written and oral communication, teamwork, personal habits, and so forth. Try to assess the skills of a professional you admire and attempt to develop similar skills.

Critical Thinking Question

- Aside from a company's mission statement, what other organizational factors can you identify that would affect cost estimating?

List these in terms of favorable or unfavorable influences.

Cost-Estimating Policies

A company's cost-estimating policies include a checklist that identifies who is involved in the estimating process, and how each potential project is categorized. Cost-estimating policies include information on the following:

- **Who.** This includes information on who creates the cost estimate, who reviews the work, and who is responsible for the outcome.

- **Travel.** This includes information on additional charges for each project based upon travel requirements.
- **Project Management Factor.** This is a key indicator that allows the overall project management time to be increased or decreased based upon the anticipated difficulty of the project. We can think of the project management factor as a multiple of the total project hours. For an "easy to implement" project that number can be as low as 15 percent, and as high as 25 percent on "difficult to implement" projects.
- **House Factors.** These include a difficulty factor for running the cables, the average cable length, and the house size.

It is important to formulate policies with respect to cost estimating to ensure each estimate generated is consistent with prior proposals. Let's look at the average cable length factor, for example. This factor helps determine the cost of the cabling. ABC Integrations determines the average cable length and uses a standardized formula for costing. The formula, as shown in Table 8–2, determines the actual cable length and time required to install. Cost-estimating policies such as this one are included in the scope management plan (see documents PM07, PM17, and PM12), which we will discuss in Chapter 9, The Scope Management Plan.

Cost-Estimating Templates

Cost-estimating templates (see document PM10 in Appendix A) are created as an aid to rapid development of an estimate. The templates are continually improved to provide a highly effective estimating tool. As mentioned earlier, a sales engineer can start developing cost-estimating templates using information from a commercial database. Later the sales engineer can modify the templates to meet the changing needs of the organization.

Templates have both a positive and negative impact on the cost-estimating process. Templates allow the sales team to streamline the sales process by predetermining the pricing for products and service prior to meeting with the customer. This allows the team to rapidly provide a thorough scope statement and cost estimate for the customer in short order. On the other hand, templates do not allow for much customization and can have a canned look and feel to the

TABLE 8–2
Required Cabling Time for a Telephone and Networking Wallplate for Various Size Homes

House Size	*Required Cabling Time*
3000 sq. ft.	30 mins.
5000 sq. ft.	40 mins.
7500 sq. ft.	60 mins.
10000 sq. ft.	85 mins.

customer that could result in a less personal appearance. Blending some level of customization as part of the cost-estimate and scope-statement templates can offset this affect. For example, the sales representative could offer several models of speakers, but the cabling would be the same in all cases.

Historical Information

Residential integration companies use historical information to formulate the costs of future project activity. Clearly, the longer the history of information, the more refined this process becomes. This information can be gathered from commercial databases and from a company's project files, which house previous project performance. Maintaining a "lessons learned" knowledge base can help minimize project risks.

In Chapter 16, Project Execution, we look at the final BOM (bill-of-materials) for a project and the project's actual costs. This information should be saved from project to project as historical information. It contains critical data on the actual quantities required for each work package, as well as final prices of materials and services. The information can then be used to tweak the cost-estimating model for an organization. As more data are included, the model becomes more refined and accurate.

Consider the ABC Integrations contract to install a telephone and networking wallplate. In its cost estimate to the customer, ABC Integrations must determine how much time is required to cable the wallplate based upon the size of the house. If ABC Integrations has never installed the wallplate cables before, then it will be only a best guess estimate on how long it takes to install the cabling. To mitigate this, the company decides to use a commercial database, which provides actual data from the financial analysis of the cabling phase for many other completed projects to pinpoint the cost of labor for the upcoming project. (Information from an actual database was used to formulate Table 8–2, which shows averages for various size homes.) Now, with this critical historical information in hand, ABC Integrations can easily determine how long it takes to cable the telephone and networking wallplate. The number of labor hours can be determined and multiplied by the hourly rate per installer to determine the cost estimate for that particular project. The same process can be used in developing costs for other company products. In this way, a company can reduce the risk of error and end up with a more accurate cost estimate.

Project Team Knowledge

Project team knowledge of the residential integration industry provides an organization with a significant data on which to base its cost estimates. The cost-estimation process is aided by how long each team member has been working in the industry and how much individual experience team members bring to the project.

Companies can also benefit from project team members who have worked on other projects for other organizations because each member brings their own personal expertise to the new company. To illustrate, ABC Integrations has never installed a lighting control system, yet they want to start installing these systems in the future. A great way to gain the knowledge of how to cost estimate and install these systems would be to hire a lighting control expert into the organization. The idea is that the experienced lighting control expert could provide the necessary information for what needed to be procured (both goods and services) and what it would cost.

Let's look at how the experience level of the installation team affects the cost estimate. It is not hard to imagine that an experienced installer will perform his duties faster than that of an inexperienced installer. The question for the cost estimator is, on whom do you base the estimate—the experienced or the inexperienced installer? The answer depends on the marketplace. Ultimately, the market limits the price of the goods and services provided as part of the cost estimate. Therefore, if the going rate for calibrating a projector is $600, then the sales representative will have a difficult time justifying a cost to the customer greater than $600. It is therefore advantageous to the residential integration company that it maintains a staff of installers that are trained at least to the level of the current market.

Study/Career Tip

Use the summary at the end of each chapter of this textbook as an outline for study. Take each bullet point and write out as much information as you can about that bullet point. Use these notes as a study tool.

Critical Thinking Question

- What are some other reasons for leveraging project team knowledge?

Be specific as it pertains to completing the job.

Project Scope Factors Affect Cost Estimating

As we have discussed, the project scope statement and the project cost estimate are integral parts of a successful custom installation. In fact, these two documents must work in concert with one another so that all project deliverables are identified and costed out correctly. Properly designed templates that allow for standardization, yet customization, allow the cost estimator or sales representative the freedom to design a system that meets customer needs and the company's bottom line.

See the case in point on page 157.

Causes of Poor Cost Estimating

Cost estimation problems have been with humankind from the beginning. No doubt there were cost overruns during the construction of the Great Pyramids of ancient Egypt, and the building of Rome and the

A CASE IN POINT

Speak(er) to Me

Let's review the architectural speakers section of ABC Integrations' project scope-statement template (see Figure 8–1). It describes, in generic terms, the types of speakers that can be installed in a home as part of a house-wide music system. As you'll note, the music system speakers have been grouped into three different categories: the entry-level speakers, the background speakers, and the hi-fidelity speakers. In the case of this project scope statement template, the specific model and brand are not listed; rather, a host of speakers can be selected at each level and those choices will be reflected in the cost estimate.

ABC Integrations can then set a reasonable selling price for each of these levels, say $300, $400, and $500 per speaker pair, and a reasonable margin, say 50 points. These amounts can be used as part of the cost estimate, and the sales representative can then select from a list of speakers at each price point. The project scope statement template provides three ranges that the cost estimator can use without knowing the exact model of speaker to be selected by the customer.

Project Management Plan
The Scope Statement PM05
Architectural Speakers

Architectural speakers are designed to fit into the décor of your home. They consist of models that fit into the wall, ceiling, and even the outdoors. Today's technology allows you to enjoy high quality sound with very little visible presence of equipment in your home. We have divided these architectural speakers into three different value points:

Entry Level Speakers – Select these speakers for areas where low-level music is desired. They're generally located in the kids' playroom, laundry room, and garage.

Background Speakers – Select these speakers for areas where good background music is desired. They're generally located in the kitchen, breakfast area, and the master bathroom.

Hi-Fidelity Speakers – Select these speakers for areas where the quality of the sound is more important. They're generally located in the dining room, living room, and master bedroom.

FIGURE 8–1 Project Management Plan—The Scope Statement PM05 Architectural Speakers

Great Wall of China. In examining the reasons for poor cost estimations, it is certain there is no single cause, but in fact several causes, all conspiring together to undermine the best guess of project costs, duration, scope, and final quality. Let's examine some of these causes.

Inexperienced Estimators

Inexperienced estimators cause a fair percentage of poor cost estimates. They are more likely to underestimate a project than to set a price that is too high. They also make a number of mistakes that compound the inaccuracies of their estimates.

Study/Career Tip

Use the Web to research information in this chapter in more detail. Be sure to critically analyze the Web sites, however. Ensure that the information is credible and accurate.

Critical Thinking Question

- Before reading any farther, what do you think are some causes of poor cost estimating?

Let's look at how an inexperienced estimator exacerbates an error in estimating the cost of a telephone and networking wallplate. The cost estimator must not only determine the amount of cable required for the wallplate, but also the amount of time required to install the cable. If the cost estimator underestimates the amount of cable required, then it would follow that the time required for installation also will be miscalculated. Underestimating a small piece of a project also compounds itself when that piece of the project is used many times—as in a telephone and networking wallplate. To illustrate, a customer may request that a particular wallplate be installed throughout the home. Each time the wallplate is installed at the underestimated price, the price of the total job is even more skewed. Miscalculating one wallplate by $10 can end up costing the company 10 or 20 times that when the project is completed.

Inexperienced estimators are inclined to have similar issues in common. They tend to be:

- **Overly optimistic** about what is required to accomplish a successful project from start to finish. Failure to take into account *all* the factors that go into completing a project is a recipe for financial mayhem.
- **Prone to ignore or leave out small parts when estimating.** For example, an inexperienced cost estimator could assume that the only parts required for a plasma TV installation are the television and a mounting bracket. That is simply not the case.
- **Inconsistent in their methodology.** Inexperienced cost estimators are often inconsistent in their costing methodology, which can come from being too attached to a particular project problem rather than establishing overall company standards. By focusing too much on the issue at hand and too little on the overall project requirements, it is easy to lose sight of good business practices. Also, inexperienced estimators may try to "improve" the way they make estimates on a per project basis and again stray from established norms. By operating in this fashion there can be no consistent parameters to help in creating a better cost estimate for the next time around.

Here are three cases in point:

A CASE IN POINT
The Optimist

ABC Integrations assigns Cost Estimator Frank the job of calculating the amount of time needed to mount a plasma TV on a bedroom wall. Being inexperienced at this job, yet knowing the good reputation of ABC Integrations and its installers, Frank reasons the job can be completed in an hour. After all, it's simply a matter of getting the equipment to the site, mounting the TV, and making sure it works, right?

Here's what Cost Estimator Frank failed to consider in terms of time:

- Having the installer ensure that all hardware is present in the mounting kit before leaving the shop
- Verifying that all holes on the brackets match with the mounting bolts on the TV; testing the TV in the shop, loading the truck, driving to the job site, and unloading the truck
- Having the installer work with the frame crew to ensure that the wall has been adequately prepared for the load of the plasma TV (about 100 lbs.); making sure the electrician works with Project Manager Owen to verify the outlet is placed in the exact position for the given TV
- Providing customer training.

The end result? Cost overruns for additional labor and project management involvement.

A CASE IN POINT
It's a Small World After All

On top of failing to estimate all of the time factors involved in mounting that plasma TV, Cost Estimator Frank also forgot to add in the cost for audio/video cables from the television to the home theater system, tie wraps to secure the cables, and spare mounting hardware. There are likely a few miscellaneous items particular to this job as well that have been forgotten.

The end result? Cost overruns for material not included in the cost estimate.

A CASE IN POINT
No Method to the Cost Estimating Madness

Inexperienced Cost Estimator Jim is calculating the installation costs of the same type of plasma TV in three different scenarios. Here's how he handles each one. For one job he lumps all of the installation time in with the installation of the whole-house systems. Another time he is only estimating the cost of the TV, so he adds the installation cost in with the price of the set. And on a third occasion, when the customer appears price sensitive to the installation price, he hides that cost into the cost of the mounting brackets.

The end result? Ambiguous estimating methodologies leading to inconsistent calculations and cost overruns.

Managing an Inexperienced Estimator

Despite what may sound like all gloom and doom when working with inexperienced cost estimators, there are methods companies can employ to help estimators get up to speed more quickly and to aid them in making the correct decisions.

To illustrate, management can develop competent estimators through training in all aspects of home integration. High quality training is available from manufacturers, distributors, trade organizations such as CEDIA, NSCA, and CompTIA, and trade schools. Societies such as the Association for the Advancement of Cost Engineering (AACE) focus on cost estimating.

Study/Career Tip

The Society for Cost Estimating and Analysis (**www.sceaonline.net**) is a nonprofit organization dedicated to improving cost estimating and analysis in government and industry and enhancing the professional competence and achievements of its members. Check them out!

Critical Thinking Question

- What are some other methods of managing an inexperienced cost estimator?

List at least three ways.

As we discussed earlier, developing standard methods and procedures within an organization helps to support the estimating process. Another support mechanism is to provide simple forms to guide an inexperienced estimator through the process. A wise management team provides inexperienced estimators with a checklist that spells out the approved procedures and includes standard items to be integrated within each project.

Let's take a closer look at one company standard. The amount of project management hours included in a bid on every job should follow Table 8–3. The number of project management hours equals the multiplier times the total installation hours on the project. (See Table 8–3.) Establishing standards for other areas helps reduce the risk of errors in the cost estimate.

TABLE 8–3
Project Management Cost Multiplier

Project Type	*Multiplier*
Total Integrated Home	.20
Media Room	.15
Dedicated Theater	.25
Developer Cabling	.05

TABLE 8–4
Estimated Project Profit

	Estimated	*Actual*
Project Value	$ 50,000.00	$ 50,000.00
Labor Sales	$ 15,000.00	$ 15,000.00
Project Hours	300	330
Labor Rate	$ 50.00	$ 50.00
Cost of Labor	$ 40.00	$ 44.00
Labor Cost	$ 12,000.00	$ 14,520.00
Labor Profit	$ 3,000.00	$ 480.00
Product Sales	$ 35,000.00	$ 35,000.00
Product Margin	45%	41%
Product Costs	$ 19,250.00	$ 20,825.00
Product Cost Overages	$ -0-	$ 2,082.50
Product Profit	$ 15,750.00	$ 12,092.50
Gross Project Profit	$ 18,750.00	$ 12,572.50
Gross Margin	54%	36%

The 10 Percent Optimist

As we have learned, the role of the cost estimator is to accurately determine the resource requirements throughout the project life cycle. If the cost estimator is consistently too optimistic or confident about project costs and the time required for completion, even if only by a small amount, the overall impact is tremendous.

Let's look at an estimate of project costs for a home theater project shown in Table 8–4. Early in his career, Cost Estimator Frank estimates a particular home theater project will take 300 labor hours of time to complete. However, the actual time comes in at 330 labor hours. The cost of labor is estimated at $40 per hour, but in fact it is $44 per hour. The profit margin derived from the sale of the product is assumed at 45 percent. However, this assumption is calculated using the best payment terms and does not take into account shipping costs. So the actual profit margin for the project is only 41 percent. The final mistake Frank makes is to ignore 10 percent of the parts required for the project. Parts such as brackets, interconnects, connectors, tape, wall boxes, and infrared repeating devices are often missed by the optimistic cost estimator. Frank's estimate results in $2,082.50 in product cost overages for the company.

Table 8–4 indicates that Frank is 10 percent optimistic on the project in four major areas (Project Hours, Cost of Labor, Product Margin, and Product Cost Overages), which results in $6,177.50 in lost project profit. Because the organization has fixed expenses such as advertising, non-project related payroll expenses, rent, utilities, and much more, each project must maintain the net margin set by the executive team. If this team determines that each project requires a 50 percent gross margin as a minimum in order to meet overhead expenses, it won't take long for this company to go out of business!

Every aspect of the cost estimate must be monitored closely. By continually reviewing the margins of products and actual labor hours, the cost estimator can ensure an accurate cost picture for the project.

Lowballing

Lowballing occurs when the organization makes a conscious decision to do a job for less than it will actually cost the company to complete the work. We all know that if the organization adopts this corporate practice they won't stay in business for long, so why do companies engage in this habit?

Sometimes projects that are sent out to bid are awarded solely on price. Knowing this selection criterion, some residential integration firms may choose to bid on those projects at a financial deficit in order to hopefully make a profit on subsequent change orders. This certainly is not good business acumen, but it does happen in the industry.

In addition to possible fiscal mismanagement by the company, lowballing can create an unfavorable image for an organization; it might be viewed as being dishonest. A company using such practices also takes a huge risk that the customer will actually *want* and *approve* project changes.

Here's a case in point.

A CASE IN POINT

The High and Low of It

John is a builder of mid-size homes. He approaches ABC Integrations to cable one home currently under construction within a new development. ABC Integrations decides to lowball the cabling cost for this one home in the hope that they will be awarded the remaining cabling jobs in the development.

Sure enough, the company is awarded that particular job. As it turns out, John is so thrilled with the price point that he sends along a purchase order for cabling the remaining homes in the development at the price of the first home.

Oops! ABC Integrations is now in a financial pickle. The company can choose to raise the price of cabling on the remaining homes, but that could cause John to open up the bidding process again. And, of course, if ABC Integrations decides to honor the purchase order, they are losing money for a long time to come.

The idea of bidding low now in the hope of later work at a better price has never been a good one.

A better business approach is that ABC Integrations could negotiate to cable all the homes in the development at a reduced pricing, while the developers agrees to provide an opportunity to sell integration systems to the people who buy John's homes. That way, ABC is creating an opportunity to sell its higher priced products. This type of marketing is not considered lowballing.

Study/Career Tip

When considering this information, try to explain it to someone who knows nothing about cost-estimating procedures. If you can do this effectively, you really know the information. Think about how you might explain it to a client.

Critical Thinking Questions

- Do you think the practice of "lowballing" is ethical?
- If so, why?
- If not, why not?

Approaches to Cost Estimating

In this textbook, we confine ourselves to two different methods of cost estimating: bottom-up, and top-down. Both of these methods are used in the residential integration industry; they are derived from a detailed Work Breakdown Schedule (WBS) (see document PM10 in Appendix A). In fact, after the WBS has been completed, the work of determining cost is rather straightforward. Simply fill in quantities and pricing, and sum (add together) to create the cost estimate.

Bottom-Up Cost Estimating

As the name implies, the **bottom-up estimating** involves starting at the bottom and working our way up to the finished project. Table 8–5 details the costs for a work package for a DVD player being installed as part of a home theater system.

The DVD player work package cost estimate can then be added to the overall home theater system estimate and stored in the WBS dictionary for later retrieval for future estimates on other projects.

It's important to note that the details of the Work Breakdown Structure do not necessarily need to be presented to the customer. In fact, too much detail may be confusing, or it may provide the customer with data to negotiate the price downward. Table 8–6 shows a completed cost estimate for our home theater without revealing the details of the individual work packages.

When to Choose the Bottom-Up Cost Estimate

The bottom-up cost estimate provides a very accurate measure of the actual costs involved in a project. By using historical information, the cost estimate model of each piece of equipment (work package) can be continually improved, and information on a wide range of products can be stored for use in future cost estimates. When creating a cost estimate for a very large project, such as a full-blown residential integration project, the bottom-up approach becomes the better choice. Let's look at an example.

TABLE 8–5
DVD Player Work Package Cost Estimate

Quantity	*Resource Description*	*Unit*	*Price*
1	Gamma Electronics: DPS-7.3—DVD Player	$ 799.00	$ 799.00
1	Upsilon Cables: CCV-2M—2 Meter Component Video Interconnect	69.99	69.99
1	Upsilon Cables: MDC-2M—2 Meter Digital Coax Interconnect	34.99	34.99
1	Zeta IR: 283M—Blink-IR-Mouse Emitter	13.99	13.99
30	Installation, Finish Phase, minutes	1.25	37.50
60	Production, Home-Wide Systems, minutes	1.25	75.00
12	Project Management, minutes	1.50	18.00
	DVD Player Work Package Total		$ 1,048.47

TABLE 8–6
Home Theater Bottom-Up Estimate

Quantity	*Description*	*Amount*
1	Alpha Screens: Fifty-inch plasma display—P50X10	$12,428.02
1	Gamma Electronics: DIR 7.2 Surround Sound Receiver	$ 2,595.76
1	Gamma Electronics: DT 7.3 DVD Player	$ 1,048.47
1	Iota Controls: Integrated Controller w/RF Receiver—IC-7200	$ 2,590.45
1	Tau Acoustics: CLR7.5—LCR Front Speakers, SR585 Surround Speakers, SB10F Subwoofer	$ 2,810.68
1	Mu Pacific: Equipment Rack—RL-15	$ 1,804.15
	Total Products	$23,277.53
	Total Services	$ 3,370.58
	Sub Total	$26,648.11
	Sales Tax	$ 995.35
	Grand Total	$27,643.46

ABC Integrations is asked to prepare a scope statement and cost estimate for a project on an 8,000 square foot home to include a music system, lighting control system, structured cabling, and integration with HVAC and security systems. The project's total cost to the customer is $125,000 and includes more than 700 parts and labor categories. With this type of project, predetermining the cost of each system and item within the system greatly enhances the cost estimator's ability to quickly and accurately determine the total project costs. In addition, when a change order is required, say for an additional wallplate, a unit cost for the wallplate has already been established within the original cost estimate, which includes an appropriate amount of cable, connections, termination time, cabling time, project management, and so on. The number of wallplates can be easily changed, and by tracking the resources required for the additional wallplate, the bill-of-materials (BOM) can be easily updated.

Using Formulas as Part of the Bottom-Up Cost Estimate

The bottom-up approach can include formulas as part of each item. The formulas make it easier to determine various pieces in calculating the final estimate. One such formula can be seen in Table 8–7.

By using formulas as part of the cost estimates and by keeping each item in a WBS dictionary, the item can be used across many projects by the organization.

Let's look again at the telephone and networking wallplate. In Table 8–7 we can see that the amount of time reserved for cabling each wallplate is dependent upon the size of the home, which makes sense.

TABLE 8–7
The telephone and networking wallplate as an item in the WBS dictionary and its associated resources.
(Note: The letter A represents the average cable length in the home.)

Formula	*Quantity 3000 sq. ft.*	*Quantity 5000 sq. ft.*	*Quantity 8000 sq. ft.*	*Description*
0.2 * A + 30	47	54	63	Installation, Cabling Phase, minutes
0.2 * A	17	24	33	Installation, Termination Phase, minutes
0.1 * A	9	12	17	Installation, Move-In Phase, minutes
0.05 * (Total Service Time)	4	5	6	Project Management, minutes
1	1	1	1	Wall Mount Telephone Wallplate, each
1	1	1	1	Modular Wallplate, Two Port, each
2	2	2	2	Modular Connector, RJ45 Each
10 + A	95	130	174	Wire Cat/5e: Cable 24/4 pr. Ft.

TABLE 8–8
Home Theater Top-Down Estimate

Materials				
Quantity	***Category***	***Description***	***Unit***	***Price***
1	Display	Alpha Screens: P50X10—50″ Plasma Monitor, 1366×768	$10,999.00	$10,999.00
1	DVD	Gamma Electronics: DPS-7.3—DVD Player, RS232	$ 799.00	$ 799.00
1	Processor	Gamma Electronics: DIR 7.2—Surround Receiver, 7×100, RS232	$ 1,399.00	$ 1,399.00
1	Controller	Iota Controls: RFX6000—RF Extender for Pronto Remote, Black	$ 149.00	$ 149.00
1	Controller	Iota Controls: TSU7000—Pronto Remote, Color, Black	$ 999.00	$ 999.00
3	Speakers	Tau Acoustics: CLR7.5—Bookshelf Speaker, Black	$ 150.00	$ 450.00
1	Speaker	Tau Acoustics: SR585—8″ In-Ceiling Speaker, White	$ 450.00	$ 450.00
1	Speaker	Tau Acoustics: SB10F—Subwoofer Amplifier, Black	$ 600.00	$ 600.00
4	Equipment Rack	Mu Pacific: EVT-1 Vent Panel, 1 Space, EA	$ 10.99	$ 43.96
1	Equipment Rack	Mu Pacific: RK-16 16 Space Black Laminate KD Rack, Black, EA	$ 144.00	$ 144.00
1	Equipment Rack	Mu Pacific: RKW Casters for RK Racks, Set of 4	$ 44.50	$ 44.50
2	Equipment Rack	Mu Pacific: RSH-4S Custom Rack Shelf (See Rack Drawings), EA	$ 279.90	$ 559.80
1	Equipment Rack	Phi Power: M4310—8 Outlets (4 Switchable), Plus front panel convenience outlet	$ 249.00	$ 249.00
1	Cables and Connectors	Upsilon Cables: CCV-10M—10 Meter Component Video Interconnect	$ 199.99	$ 199.99
1	Cables and Connectors	Upsilon Cables: CCV-2M—2 Meter Component Video Interconnect	$ 69.99	$ 69.99
1	Cables and Connectors	Upsilon Cables: MDC-2M—2 Meter Digital Coax Interconnect	$ 34.99	$ 34.99
2	Cables and Connectors	Zeta IR: 283M—Blink-IR Mouse Emitter	$ 13.99	$ 27.98
5	Cables and Connectors	Upsilon Cables: Double Banana Plug	$ 3.50	$ 17.50
250	Cables and Connectors	Upsilon Cables: CL412—14 Guage Speaker Cable	$ 0.80	$ 200.00
1	Miscellaneous	Alpha Screens: PST-2000—Standard Plasma Wallmount Bracket	$ 199.99	$ 199.99
1	Miscellaneous	Tau Acoustics: NCBR—New Construction Bracket	$ 39.99	$ 39.99
			Total Equipment	$17,676.69
			Sales Tax	$ 883.83
Services				
Quantity	***Category***	***Description***	***Unit***	***Price***
10	Equipment Build		$ 75.00	$ 750.00
10	Equipment Installation		$ 75.00	$ 750.00
6	Project Management		$ 90.00	$ 540.00
6	Programming		$ 90.00	$ 540.00
			Total Services	$ 2,580.00

By adjusting the average cable length (the average length of a cable run within the home), the overall resource quantities are changed. In other words, as the size of the home increases, more time is required to install the wallplate. In addition, more time has been added to the termination and move-in phases, as it takes longer to walk around in a larger home during the installation and other phases for these items. Using this technique and monitoring the formulas for a long period of time allows the project manager to create a very accurate WBS dictionary for the purposes of cost estimating.

Top-Down Cost Estimating

The top-down estimate follows the top-down WBS. In this type of cost estimate we break out all of the major components and list them in the estimate. In addition, we list each of the service areas and estimate the amount of time required to complete the system.

Top-down estimating is a good practice for small systems, where the resources required are readily accounted for. Table 8–8 shows an example of a simple home theater system based on a top-down cost-estimating approach. In this example, the home theater is built by first listing the major components, such as the DVD player, plasma monitor, and surround receiver. At the bottom of the equipment list is the services list. Because this estimate involves a single system and is installed in a relatively short time period, it is realistic to estimate the time required for installation and project management based on the project as a total. When change orders arise, the equipment and services will be adjusted as the change requires.

While this is not a good fit for large, complex systems, the top-down approach can be used as a check against the bottom-up estimate. By creating a top-down estimate and then comparing the two BOM lists, potential mistakes can be found in the bottom-up approach. Perhaps some of the key factors used in the formulas will need to be adjusted, such as an increase in the overall project management time. The top-down estimate can then be used to go back to the bottom-up approach and tweak the final numbers for a more accurate picture of project costs.

Summary

In Chapter 8, we examined cost estimating, factors that influence cost estimating, the people who typically perform the cost estimate, causes of poor estimating, and the top-down and bottom-up cost-estimating strategies.

Important points in Chapter 8 are:

- Information in the scope statement and Work Breakdown Structure (WBS) is used to help determine a cost estimate.
- The scope statement provides the input for the cost estimate, and the bill-of-materials (the quantity and cost of each resource needed for the project) is the output.
- The size of a company usually determines who performs the cost estimate. Generally, sales representatives or sales engineers are responsible for developing cost estimates.

- Industry, organizational, and scope factors affect a cost estimate.
- Inexperienced and optimistic estimators can cause poor cost estimates. Another cause can be asking the same person to both sell the project and estimate its cost.
- Lowballing is a cost-estimation technique that is sometimes used in the industry. This practice can create an unfavorable image for an organization and might assume a huge risk that the customer will actually want and approve project changes.
- Bottom-up estimating starts at the bottom (separate individual items in a project) and works up to the finished project (everything added together). It is used on larger cost estimates, and can be used as a check against top-down cost estimating.
- Top-down estimating uses the cost of the major components and the amount of time required to complete the system to determine an overall estimate. It is used on smaller cost estimates, and can be used as a check of bottom-up estimating.

Key Terms

Bill-of-materials (BOM) A document that is a formal hierarchical tabulation of the physical assemblies, subassemblies, and components needed to fabricate a product.

Bottom-up estimating Bottom-up estimating starts at the bottom (separate individual items in a project) and works up to the finished project (everything added together).

Commercial database Contains information pertaining to cost, installation time, and other key product information that is available through third-party resources.

Cost estimating The process of developing an approximation of the costs of the resources needed to complete project activities.

Enterprise environmental factors Any or all external environmental factors and internal organizational environmental factors that surround or influence the project's success. These factors are from any or all of the enterprises involved in the project and include organizational culture and structure, infrastructure, existing resources, commercial databases, market conditions, and project management software.

Lowball The practice of estimating a project below the cost of delivering that project, in hopes of creating profit with additional project changes.

Market conditions Refers to the strength of the market or a market segment.

Top-down estimating Top-down estimating uses the cost of the major components and the time required to complete the system to determine an overall estimate.

Review Questions

1. Define a bill-of-materials (BOM).
2. What factors may influence who completes a cost estimate at a particular company?
3. What is one potential drawback of having the sales representative also complete the cost estimate?
4. Why are sales engineers considered particularly good candidates for completing cost estimates?
5. What are enterprise environmental factors?
6. What are commercial databases used for in the residential integration industry?
7. What information is contained in cost-estimating policies?
8. What is one drawback of using cost-estimating templates?
9. Why is project team knowledge important?
10. What is one reason for poor cost estimates?

Planning the Project

Introduction

Project planning is a proposed or intended method of getting from one set of circumstances to another set of circumstances. The circumstances often are used to move from the present situation toward the achievement of one or more objectives or goals. In this section, we build the processes used by ABC Integrations to develop its plans to lead to a successful project. Developed by the ABC Integrations project manager, Figure S3–1 shows the flow and relationships of these processes to each other. They are activity definition, activity response estimating, activity duration estimation, activity sequencing, schedule development, human resource planning, risk identification, qualitative analysis, quantitative analysis, risk response planning, plan purchases and acquisitions, plan contracting, communications planning, quality planning, and cost planning.

Activity Definition

Activity Definition is the process created and used to identify specific schedule activities that the project team carries out in the process of producing the project deliverables. At ABC Integrations, the project manager builds this process using the WBS to determine the schedule activities or tasks.

Activity Duration Estimation

The Activity Duration Estimation is the process created and used to determine the time in workdays or workweeks needed to complete a

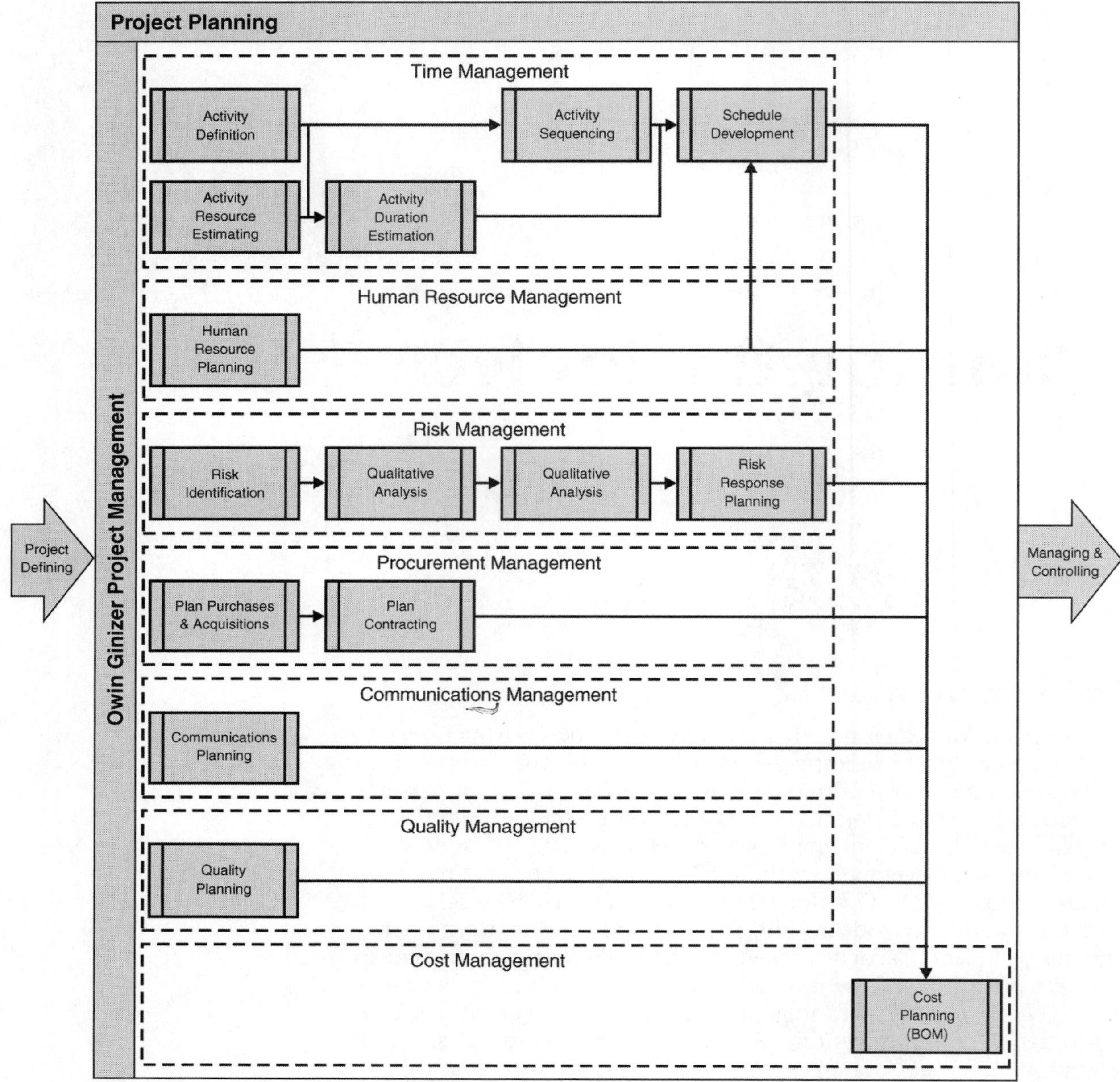

FIGURE S3–1 Project Planning

schedule activity as defined in the WBS. The work periods do not include holidays or other nonworking periods.

Activity Sequencing

Activity Sequencing is the process created and used to identify and document dependencies among schedule activities. The project manager uses the process to identify the order in which these activities take place.

Schedule Development

Schedule Development is the process created and used to incorporate the activity definition, activity sequence, activity resource estimating, and activity duration estimation into a preliminary schedule. At ABC Integrations, the project manager analyzes activity sequences, durations, resource requirements, and schedule constraints to create the project schedule.

Human Resource Planning

Human Resource Planning is the process created and used to manage and organize the project team. The project team is comprised of the people who have assigned roles and responsibilities for completing the project. Project team members can be referred to as the project's staff. Project Human Resource Management processes include human resource planning, acquiring the team, developing the team, and managing the team. Administrative resource planning also is part of human resource planning but beyond the scope of this text.

Risk Identification

Risk Identification is the process created and used to determine which risks are likely to affect the project and to document the characteristics of each.

Qualitative Risk Analysis

Qualitative Risk Analysis is the process created and used to set priorities for identified risks so that the probability of whether they will occur and their potential impact on a project can be assessed. The analysis also determines what action, if any, should be taken.

Quantitative Risk Analysis

Quantitative Risk Analysis is the process created and used to analyze numerically the effect on the overall project objectives of the identified risks.

Risk Response Planning

Risk Response Planning is the process created and used to develop options and actions to enhance opportunities (positive events) and to reduce threats (negative events) to project objectives. There are five ways to respond to risk: avoidance, mitigation, assumption, transference, and prevention.

Plan Purchases and Acquisitions

Planning Purchases and Acquisitions is the process created and used to decide what to purchase or acquire. It also determines when and how to execute those purchases and acquisitions.

Plan Contracting

Plan Contracting is the process created and used to document products, services, and outcomes that are required. It also identifies potential sellers.

Communications Planning

Communications Planning is the process created and used to determine the information and communications needs of the project stakeholders.

Quality Planning

Quality Planning is the process created and used to identify which quality standards are relevant to the project and to determine how to meet those standards.

The Scope Management Plan

OBJECTIVES

After studying this chapter, you should be able to:

- Define and explain the terms project plan and scope management plan.
- Discuss how a company's business structure affects the project management plan.
- Explain the importance of developing a project management plan's policies and procedures.
- Define a project charter.
- Explain how the scope management plan influences gathering information, the scope statement, Work Breakdown Structure, and cost estimate for a specific project.
- Discuss the importance of creating naming conventions for project files and documentation.

OUTLINE

Introduction

Developing the Scope Management Plan

Developing Scope Management Plan Policies and Procedures

Introduction

Residential integration is inherently complex, and as such, expectations can vary greatly from customer to customer. While custom installation companies seek to personalize each project, regardless of size or dollar amount, they must formalize their design process so as to maximize profits while maintaining an exceptional level of customer service. This chapter addresses the need for the development of the scope management plan that helps guide residential integration firms through this design process.

Study/Career Tip

Put as much effort into homework assignments as you would into an assignment from your employer. Consider the quality of your work, including how it is presented and the accuracy and detail of the information. Get into the habit of always submitting professional, quality work. This practice will no doubt pay valuable dividends in the long run.

Critical Thinking Question

Before you read on, try to define the term "scope management plan."

- What do you think it means?

Developing the Scope Management Plan

One surefire way for residential integration firms to adopt smart business practices is to employ a standardized document known as the **project scope statement** (see document PM05 in Appendix A). Chapter 6 outlined the specifics of this document and its importance. This document is written in language that the customer can easily understand, and it helps every project stakeholder recognize his own role in the project.

In addition to the project scope statement, companies also make use of information gathered by the sales representative from the customer (Chapter 5), as well as the use of two other important design documents: the Work Breakdown Structure (Chapter 7) and the cost estimate (Chapter 8). The design of these core documents, which are tailored to the specific project, is based upon an overarching project management approach known as the **project scope management plan** (see document PM08 in Appendix A).

The main purpose of the scope management plan is to define how the project is to progress from its beginning to completion. It is considered a supplementary element of the overall **project plan.** Specifically it addresses the following:

- It describes how the project scope will be managed.
- It describes how scope changes will be integrated into the project.
- It generally contains an assessment of the expected stability of the project scope (i. e., how likely it is to change, how frequently, and by how much).
- It should include a clear description of how scope changes will be identified and classified. This is especially important when the product characteristics are still being defined.

In short, the scope management plan provides the high-level blueprint for how the project moves through its life cycle. Developing the scope management plan includes coordinating the development of additional or subsidiary plans and incorporating them into the complete project plan.

There are many potential subsidiary components that make up the overall scope management plan. These subsidiary plans provide the specific details for managing each aspect of the project from initiation through closure. The subsidiary project management plans could include the:

- Time management plan
- Risk management plan
- Procurement management plan
- Communication management plan
- Quality management plan
- Cost management plan.

Study/Career Tip

If you don't have the skills already, start now to fully develop your Internet research skills. It is essential to be able to efficiently research information on the Internet and then critically evaluate that information to determine if you should use it. The guiding principles of project management rarely change, but the residential integration industry is in a constant state of flux. Stay on top of those changes!

Critical Thinking Question

- What other subsidiary plans might be a part of a scope management plan?

These plans (and others as warranted by the complexity of the project) work in concert with the scope management plan to provide structure to the overall project plan. We will take a closer look at each of these subsidiary plans in subsequent chapters of this textbook.

Remember that the scope management plan is very different from the project scope statement. The scope statement is an external document given to the primary stakeholder, generally the customer. That is why, in the residential integration industry, it is often called the customer proposal. As we have learned, the scope management plan is an internal document that describes how scope changes are evaluated, processed, and integrated into the project. Both the project scope statement and the scope management plan are outputs of what is known as **scope planning.**

Company Business Structure

Each company develops its own scope management plan based on its deliverables, its customer demographics, and its competition. Typically, the **executive team** collaborates to create and define company standards.

And, if applicable, the team works together to create the set of templates used by various members of the company staff to complete information gathering, a scope statement, a Work Breakdown Structure (WBS), and a cost estimate (see documents PM02, PM03, PM04, PM05, PM08, and PM10 in Appendix A). It is the project team's responsibility to focus on applying these company standards to each customer assignment.

The experiences and vision of people within the organization weigh heavily on the company's scope management plan. Company executives need to decide on common issues. Will the organization focus on very large custom homes providing complete integrated systems, or will it focus on small home theater systems for the average homebuyer? This decision determines whether the systems are designed as highly repeatable or highly custom. Each will have its own requirements.

For example, if the company focuses on large custom homes with complete integration systems, these systems are installed over many months. As a result, the company may organize separate teams that work on the project, perhaps one team per phase of the project. In addition, the project is treated like several mini-projects, one for each phase. However, if the company focuses on small home theater systems, these would be handled differently. Each project would be treated as a single phase with only a single installation team. We can see that the company is organized around the type of work that is performed.

As experience is gained, a company could change or expand its focus. Company employees can gain a wealth of experience through networking, length of time in the trade, trade magazines, and interviews with customers, builders, architects, and those in other related fields. If it is the first time the organization has installed a telephone system, a project team member who has performed these installations as part of prior job can help minimize the **learning curve.** An experienced installer can show others on the installation team how to install the telephone system. As a result, next time the installation is likely to be accomplished faster and more efficiently. Ultimately the installation team will be capable of installing many telephone systems in a short period of time, allowing the company to expand into additional areas. That means the company will grow and become more profitable.

Study/Career Tip

Consider reading this text more than once or twice. The first reading is a preview. The second will be effective learning. The third reading will be an important review. Also read the text in relation to the homework you are assigned. Using insights you gain from the homework and classroom discussion, annotate the margins of the text with what you've learned as a reminder for the next time you read the text.

Critical Thinking Questions

- How do you manage your learning curve when it comes to the residential integration industry?
- Where do you find answers to your questions?

Developing Scope Management Plan Policies and Procedures

Many factors influence the development of the scope management plan. Understanding the residential integration industry, the company organization, its customers, and its strengths and weakness are a few common traits that are detailed in preparing this document. To that end, companies need to establish policies and procedures for the development of the following areas in the scope management plan:

- Project Charter
- Gathering Information
- Project Scope Statement
- Work Breakdown Structure
- Cost Estimate
- Project Files and Documentation

Project Charter

The first step in developing a project plan is to create the **project charter.** The project charter is the document that formally authorizes a project. In some situations, the project charter is issued by the project sponsor, the customer. It might take the form of a set of plans from the architect, or it might even be verbal communication from the customer describing his wants and needs for the home.

The intent of a project charter is to notify stakeholders of the new project and the new project manager and to demonstrate upper management support for the project. The sponsor uses the project charter to provide a broad direction to the project manager. The charter should precede the other project documents as it establishes the project manager's authority that, in turn, is necessary to create the stakeholder written agreements (the scope statement).

The charter should be made available to all project stakeholders—everyone who may be associated with the project—reaching as wide an audience as practical. The charter is usually written by the sponsor or the project leader and should be approved by the sponsor, the customer, and the functional manager(s) who provide company policy and resources. When a project is performed under contract, which is generally the case in the residential integration industry, the signed contract will serve as the project charter for the seller.

For ABC Integrations, the project charter is created by the executive team and is a managerial overview of the scope statement template. ABC Integrations' project charter includes an overview of the system requirements, and defines the key technologies and standards to be utilized within each system. This document is designed for the demographic rather than the individual customer. It looks at the common needs and requirements of a broad range of customers and formulates them into a single charter (see Figure 9–1).

Project Management Plan
Project Charter PM05

Key Technologies

Our structured cabling system must have the capability of distributing the following technologies:

- Broadband Internet
- Multiple telephone lines
- High-definition signals for off-air, satellite, and cable
- Home-wide music system
- Shared files, printers, and on-line resources on all computers in the home

Structured cabling is limited to communications within the home and does not include lighting, HVAC, or security.

FIGURE 9–1 Project Management Plan—Project Charter PM05

Study/Career Tip

Use this text carefully as a learning tool. What if you took it to the workplace as a reference? How would you make notes, underline, highlight, use sticky notes, and so forth? Would this differ from what you are doing now?

Critical Thinking Questions

Think about a project that you have worked on or are currently working on, whether for school, work, or pleasure. Try writing out a project charter as it specifically relates to that project.

- Who are the stakeholders?
- Are you the project sponsor?
- How does this provide structure for the completion of the project?

For example, our case study lists key technologies, such as broadband Internet and multiple telephone lines. Not every customer requires all of these specific technologies; however, the project charter describes the foundation that is to be installed for every system so that any of these key technologies can be installed or easily added later to the home.

Key information used to formulate the project charter includes the company mission statement, company strengths, weaknesses, knowledge base, and availability of outside consultants. The project charter drafted for ABC Integrations can be expanded into each area of technology provided by the organization. These areas could include music, lighting, home theater, HVAC, security, and other areas of integration.

Gathering Information

As you recall from Chapter 6, The Scope Statement, customer information needs to be gathered by the sales representative in order to feed that information into the design of the scope statement. The scope management plan states how that customer analysis is written. For ABC Integrations, that means stating in the project plan that company templates will be used to gather customer information, site information, and system information. The scope management plan provides details on how each template is used and who uses each one. For example, it will state that the sales representative will develop the scope statement from information recorded on the site survey template, the customer survey template, and the structured cabling survey (see documents PM02, PM03, and PM04 in Appendix A). In this way, the scope statement will be developed for each deliverable offered by ABC Integrations. The scope management plan will also provide information on how to present the pricing of deliverables in the scope statement for the customer.

Study/Career Tip

When completing the end-of-chapter Review Questions, find the part in the text to which the question is related and reread that section of the text. Make notes in the margins as necessary, and perhaps write the question itself in the margin to help you remember the importance of the material.

Critical Thinking Question

Think back to Chapter 5 of this text, Information Gathering.

- What types of questions should be asked when conducting a site survey, and a customer survey, and a structured cabling survey?

Figure 9–2 shows an excerpt from this plan. In this section, we show one of the steps involved in creating the scope statement. It specifies how information is gathered from a customer and what forms are required. Of course, different companies may have different policies for creating this portion of their scope statement. A larger organization might specify how the information is stored in the organization's intranet systems, and smaller organizations might base their documentation on a paper and folder-based system.

The Project Scope Statement

The next step in developing the project scope statement is reviewing the collected data and presenting it to the customer. The scope management plan identifies who in the organization reviews the plan and how and who presents it to the customer. For example, ABC Integrations Sales Representative Sal Moore, Engineer Maggie Pi, and Project Manager Owen Ginizer all review the scope statement for accuracy, resource availability, company capability, and company mission. Both

Project Management Plan
Project Scope Management Plan PM08

Step 1 Gather all of the required information

One or more customer interviews will be required to gather the information. The following forms must be completed.

- Site Survey
- Customer Survey
- Structured Wiring Survey

FIGURE 9–2 Project Management Plan—Project Scope Management Plan PM08

Sales Representative Sal and Project Manager Owen may present it to the customer, or depending on the size of the project, Sales Representative Sal may present it to the customer alone.

The process of negotiating a final scope statement is also outlined in the scope management plan. For ABC Integrations, Sales Representative Sal has the authority to make minimal changes requested by the customer, but substantial changes are to be reviewed by Project Manager Owen and Engineer Maggie, especially changes to key technologies. That way the company is assured the scope statement does not promise more than it can deliver.

For example, if the customer base is mainly comprised of owners of tract homes, the systems installed must be highly repeatable. As a result, the scope management plan creates policies that encourage repeatable business and discourage one-of-a-kind systems. Let's look at a sample policy, seen in Figure 9–3.

Project Scope Statement Template

As you learned earlier, the executive team, in concert with others, develops a set of templates that guides the design of the various core

Sample Scope Management Plan Policy

Each scope statement is to be created from templates released by the executive team. Additional features may be added to the template but these new features must go through the engineering evaluation process described below.

New Product Evaluation Process

1. All new products must be brought in-house for evaluation, using the in-house evaluation form.
2. Once evaluated and accepted, the cost estimate template (see documents PM10 and PM11) and scope statement template (see document PM05) should be updated and signed off by the executive team.

FIGURE 9–3 Sample Scope Management Plan Policy

documents used in the overall project plan. One such **template** is the scope statement template (see PM05 in Appendix A). This template must answer the following questions:

- What are the project objectives?
- What are the project scope definitions?
- What are the project requirements?
- What are the project boundaries?
- What are the project acceptance criteria?
- What are the project constraints?
- What are the project assumptions?
- What is the initial project organization, i. e., the project team?
- What are the initial defined risks?
- What are the schedule milestones?
- What are the configuration management constraints (administrative constraints)?
- What are the project specifications?
- What are the approval requirements?

As the executive team answers these questions, it is providing the elements needed to create a comprehensive scope statement for the project manager, the customer, and members of the project team.

Study/Career Tip

Conduct an Internet search on the phrase "scope management plan." Here you will find many articles from trade magazines and white papers that discuss this topic. This will aid you in broadening your view of the subject.

Critical Thinking Questions

Review the definition of the word "template" in the Glossary.

- What documents do you use in your daily life that can be considered templates?
- Have you ever created a template?
- If so, what was its use?

Verifying the Project Scope Statement

Verifying the information contained in the scope statement is a key element of the scope management plan. Nothing turns off a customer more than finding out the wrong products have been installed, whether it's the cabling structure or the home theater screen. Each system to be installed in the home must be verified before the project begins. The scope management plan includes this information so that each member of the project team is aware of what their responsibility is in verifying the various systems. As discussed in Chapter 6, The Scope Statement, a common way to verify the structured cabling system is "stickering" the house (the use of sticky notes to identify where the cabling should be run). The information that needs to be included in the scope management plan is to identify who is responsible for this verification process and how to accomplish it. At ABC Integrations, the "stickering" is accomplished by the project manager and the sales representative. The "stickering" is then confirmed by the project sponsor.

It's important to verify the systems that are to be installed, not only to satisfy the customer, but also to provide a smooth and efficient installation. The scope management plan needs to include a verification policy and procedure for the various aspects of the scope statement (see Table 9–1). These are the same aspects considered in developing a scope statement template. We have provided sample verification policies and processes for each aspect in Table 9–1.

Of course, each company will have different verification policies and processes based upon their business needs and alignment. Those noted in the table are intended to be brief and simple. A large company with many employees and products is likely to have complex verification processes outlined in its scope management plan. A growing company, on the other hand, can revise its policies and processes to match its current status. The important point is to have a policy and a standard written for what procedures project team members are to follow. The verification process confirms the aspects of the scope statement and provides accountability for all concerned parties.

Study/Career Tip

Even though you feel confident that you have mastered an idea, keep reviewing it. The information will become more firmly embedded in your memory and will remain there for a longer period of time. In addition, the deeper you understand a concept, the easier it is to expand it.

Critical Thinking Question

- What might be some consequences of not installing a verification process into the scope management plan?

List as many as you can.

The Work Breakdown Structure

In addition to identifying the key components of the scope statement, the scope management plan describes how the company creates its Work Breakdown Structure (WBS). As we discussed in Chapter 7, there are two basic approaches to preparing a WBS—the bottom-up WBS and the top-down WBS. The scope management plan states who within the organization creates the WBS, which approach is to be used, and how the document will be formatted (either in flowchart or tabular form).

TABLE 9–1
ABC Integrations Scope Statement Verification Policy and Process

Aspect	*Verification Policy*	*Verification Process*
Project Objectives	All project objectives are to be reviewed by the sales representative with the customer for verification.	Sales representative presents final scope statement to the customer. Customer signs.
Project Scope Definition	The project scope definition is to be reviewed by the sales representative, project manager, lead technician, and installer to verify.	Sales representative, project manager, lead technician, and installer review the scope definition and sign off.
Project Requirements	Project requirements are to be reviewed by the project manager and engineer.	Project manager and engineer review project requirements and sign off.
Project Boundaries	All project boundaries are to be included in the scope statement.	The project manager reviews the project boundaries section to ensure each is reasonable and all boundaries have been included.
Project Deliverables	Project deliverables are reviewed with the homeowner.	Homeowner reviews project deliverables and signs contract.
Project Constraints	Project constraints are to be reviewed by the project manager and compared to survey documents.	Project manager confirms using original survey documents and signs off.
Project Assumptions	Project assumptions are to be reviewed by the project manager and the lead installer.	Project manager reviews with lead installer and both sign off.
Initial Project Organization	The initial project organization is to be reviewed by the project manager to verify resource availability.	Project manager confirms resource availability and signs off.
Initial Defined Risks	Initial defined risks are to be confirmed by the project manager and the installer.	Project manager reviews with lead installer. Both sign off.
Schedule Milestones	Schedule milestones are to be confirmed by the project manager with assistance from the lead technician and lead installer.	Project manager reviews with lead technician and lead installer, and writes memo to customer to confirm. All sign off.
Fund Limitation	The fund limitation is to be confirmed with the customer.	Sales representative sends memo to the customer to confirm. Customer signs and returns to company.
Project Configuration Management Requirement	Project Configuration Management Requirement is to be confirmed by the project manager and the administrative assistant.	Project manager reviews with administrative assistant.
Project Specifications	Project specifications are confirmed by the customer on a review of the field located stickers, along with the sales representative or project manager.	Homeowner, builder or architect walks through home to confirm "stickering." Customer signs off.
Approval Requirements	The standard approval requirement section is to be included within the scope document.	The project manager ensures the approval requirement has been included.

ABC Integrations' Preparation Policy (PM05)

1. The Project Manager and the engineer work together to prepare a bottom-up WBS for each project.
2. The Project Manager uses the current, approved scope statement to determine what systems are to be installed in the customer's home.
3. The Project Manager prepares a preliminary bottom-up WBS, using the company's WBS template (both flowchart and tabular format). The template includes pre-determined, detailed information on each system that the company offers. The Project Manager deletes the systems that are not to be installed for a given project.
4. The Project Manager forwards the preliminary WBS to the engineer.
5. The engineer reviews the preliminary WBS, adding any necessary pieces to the WBS, and returns it to the Project Manager.
6. The Project Manager reviews and approves the WBS.
7. The Project Manager stores the approved WBS in a computer file, using the company's file naming protocol.
8. The Project Manager uses a second WBS template to prepare a top-down WBS, reviews it and stores the file in company computer, using the company's file naming protocol.
9. The Project Manager also prepares a top-down WBS to assist in the cost estimate process.
10. The cost estimator, who can easily access the most recent and approved documents, uses both the bottom-up and top-down WBS to prepare the cost estimate.

FIGURE 9–4 ABC Integrations' WBS Preparation Policy (PM08)

To illustrate, to prepare the WBS, the ABC Integrations project manager could develop the following procedure and add it to the scope management plan (see Figure 9–4).

Each residential integration company may have a different procedure for preparing the WBS. Today, most companies have computer systems to aid in the development of the WBS, rather than conventional file drawers, and the project manager can easily communicate with other staff members via e-mail to garner information. Companies are likely to use different file naming systems as well. Regardless of the methods used, the issue for organization is consistency. That way each person in the company can quickly look at a WBS and understand the tasks ahead.

Work Breakdown Structure Template

Just as with the scope statement, the executive team will develop a Work Breakdown Structure (WBS) template. The following questions must be addressed in this template. The answers to these questions will then be identified during the WBS verification process.

- Who prepares the WBS?
- Who verifies the WBS?
- How frequently are the WBS templates revised?
- What approach (top-down or bottom-up) is used to prepare the WBS?

- What format for the WBS will be used (flowchart or tabular)?
- What protocol is used in naming the WBS files?
- Are computer files and hard copy files to be stored?
- How long are the WBS files to be stored?
- How is approval given to the WBS?
- Who gives approval to the WBS?

Verifying the Work Breakdown Structure

An important part of any scope management plan is to verify the work that has already been completed in order for the project to commence. To verify the Work Breakdown Structure, the scope management plan must include a general policy on verifying the WBS. A policy needs to be developed to verify each point of the WBS template. Verifying these and other pertinent questions is how the executive team formulates the policy section of the WBS as part of the scope management plan. Again, the complexity of the plan depends on the size of the company.

The Cost Estimate

As we discussed in Chapter 8, Cost Estimating, a number of people in the organization can prepare the cost estimate. The scope management plan dictates the person responsible for preparing this document and it details a procedure for preparing the estimate (see document PM10 in Appendix A).

At ABC Integrations, the cost estimate is prepared by the sales engineer and the sales representative, with sign-off required by the project manager, which is stated in the scope management plan. The plan also states that the cost estimate is prepared using the bottom-up WBS and the company's cost estimate template. As with the WBS, the template provides consistency from customer to customer.

The Cost Estimate Template

A number of factors are expressed in the scope management plan about preparing a cost estimate. Some residential integration companies have few details in the scope management plan, while others are fairly precise about the method. Some of the issues that need to be considered, although they may not all be included, are deciding:

- Who prepares the cost estimate, who reviews the work, and who is responsible for the outcome?
- Will a template be used? How frequently will it be updated?
- Will the WBS be used in preparing the cost estimate? If so, which type?
- How will the cost estimate be presented to the customer?
- Is the cost estimate included in the scope statement or prepared as a separate document?
- Will the cost estimate be top-down or bottom-up?
- Will both approaches be used?
- What financial resources will be used?

- What commercial databases will be used, if any?
- How will the cost estimate be verified?
- What travel charges will be added for each project?
- How will project management time be charged?
- How will house factors, such as difficulty of cabling and size of house, be addressed in the cost estimate?

The executive team answers these and other questions pertinent to its organization in the scope management plan.

Verifying the Cost Estimate

Verifying the cost estimate is another important element to include in the scope management plan. Without an accurate cost estimate, company profits can easily disappear. Slight errors in estimating can reach major dollar amounts when carried through an entire project. A simple, yet practical method of verifying the cost estimate is to compare a top-down cost estimate with a bottom-up estimate. The final figures will not be identical, but they will provide a reasonable accuracy. The process is similar to the process mathematicians use to verify totals in a long column of figures. First they add the figures from the top to the bottom. And to check their total, they add the figures from the bottom to the top. A policy for the cost estimate verification process may be included in the scope management plan and will ask the following:

- Who performs the cost estimate?
- Who verifies the cost estimate?
- Are templates used to prepare the cost estimate?
- If so, how frequently are the templates revised?
- How are the cost estimates presented to the customer?
- What format is used to present the cost estimate?
- What protocol is used in naming the cost estimate files?
- How is approval given to the cost estimate?
- Who gives approval to the cost estimate?
- Are computer files and hard copy files of the cost estimate to be stored?
- How long are the cost estimate files to be stored?

Again, the verification process for the cost estimate cannot be underestimated. Failure to validate figures prior to the beginning of the project spells financial woes for the company.

Project Files and Documentation

As we have seen, the project manager uses the scope management plan to change, update, and improve the processes used within the company. By continually adding to and modifying the processes and procedures in the scope management plan, the project manager ensures a consistent business methodology. Without a set of rules for the development and implementation of the scope management plan, every project would be unique. And uniqueness costs the organization money. Even though the focus of project management is to provide project uniqueness, a successful project manager knows how to transfer successful ideas from project to project to improve the process and procedures, allowing each project to run as smoothly as possible.

Study/Career Tip

Do you have a naming convention in place for your digital files? If so, you are ahead of the game. If not, give strong consideration to thinking through the process and establishing standards for the various folders on your hard drive. You never know when you'll need to go back and research something or otherwise retrieve a file.

An example of a good process, shown in the sample scope management plan for ABC Integrations (see Figure 9–5), is a file-naming convention for all project files. By adapting standard practices and

File Naming Conventions (PM08)

All of the files must be completed in a softcopy and follow company standard naming conventions. Each project has many documents: a WORD document for the scope, an EXCEL document for bid and estimation, VISIO files for schematics, and so forth.

Obtain the proposal number from the company proposal registry. For example, let us assume the scope number is 101, the customer name is Douglas Adams, (ADAMD—first four letters of the last name and first letter of the first name). It is the first version of the proposal (V01) and it was released (rel) on February 5, 2005 (020505).

0101ADAMDV01 Proposal rel020505.doc

"V01" represents the first version or the original document. If the customer requests changes to the proposal, then the version number becomes 02, 03, and increases until the customer accepts the proposal. Once the proposal is accepted, the first number is changed to one (1) to indicate the first major release of the document.

When minor changes to the document are requested, the second number reflects the change requests. Therefore, the first version with the first requested change would be 11.

As additional minor changes are requested, minor releases are created with the number increasing incrementally as 12, 13, 14, and so on, until formally accepted.

0101ADAMDV23 would be the second version of the proposal with three change requests.

Finally, your project team will create many other files. Let's use the above example and assume the customer signed the third version of the contract. Here's how the files would be named.

0101ADAMDV03 Proposal rel020505.doc

0101ADAMDV03 Estimate rel020505.xls

0101ADAMDV03 Riser Diagrams rel020505.vsd

0101ADAMDV03 Work Order Cabling rel020505.doc

FIGURE 9–5 File Naming Conventions (PM08)

keeping to those practices, the project team enhances its productivity and minimizes unwarranted management time.

In addition to the file-naming conventions established in the scope management plan, all other plans and procedures that are developed for the company should follow the company's document release policy. This ensures consistency across all projects, reduces risks and miscues, and helps to minimize extra project management time.

Summary

In this chapter, we defined the project plan and the scope management plan, and we discussed how the executive team creates the scope management plan. We also discussed how the scope management plan is used to guide the development of information gathering techniques, the scope statement, Work Breakdown Structure, and cost estimate using company templates.

Important points in this chapter are:

- A scope management plan contains all of the internal policies and procedures required to carry out the creation of the information-gathering techniques, the scope statement, the Work Breakdown Structure (WBS), and the cost estimate.
- The company's business structure influences how a scope management plan is defined and written.
- A scope management plan includes many components to define, document, verify, and manage the scope statement, otherwise known as the customer proposal.
- The first step in developing a scope statement is to create the project charter. The project charter formally authorizes a project. A company can write its own charter or, in some situations, the project charter is issued by the project sponsor, otherwise known as the customer.
- The project manager uses templates created by the executive team to develop the needs analysis (information gathering), the scope statement, the Work Breakdown Structure, and the cost estimate for each individual project.
- Having a standard naming convention for project files and related documentation is important.

Key Terms

Executive team The executive team is the governing branch of the organization. In a small company, this may be the owner, and in larger companies, it may be the director of operations in conjunction with the heads of each of the departments—engineering, sales, project management, installation, and administration.

Learning curve The length of time it takes to learn successfully any given task. When many items are produced repeatedly, the unit cost of those items normally decreases in a regular pattern as more units are produced.

Project charter A document issued by the project initiator or sponsor that formally authorizes the existence of a project, and provides the project manager with the authority to apply organizational resources to project activities.

Project plan A formal, approved document used to guide both project execution and project control. The primary uses of the project plan are to document planning ssumptions and decisions, facilitate communication among stakeholders, and document approved scope, cost, and schedule baselines. A project plan may be summary or detailed.

Project scope statement The narrative description of the project scope, including major deliverables, project objectives, project assumptions, project constraints, and a statement of work, that provides a documented basis for making future project decisions and for confirming or developing a common understanding of the project scope among the stakeholders. The definition of the project scope—What needs to be accomplished.

Project scope management plan The document that describes how the project scope will be defined, developed, and verified and how the work breakdown structure will be created and defined. It provides guidance on how the project scope will be managed and controlled by the project management team. It is contained in or is a subsidiary plan of the project management plan. The project scope management plan can be informal or broadly framed, or formal and highly detailed, based on the needs of the project.

Scope planning The process of progressively elaborating the work of the project, which includes developing a written scope statement that includes the project justification, the major deliverables, and the project objectives.

Template A partially complete document in a predefined format that provides a defined structure for collecting, organizing, and presenting information and data. Templates are often based upon documents created during prior projects. Templates can reduce the effort needed to perform work and increase the consistency of the results.

Review Questions

1. What is the main purpose of the scope management plan?
2. Name at least two things the scope management plan addresses.
3. What is the difference between the scope statement and the scope management plan?
4. Who develops company project management standards? What methodology is often used to create these standards?
5. What is a project charter?
6. What is the intent of a project charter?
7. Why is it important to create a scope statement template?
8. Why is the scope statement verification process important?
9. What is one way to verify a cost estimate?
10. Why is developing a naming convention for all project files important?

Time Management Planning

OBJECTIVES

After studying this chapter, you should be able to:

- Discuss time management planning.
- Define and explain activity definition, activity resource estimation, activity sequencing, and activity duration estimation.
- Define outsourcing and explain its use in the residential integration industry.
- List and explain the three dependencies in activity sequencing.
- List the types of resources that a project manager considers in planning a project.
- Define and explain the terms critical path and critical path method.
- Discuss schedule development, critical path activities, free float, and total float in a critical path.
- Explain the concepts of forward pass and backward pass.
- Explain fast tracking.

OUTLINE

Introduction

Remember the triple constraints we discussed in Chapter 1, Introduction to Project Management? Projects always are evaluated and measured against the three constraints of cost, scope, and time. These three project constraints are interrelated trade-offs, pulling in different directions

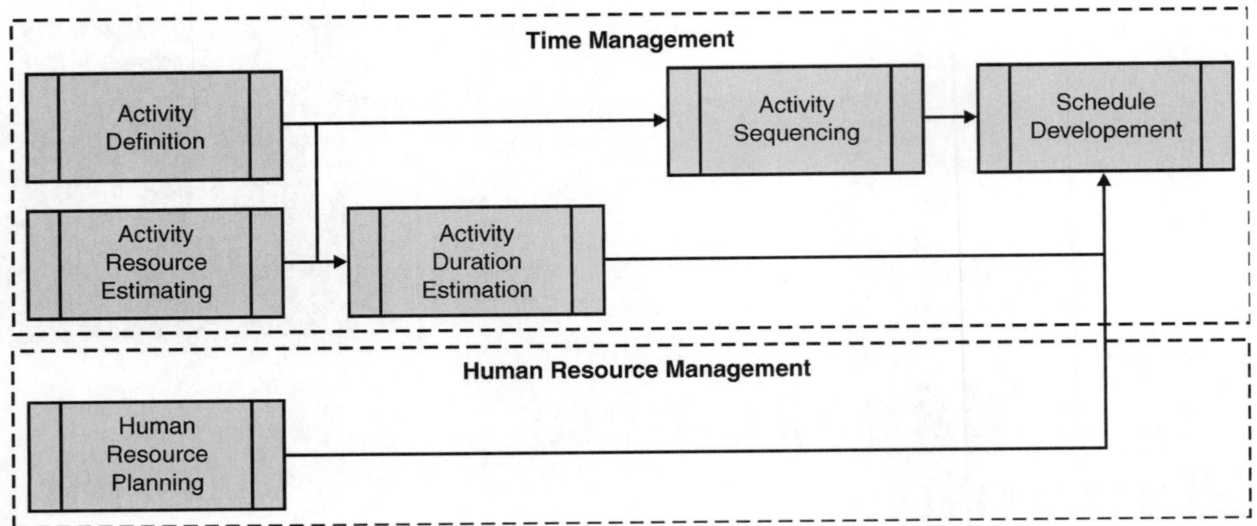

FIGURE 10–1 Time Management Planning Processes

throughout the project. As we have seen, the project manager first defines the scope of the project, thus creating the project scope statement, then the project timeline is developed through the use of a Work Breakdown Structure (WBS), and finally the cost estimate is determined. These same project management steps correspond to the planning of the project. The activity definition is the scope, the activity sequencing is the time, and the activity duration corresponds to the cost of the activity. In this chapter we discuss the processes involved with time management planning as shown in Figure 10–1.

What Is Time Management Planning?

In its most fundamental form, **project time management** is the process by which projects are completed in a timely manner. Simple enough. And yet there is so much more to the efficient planning and execution of a project to ensure that it is completed on time and under budget, and that it meets customer expectations.

Time management planning is comprised of five major processes as depicted in Figure 10–1.

These processes include:

- Activity Definition
- Activity Resource Estimation
- Activity Sequencing
- Activity Duration Estimation
- Schedule Development

Let's review each in greater detail.

Study/Career Tip

Time management is a critical cog in the overall project management machine. Be sure that you understand the key terms associated with time management planning and how they relate to the development of specific forms used in project management, such as the scope statement, the WBS, and the cost estimate.

Critical Thinking Questions

- How do you practice time management in your daily life?
- Do you use a day planner, a PDA, a bunch of sticky notes?
- What's another way you could plan your daily schedule?

Activity Definition

As we have learned, the project schedule is developed from the schedule milestones outlined in the scope statement document. If the preliminary project schedule differs greatly from the scope statement's milestones, then the project manager will need to negotiate with the sales representative on the timing of the project. The purpose of the preliminary schedule is to make the project manager aware that key decisions have to be made.

To aid in developing the preliminary project schedule, project managers use a technique called **activity definition.** Activity definition is the step that identifies the specific activities that the project team carries out in the process of producing the project deliverables. An activity or task is the basic element of measure that is found in the WBS.

In Chapter 7, Work Breakdown Structure, we developed a WBS for a home theater system, shown in Figure 10–2. Let's look at how this WBS translates into activity definition.

If we focus on the activities shown in Figure 10–2, we can define common activities for each work package. These common activities are:

- **Electronics Build.** This activity involves connecting and testing all of the electronics in the shop before delivery to the customer.
- **Electronics Install.** This activity involves installing the electronic components at the job site.
- **Project Management.** The focus of our textbook, this activity involves the steps required to manage the installation of the system.
- **Electronics Programming.** This activity involves programming each piece of electronics.
- **Cable Install.** This activity involves installing cabling at the job site to accommodate speakers, wallplates, and electronic components.

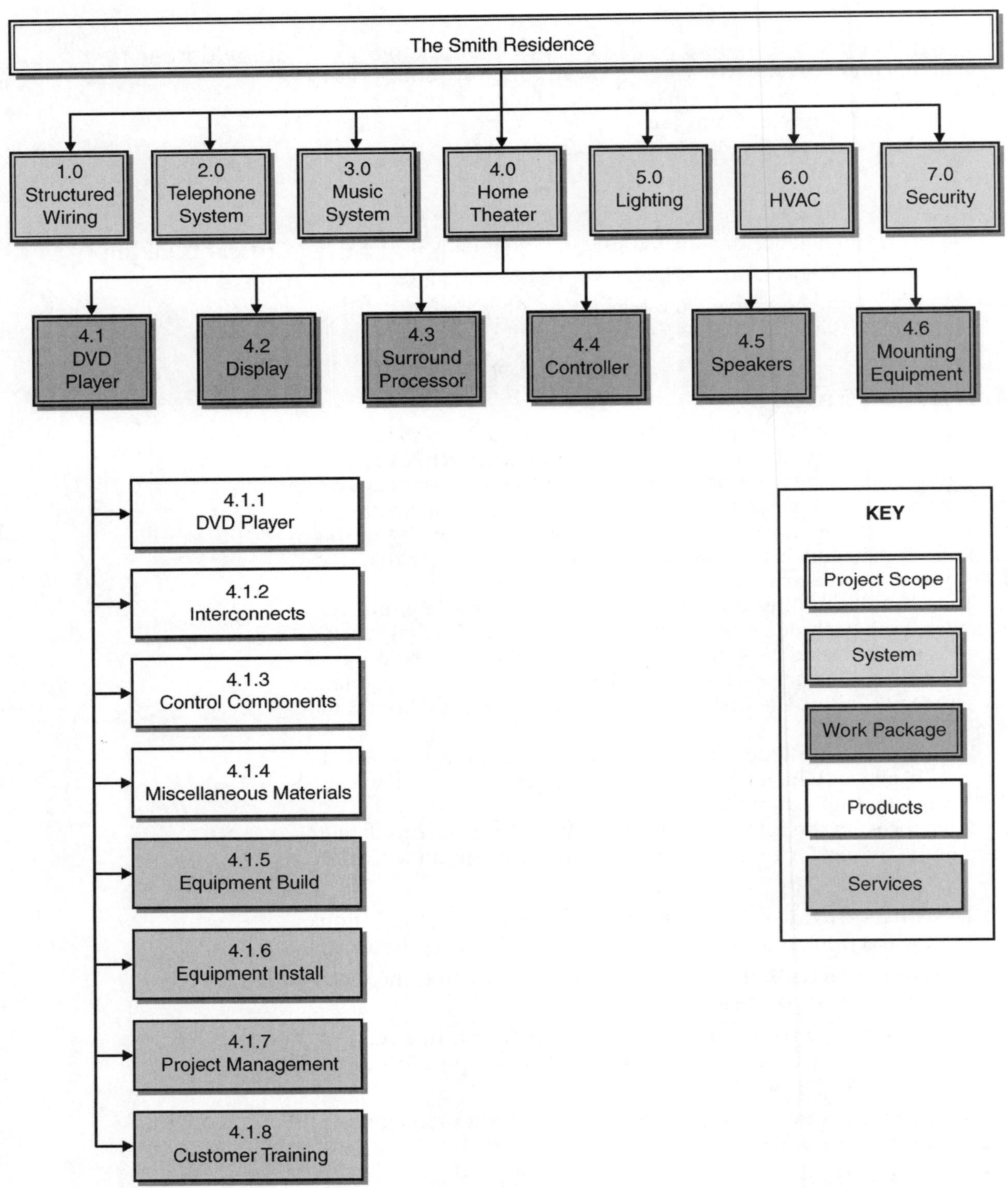

FIGURE 10–2 Work Breakdown Structure

Study/Career Tip

Time management planning is much more involved than what is presented in this chapter. To learn more, conduct an Internet search to garner additional information. Consider either printing this information or storing it electronically. You should be continuing to build your own personal library of related industry terms and concepts.

Critical Thinking Questions

- Why do you think it's important to have "project management" listed as one of the activities in the WBS depicted in Figure 10–2?
- Isn't project management considered the overall thrust of the project?
- If that's the case, why bother to include it at this level of detail?

The project manager starts with the WBS and breaks the work down further for the purpose of developing a detailed schedule. This is done to more readily track the progress of a large project. A smaller project may only have a single activity definition for project management, while a larger project will break the project management tasks into greater detail.

It is important to understand that a WBS can be designed for the purpose of cost, scope, or time. In subsequent sections we look at how the project is broken down for the purpose of time planning. Any way you break out the project, the total amount of project management time, or installation time, will be the same.

Oftentimes it is difficult to know everything about a project during the planning stage, especially if the project is complex. It is common to learn more about the project as you work through the project life cycle. As discussed in Chapter 1, this is called progressive elaboration and it affects the planning process. If you don't know everything about the project at the outset, you can't plan the whole project to the level of detail necessary in order to complete the project. For large projects, it is common to plan the entire project at a high level. The project starts with detailed plans in place for the work packages that are near the beginning of the work. As the time draws near to begin additional phases, the more detailed, low-level plans for those work packages are added to the project plan. The project manager should revisit the planning process on a recurring basis to ensure that the detailed plans contain the latest information known about the project.

Activity Resource Estimation

In activity resource estimation, the project manager prepares a list of internal and external resources that have been identified in order to meet the established project milestones, which is a significant point in the project life cycle. This process lists each type and amount, or quantity, of each required resource. What are considered resources? Activity resources *can* include:

- Equipment
- Materials and supplies
- Money
- People

As we discussed earlier, one of the best ways to approximate resources is to use the bottom-up estimating method. As you recall, one of the purposes of creating the WBS is to decompose project work into work packages that are small enough to reliably estimate activity duration and resource requirements. Using the WBS, the project manager can provide estimates for mid- and high-level work by totaling the estimates for the work packages that make up the required work. This type of estimating tends to be fairly accurate because the estimates come from the people doing the actual work.

Outsourcing

An important trend in recent times has been the growing use of outside contractors to carry out an organization's business, a practice known as **outsourcing.** This practice results in a number of benefits for the company and is quite common in the residential integration industry. Here are a few reasons why. We discuss additional aspects of outsourcing in Chapter 12, Procurement Management Planning.

- The company saves money because company benefits are not provided to outsourced contractors. These benefits include health and dental insurance, vacation and holiday pay, and retirement and pension programs.
- The organization can provide a highly specialized service that would be impossible to maintain by hiring a direct employee. For example, an organization can outsource the certification of networking cables. This practice allows an organization to provide a service that would be too expensive for most companies to take on internally because the test equipment for certification is approximately $6,000.
- Outsourcing allows a company to quickly expand and contract its workforce. As you will see, ABC Integrations' cabling team utilizes outsourcing to expand from two to four members based upon the workload. This can be accomplished without hiring additional direct employees.

Study/Career Tip

Easily distracted when you are studying or doing homework? One way to improve your concentration is by reading actively: try using a highlighter, making notes of significant points in a notebook, recording questions about items you don't understand, trying to predict what will be on the next page, and connecting what you're reading with other things you've read.

Critical Thinking Questions

- Are you familiar with the term *outsourcing*?
- Do you work for an organization that has this business practice?
- What are some advantages to outsourcing?
- Disadvantages?

Outsourcing can be used in many ways. A company may hire someone temporarily to perform a specific task or a company can repeatedly hire the same subcontractor to perform the lion's share of the workload. With judicious management, a company may outsource all of the installation and engineering work. This is how a management-based company functions. The company focuses on completing the project to the customer's specifications by using many different low-voltage contracting companies to build a unified, integrated system. One company may provide the cabling, terminate, and move-in phases, while an audio/video company might be hired to install the music and theater systems, a telephone company contracted to provide the telephone system, and other companies retained for additional systems. While this may be an extreme example, it demonstrates how a residential integration company may employ outsource companies to expand their offerings.

Contracts for outsourcing are the same as the contracts with the customer. The various types of contracts are discussed in Chapter 1, Introduction to Project Management and in Chapter 12, Procurement Management Planning.

Table 10–1 represents three different ways to resource the cabling of a project, based upon the size of the project. In the first scenario, the resources assigned to the task are the two internal

TABLE 10–1
Cabling Team Requirements

Scenario	*Bid Hours*	*Team*	*On-Site Days*
1	16	1 Lead Installer 1 Installer	1
2	64	1 Lead Installer 1 Installer 2 Outside Contractors	2
3	128	1 Lead Installer 1 Installer 2 Outside Contractors	4

installers working for ABC Integrations. The second and third scenarios allow two additional outside contractors to aid in staffing the cabling task. This allows the task to be performed in a shorter duration than would be accomplished with using just the internal resources. A project manager can look at the chart and quickly compare the need of the project with the availability of in-house staff. Based upon that comparison the project manager can then outsource the additional installers as needed to complete the cabling by the project milestone. Here's a case in point.

A CASE IN POINT

From the (Out) Source

A customer contracts with ABC Integrations to install a satellite dish on his home. Project Manager Owen Ginizer starts his cost estimation process by assessing the cost of tools and training. Then he looks at the overhead costs associated with the task. In the case of satellite dishes, which typically are installed on a roof, the insurance costs are significantly higher than they would be otherwise. Even if the ABC Integrations installer only installs a single dish in a given month, the entire month must be covered at the higher insurance rate. For this reason, Project Manager Owen decides to outsource this task.

The following is an excerpt from a Letter of Agreement between ABC Integrations and the outsourced contractor to install the satellite dish:

Letter of Agreement between ABC Integrations and The Satellite Company

Please follow the instructions listed below.

1. Install the dish at the home in the area least visible from the front of the home.
2. Install four (4) RG6–QD cables from the dish to a ground block mounted in the attic. We will provide and install the ground block.
3. Install a #10 ground wire from the dish to the ground block located in the attic.
4. Test the installed dish for the minimum acceptable signal strength of 87, as measured at the ground block in the attic. If you cannot attain a signal of this level or higher, please contact me directly to arrange for an acceptable alternative. Measurements should pass on the following satellites; Satellite A, 101—with transponders #1 to #32, Satellite B, 119—with transponders #22 to #32, and Satellite C, 110—with converted transponders #8, #10, #12.

It should be noted that this document serves as the scope document for the single task of installing a satellite dish, as well as a contract for the service.

Activity Sequencing

After defining the project tasks and determining what resources will be required to handle these tasks, the next step is to identify in what order these tasks take place; this is known as **activity sequencing.** In activity sequencing, the project manager reviews the activities in the detailed WBS and makes sure he understands the product deliverables, assumptions, and project constraints. Each task has a **logical relationship** with the other activities. These relationships or dependencies can be divided into three categories.

- **Mandatory dependencies** are inherent in the nature of the work. For example, we cannot program the equipment until it has been built. By built, we mean assembling the equipment and making the system-level connections. This may take the form of installing the electronics in an equipment rack and connecting each piece of electronics and testing to ensure each is working properly.
- **Discretionary dependencies** are defined within the organization. These dependencies include best practice policies, and the occasional exception. For example, the implementation segment of the cabling phase requires the planning segment to be completed prior to starting. However, due to time constraints the project manager may start the work without the low-voltage permit. While local code requires the permit first, the project manager could get the team started on site, and then obtain the permit later in the day.
- **External dependencies** involve the relationships with the other contractors on the project. To illustrate, the start of the implementation segment of the cabling phase not only requires the planning segment to be completed, it also requires the electrical, HVAC, and plumbing contractors to be finished with their work.

Study/Career Tip

Make note of the fact that the terms "logical relationship" and "dependency" are widely used interchangeably in the world of project management.

Critical Thinking Question

- What might be the advantage of having several discretionary dependencies on a project?

Aside from the example given, think of several more on your own.

In order to understand the dependencies of the activities in our cabling phase, we have created a table that lists each of the planning segment activities showing how they are related to one another and in what order activities must occur before another activity can begin. In practice, this step can be done with a white board or a stack of sticky notes that can be moved around on the white board. Table 10–2 contains an **activity list,** which represents the planning segment of the cabling phase; each has an **activity identifier** (also called the activity ID), an **activity description,** and its **predecessor activity.**

Figure 10–3 shows a network diagram of the planning segment of the cabling phase. In this diagram, we can see visually how the activities are related. It is clear that while Activities 2, 3, 4, and 5 can be performed simultaneously, Activities 3, 4, and 5 need to be finished prior to the start of Activity 6, according to the WBS review. Keep in mind that it is not a race from start to finish. Every activity must be completed. Even though Activity 2 (obtain the low-voltage permit) does not need to be completed prior to Activity 6, it must be completed prior to Activity 11, which is the project kickoff meeting.

TABLE 10–2
Cabling Phase—Planning Segment Activities

Activity ID	*Activity Description*	*Predecessor Activity ID*
1	Sales to Ops Turnover Meeting	None
2	Secure Low-Voltage Permit	1
3	Scope Verification	1
4	Risk Assessment	1
5	Preliminary Schedule	2, 3, 4
6	WBS Review	5
7	Cost Estimate Review	6
8	Resource Commitment	7
9	Critical Path Analysis	7
10	Finalized Project Schedule	9
11	Project Kickoff Meeting	2, 10

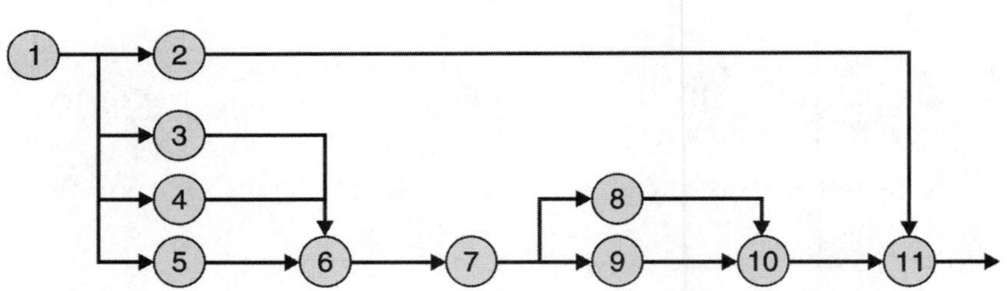

Legend:

Activity ID	*Activity Description*
1	Sales to Ops Turnover Meeting
2	Secure Low-Voltage Permit
3	Scope Verification
4	Risk Assessment
5	Preliminary Schedule
6	WBS Review
7	Cost Estimate Review
8	Resource Commitment
9	Critical Path Analysis
10	Finalized Project Schedule
11	Project Kickoff Meeting

FIGURE 10–3 Network Diagram Showing Activity Sequencing of the Planning Segment of the Cabling Phase

TABLE 10–3
Cabling Phase—Planning Segment Activities with Durations

Activity ID	*Activity Description*	*Duration (Days)*	*Predecessor Activity ID*
1	Sales to Ops Turnover Meeting	½	None
2	Secure Low-Voltage Permit	4	1
3	Scope Verification	½	1
4	Risk Assessment	½	1
5	Preliminary Schedule	½	2, 3, 4
6	WBS Review	½	5
7	Cost Estimate Review	½	6
8	Resource Commitment	½	7
9	Critical Path Analysis	½	7
10	Finalized Project Schedule	½	9
11	Project Kickoff Meeting	½	2, 10

Activity Duration Estimation

After working to define the activities and determining their sequence, the next task is to estimate the **duration** of each of the activities, also known as the **activity duration.** The activity duration estimating process assigns the number of workdays that are needed to complete schedule activities. Each estimate assumes that the necessary resources are available to be applied to the work package when they are needed. It is important to understand that the duration is not the time spent working on the activity **(effort),** but the elapsed time plus the working time.

In Table 10–3 we add the duration times for each of the activities shown in Table 10–2. For example, the second activity in the planning segment of the cabling phase is to obtain a low-voltage permit (Activity ID #2 in Table 10–2). The effort time is the 15 minutes of time it takes to fill out the form and write a check, while the duration is four days, because the form and check must be sent via mail to the local inspection office. The permit is considered in effect once received at the local inspection office with payment.

The people actually performing the work have a great amount of input into determining the duration of each activity. For example, the lead installer of the cabling team knows from experience how long it takes to complete the cabling of the home. In addition to the project team input, historical data is invaluable in this process. In the closing segment of each phase of our project, we update our original estimations, which are used to improve the estimations on the next project.

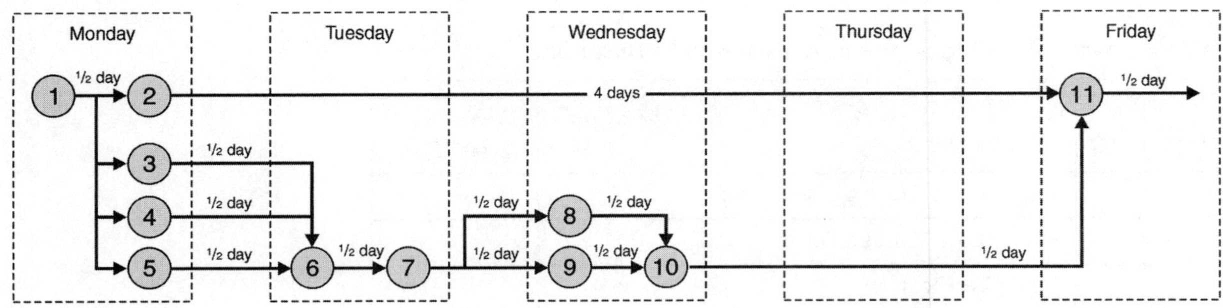

Legend:

Activity ID	Activity Description
1	Sales to Ops Turnover Meeting
2	Secure Low-Voltage Permit
3	Scope Verification
4	Risk Assessment
5	Preliminary Schedule
6	WBS Review
7	Cost Estimate Review
8	Resource Commitment
9	Critical Path Analysis
10	Finalized Project Schedule
11	Project Kickoff Meeting

FIGURE 10–4 Activity Duration of the Planning Segment of the Cabling Phase

Looking at the same network diagram of the planning segment of the cabling phase, we add duration times. These are shown in Figure 10–4. In our scenario, we are starting with Activity 1, the Sales to Operations Turnover Meeting, first thing on Monday morning. Once the meeting ends, Activities 2, 3, 4, and 5 are allowed to commence, with Activities 3, 4, and 5 each having duration of one-half of a day. That means by Tuesday morning we can expect to be ready to move onto Activity 6, the WBS Review, and subsequently to Activity 7, the Cost Estimate Review. While the actual time spent on Activities 6 and 7 may be less than a day, experience and assumptions tell us that it is best to plan to take the day to accomplish these tasks. Keep in mind that other things are taking place, so the project manager and cost engineer are not fully devoted to these tasks. If they were, the durations could be lessened.

On Wednesday, we take on Activities 8 and 9 simultaneously; these activities are the Resource Commitment and Critical Path Analysis. The second half of the day is concerned with the final schedule development. Thursday has a break in the activities. This is caused by the duration of Activity 2, Securing the Low-Voltage Permit. The project manager sent the request to the local inspection department on Monday, but the project manager probably won't receive it until late Thursday, or possibly Friday morning.

Friday morning arrives and the team holds its Kickoff Meeting, Activity 11, and by Friday afternoon the installation team is able to start on the next segment of the cabling phase, the implementation phase. During the week, many team members have played a valuable part in the project: the project manager, the sales representative, the project scheduler, and the installation team leader. The total effort for this segment is estimated to be 16 staff hours, while the duration is five days. The estimate effort is derived from the cost estimate done prior in the project, and as we see, it does not correlate to the duration time.

Study/Career Tip

For each concept, think of several examples and applications. Write these down in your notes. This strategy will assist you in recalling the information for an exam and in the workplace.

Critical Thinking Questions

- Why wouldn't a project manager simply concentrate on the activity "effort" as opposed to the duration of an activity?
- Isn't the effort the amount of time it takes to complete an activity?

Schedule Development

Schedule development is a typical issue for residential integrators. Imagine having over 50 projects running at a time, each with its own requirements for the project team, engineering, installation, programming, and project management. It only follows that a good project manager consistently reviews each project to ensure the resources are in place to carry out the project objectives.

Resources are committed by placing a team on a project team schedule. The project manager should provide a master schedule to each member of the team to view his workload.

The project manager takes the duration and sequence information from the preliminary planning and blocks time off on the schedule to commit each team. For example, the cabling team for ABC Integrations consists of two installers. If a project requires 80 hours of cabling time, then both installers would be booked for an entire week on the project schedule. In addition, the time required for each activity associated with the project will be blocked off on the team schedule.

The process of developing a project schedule for expected project activity start and finish dates is one that is continually updated and revised. The project manager reviews and revises the estimates for duration times and resource availability and commitment. The result is a project schedule that can be used to measure and track the forward progress of the project or pinpoint project delays. When project activities are completed sooner then expected, the other project activities may be moved up. On the other hand, when project activities are delayed for any reason, the remaining project activities can be moved out farther. This provides a true picture of whether or not the project can be delivered on time. If it looks like the project cannot be completed on time, then the project manager may decide to hire temporary help or pull other company workers from another project.

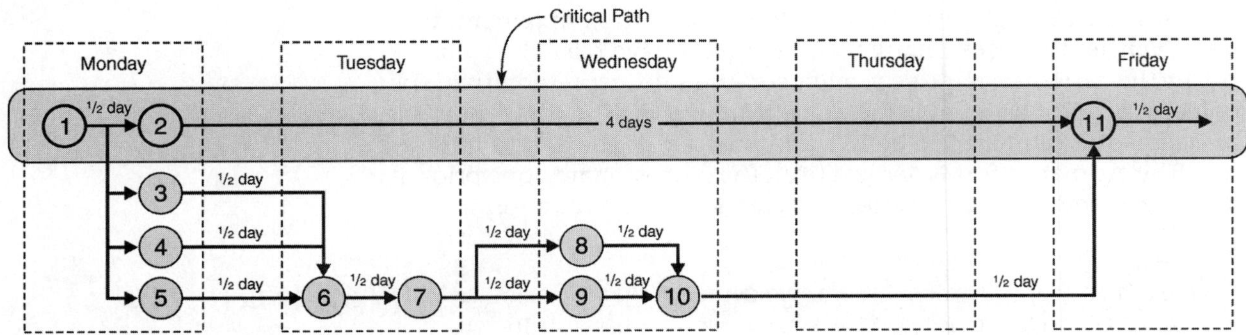

Legend:

Activity ID	***Activity Description***
1	Sales to Ops Turnover Meeting
2	Secure Low-Voltage Permit
3	Scope Verification
4	Risk Assessment
5	Preliminary Schedule
6	WBS Review
7	Cost Estimate Review
8	Resource Commitment
9	Critical Path Analysis
10	Finalized Project Schedule
11	Project Kickoff Meeting

FIGURE 10–5 Critical Path of the Planning Segment of the Cabling Phase with Duration Times and Critical Path Early Start and Finish Date

Critical Path

An important topic to understand with respect to project schedules is the **critical path.** The critical path of a project is the sequence of activities or tasks that, when followed in logical order, takes longer to complete than any other sequence in the project. A delay in the critical path of the project causes a delay in the project completion date. The critical path is represented in the network diagram as the longest path. The **critical path method** is the process of determining the longest path in the project. Reviewing the critical path helps keep the project on time.

Figure 10–5 shows the same network diagram worked out in the preliminary schedule above under Activity Duration Estimation (Figure 10–4), with the addition of the critical path. The **critical activities** have been outlined in a thick black border. In this example, we see that the shortest length, the planning segment of the cabling phase, can be completed in five days. It can also be seen by looking at Figure 10–5; Activities 3 through 10 have one day of **total float.** The amount of time that a scheduled activity can be delayed without affecting the schedule is called **free float.** That means if these activities did not start for a day, or were delayed while in progress by as much as a day, there is no impact on the duration of this segment. This is important because it gives the project manager some "wiggle room" and allows him to shift project teams to other sites where they may be needed without harming the team's current work site schedule.

Study/Career Tip

Use Post-it® notes as bookmarks. Simply put a word or two at the top to create your own reference index. This saves you time and energy when writing a paper from having to reread whole chapters. It may also aid you when studying for a quiz or exam.

Critical Thinking Questions

Try drawing out the critical path of one work (school) week.

- What activities can be moved from one day to the next before starting or finishing?
- What cannot be moved?
- Why?

The activities that lie on the critical path have no float associated with them. However, we can look at the other activities and determine for each activity an **early start date,** and an **early finish date.** The early start date represents how early an activity can start without affecting the schedule, while the early finish date represents the earliest possible date an activity can finish without affecting the schedule. These early dates are calculated by **forward passing** through the scheduled activities and stacking those items as far to the left as possible. For example, Activities 3, 4, and 5 all have an early start date of Monday afternoon, while their early finish date is the end of day Monday. Figure 10–5 shows all of the activities at their early start and finish dates. A **backward pass** through the network diagram determines the **late start dates** and the **late finish dates.** The late start date is the latest date an activity can start without affecting the critical path schedule, and the late finish date is the latest date of the finish of an activity without affecting the critical path schedule. Figure 10–6 shows each of the noncritical activities at their late start dates. Table 10–4 shows the late and early start and finish times for our project.

Fast Tracking

Why does the critical path matter? Here is a common example of how the critical path plays out in the real world. It is Saturday morning and you, as the project manager, get a call from the sales representative who informs you that he just closed the Smith project. However, in order to secure the work he promised that the cabling would start on Wednesday. Given your experience as a project manager, you know the planning segment must be completed prior to the installers arriving at the job site, and that it will take until Friday to get the plan completed. After a short laugh, you tell the sales representative you'll start a week from Monday, no sooner, and hang up the telephone.

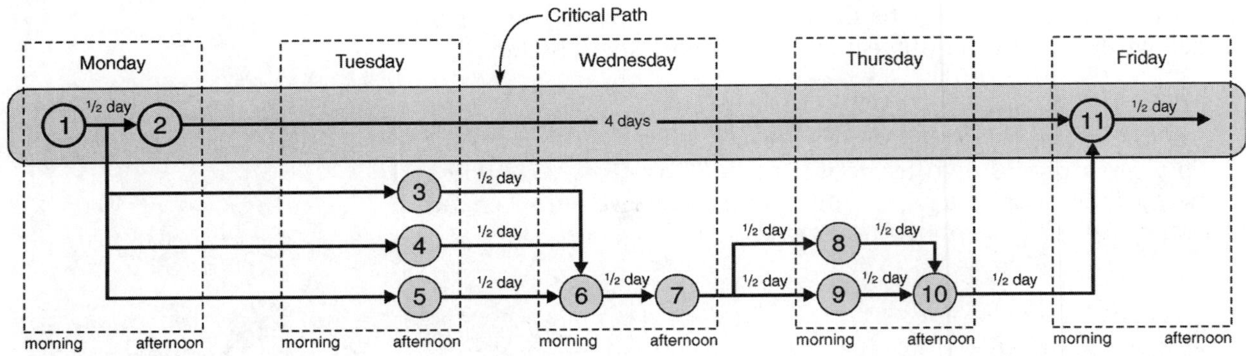

Legend:

Activity ID	*Activity Description*
1	Sales to Ops Turnover Meeting
2	Secure Low-Voltage Permit
3	Scope Verification
4	Risk Assessment
5	Preliminary Schedule
6	WBS Review
7	Cost Estimate Review
8	Resource Commitment
9	Critical Path Analysis
10	Finalized Project Schedule
11	Project Kickoff Meeting

FIGURE 10–6 Network Diagram of the Planning Segment of the Cabling Phase with Duration Times, Critical Path Late Start Dates, and Late Finish Dates

TABLE 10–4
Late and Early Start and Finish Times

Activity ID	*Start*	*Finish*	*Late Start*	*Late Finish*	*Free Float*	*Total Float*
1	5/16/05	5/16/05	5/16/05	5/16/05	0 days	0 days
2	5/16/05	5/20/05	5/16/05	5/20/05	0 days	0 days
3	5/16/05	5/16/05	5/17/05	5/17/05	0 days	0.5 days
4	5/16/05	5/16/05	5/17/05	5/17/05	0 days	0.5 days
5	5/16/05	5/16/05	5/17/05	5/17/05	0 days	0.5 days
6	5/17/05	5/17/05	5/17/05	5/17/05	0 days	0.5 days
7	5/17/05	5/17/05	5/18/05	5/18/05	0 days	0.5 days
8	5/18/05	5/18/05	5/18/05	5/18/05	0 days	0.5 days
9	5/18/05	5/18/05	5/18/05	5/18/05	0 days	0.5 days
10	5/19/05	5/19/05	5/19/05	5/19/05	0.5 days	0.5 days
11	5/20/05	5/20/05	5/20/05	5/20/05	0 days	0 days

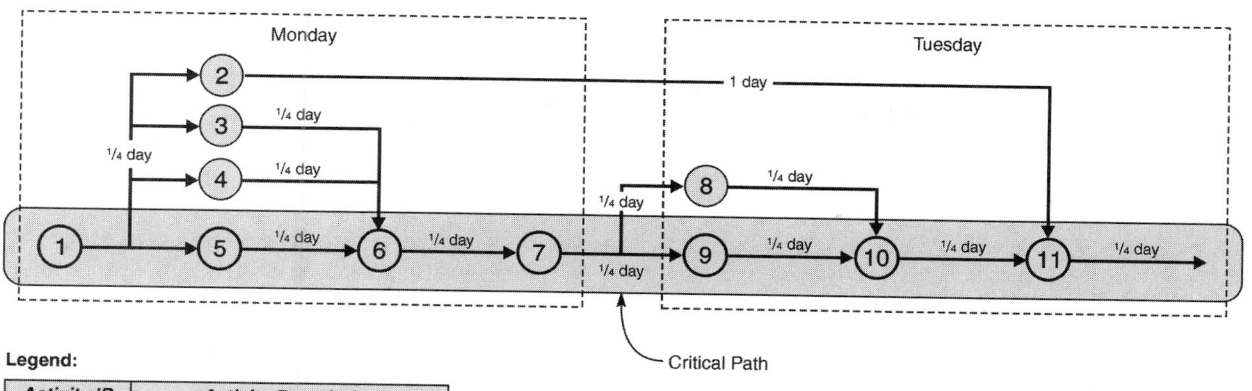

Legend:

Activity ID	Activity Description
1	Sales to Ops Turnover Meeting
2	Secure Low-Voltage Permit
3	Scope Verification
4	Risk Assessment
5	Preliminary Schedule
6	WBS Review
7	Cost Estimate Review
8	Resource Commitment
9	Critical Path Analysis
10	Finalized Project Schedule
11	Project Kickoff Meeting

FIGURE 10–7 Fast Tracked Critical Path for the Planning Segment of the Cabling Phase with Duration Times and Critical Path, Reduced to Two Days

Thirty minutes later you get a call from the owner of your company, telling you to start the cabling on Wednesday, not a week from Monday. How do you manage the challenge? Can you reduce the five-day critical path to two days? Is it possible? Remember that the plan is based upon duration rather than effort. So can we reduce the duration of the events, while keeping the effort at the same level? The surest way to do this is to make the project a priority, and put off other responsibilities in order to reduce the duration of each activity.

You inform the company owner that it can be done, but this project needs to be **fast tracked,** which is compressing the project schedule and overlapping activities that would normally be done in sequence. Fast tracking a project puts it ahead of any other project currently in progress, which will result in slippages on other project schedules. Back to our example in Figure 10–6, the longest duration item is Activity 2, obtaining the low-voltage permit. This task can be hand delivered on Monday, Day 1, and can be completed by the first day. This results in reducing the duration of the segment by one day. Next, we look at each of Activities 3 through 10 and assign them one-quarter of a day in which to accomplish the task. This is possible because everyone on the team works exclusively on this project. We redraw our network diagram, as shown in Figure 10–7.

Yes, it is possible to get the planning segment of the cabling phase completed in two days, thus laying the groundwork for arriving at the job site on Wednesday to start cabling the home. The critical path is changed from Activities 1-2-11 to Activities 1-5-6-7-9-10-11. This change happens because the low-voltage permit is hand delivered, which takes Activity 2 out of the critical path. And the sales representative gladly agrees to accomplish this task. It should also be noted that the current plan does not allow for any slippages or mistakes in the process. Every team member gives this project his full attention, resulting in time and ultimately cost overruns on other projects.

A good project manager always builds his plan for normal working conditions, thus leaving room to compress the critical path on occasion, when the political environment or other circumstances require it.

Summary

In Chapter 10, we discussed time management planning, activity definition, activity resource estimation, activity sequencing, activity duration estimation, and schedule development. We also explained the three dependencies in activity sequencing and the types of resources that a project manager considers in planning a project. We discussed schedule development and noted several reasons why companies outsource. We explained critical path, free float, and total float and their importance in a critical path. We also explained fast tracking as a means to speed up a project by allocating additional resources to it.

The important points in this chapter are:

- Time management planning involves the project constraint of time.
- The activity definition is the step that identifies the specific activities that the project team carries out in the process of producing the project deliverables.
- In activity resource estimation, the project manager prepares a list of internal and external resources that have been identified in order to meet the established milestones in the project life cycle.
- Activity sequencing is the process of identifying in what order project tasks take place.
- The three dependencies in activity sequencing are mandatory, discretionary, and external.
- The activity duration estimation is the process to estimate how long project activities will take to complete.
- In residential integration, the activity definition is the scope, the activity sequencing is the time, and the activity duration corresponds to the cost of the activity.
- In schedule development, resources are committed by placing a team on a project team schedule. The project manager should provide a master schedule to each team member to see his workload. The project manager takes the duration and sequence information to block time off the schedule to commit each team.
- In resource commitment, the project manager prepares a list of internal and external resources that have been identified in order to meet the established milestones of the project. From that list, the project manager determines which available resources will be committed to a given project.
- Resources include equipment, materials and supplies, people, and money needed to complete a project on time.
- Companies outsource to save money, to provide a highly specialized service, or to quickly expand and contract its workforce.
- The critical path of a project is the sequence of activities or tasks that, when followed in

logical order, takes longer to complete than any other sequence in the project. A delay in the critical path causes a delay in the project completion date.

- There is no effect on the duration of this segment in a project when a total float activity does not start for a day, or is delayed while in progress by as much as a day.
- The concepts of total float and free float are important because they provide flexibility in a project schedule.
- The duration is not the time spent working on the activity, the effort, but the elapsed time plus the working time.
- Ideally, the project manager first defines the scope of the project, then the timeline is developed, and finally the cost is determined.
- The process to review the WBS can be as simple as looking over the document to make sure it covers all aspects of the task or project ahead.
- The process to review the cost estimate is a simple check of the figures to make sure it matches the items called for in the WBS.
- Forward pass and backward pass techniques are used to determine early and late start dates and finish dates.
- Fast tracking is a specific project schedule compression technique.

Key Terms

Activity definition The process of identifying the specific schedule activities that need to be performed to produce the various project deliverables.

Activity description A short phrase or label for each schedule activity used in conjunction with an activity identifier to differentiate that project schedule activity from other schedule activities. The activity description normally describes the scope of work of the schedule activity.

Activity duration The time in calendar units between the start and finish of a schedule activity.

Activity identifier A short, unique numeric or text identification assigned to each scheduled activity to differentiate that project activity from other activities. Typically, unique within any one project schedule network diagram.

Activity list A documented tabulation of schedule activities that show the activity description, activity identifier, and a sufficiently detailed scope of work description so the project team members understand what work is to be performed.

Activity sequencing The process of identifying and documenting dependencies among schedule activities.

Backward pass The calculation of the late finish dates and late start dates for the uncompleted portions of all schedule activities. Determined by working backward through the schedule network logic from the project's end date. The end date may be calculated in a forward pass or set by the customer or sponsor.

Critical activity Any schedule activity on a critical path in a project schedule. Most commonly determined by using the critical path method. Although some activities are "critical," in the dictionary sense, without being on the critical path, this meaning is seldom used in the project context.

Critical path Generally, but not always, the sequence of schedule activities that determines the duration of the project. Generally, it is the longest path through the project. However, a critical path can end, as an example, on a schedule milestone that is in the middle of the project schedule and that has a finish-no-later-than imposed date schedule constraint.

Critical path method A schedule network analysis technique used to determine the amount of scheduling flexibility (the amount of float) on various logical network paths in the project schedule network, and to determine the minimum total project duration. Early start and finish dates are calculated by means of a forward pass, using specified start date. Late start and finish dates are calculated by means of a backward pass, starting from a specified completion date, which sometimes is the project early finish date determined during the forward pass calculation.

Duration The total number of work periods, not including holidays or other nonworking periods, required to complete a schedule activity or Work Breakdown Structure component. Usually expressed in workdays or workweeks. Sometimes incorrectly equated with elapsed time.

Early finish date In the critical path method, the earliest possible point in time on which the uncompleted portions of a schedule activity or the project can finish, based upon the schedule network logic, the data date, and any schedule constraints. Early finish dates can change as the project progresses and changes are made to the project management plan.

Early start date In the critical path method, the earliest possible point in time on which the uncompleted portions of a schedule activity or the project can start, based upon the schedule network logic, the data date, and any schedule constraints. Early start dates can change as the project progresses and changes are made to the project management plan.

Effort The number of labor units required to complete a schedule activity or Work Breakdown Structure component. Usually expressed as staff hours, staff days, or staff weeks.

Fast tracking A specific project schedule compression technique that changes network logic to overlap phases that would normally be done in sequence, such as the design phase and construction phase, or to perform schedule activities in parallel.

Forward pass The calculation of the early start and early finish dates for uncompleted portions of all network activities.

Free float The amount of time that a schedule activity can be delayed without delaying the early start of any immediately following schedule activities.

Late end date The latest date an activity can end without impacting the schedule.

Late start date The latest date and activity can start without impacting the schedule.

Logical relationship A dependency between two project schedule activities, or between a project schedule activity and a schedule milestone. The four possible types of logical relationships are finish-to-start, finish-to-finish, start-to-start, and start-to-finish.

Outsourcing The procuring of services or products from an outside supplier or manufacturer in order to cut costs and/or to speed up project deliverables.

Predecessor activity The schedule activity that determines when the logical successor activity begins.

Project time management Includes the process required to accomplish timely completion of the project.

Total float The total amount of time that a schedule activity may be delayed from its early start date without delaying the project finish date, or violating a schedule constraint. Calculated using the critical path method technique and determining the difference between the early finish dates and late finish dates.

Review Questions

1. List the five major processes used in time management planning.
2. For what purpose do project managers use activity definition?
3. Why should a project manager review the planning process?
4. What is activity resource estimation used for?
5. What are three reasons residential integration companies use outsourcing as a business methodology?
6. What are the three dependencies associated with activity sequencing?
7. What is the activity duration estimating process? What is the difference between it and "effort"?
8. What is the purpose of developing a project schedule?
9. What is critical path?
10. What is fast tracking?

Risk Management Planning

OBJECTIVES

After studying this chapter, you should be able to:

- Define and discuss the term project risk management.
- List and discuss the six major processes that comprise project risk management.
- Define a risk management plan and its use.
- Identify common risks in the book *A Guide to the Project Management Body of Knowledge.*
- Review an example of Qualitative Risk Analysis.
- Review an example of Quantitative Risk Analysis.
- Define the components of a risk response plan.
- Define contingency and fallback plans.
- Discuss the importance of planning meetings and analysis.

OUTLINE

Introduction

What is risk? In the residential integration industry, risk can come in many forms. A company needs to know what risks are involved in a project before deciding if it is worthwhile to start it. Failure to do so can mean cost overruns, the inability to complete the project on time, customer dissatisfaction, and a whole host of other potential setbacks. Here's a case in point.

A CASE IN POINT

Backfire

ABC Integrations took the largest project in its history. It was going to be over $300,000, and at a time when the revenue of the organization was only $300,000 per year. This project was going to double the size of the company overnight. Is there a risk in growing this fast? Yes, you bet. The project required extensive design documentation, and ABC Integrations prepared an 80-page design document which included block diagrams of each system, schematic drawings of interconnections, and a programming specification. The company took on new space and hired two new installers, adding to the two already on staff. One week prior to the start of the cabling phase, the customer cancelled the project. That's right, cancelled! The customer had provided the design document to a buddy in the car stereo business who assured the owner that he could get the same work done for less.

Yes, there were copyright and breach of contract issues that gave room for ABC Integrations to sue the owner. However, would ABC Integrations be able to stay in business long enough to carry out the lawsuit? It took 18 months for the company to rebound financially from this contract failure.

In the end the company instituted a policy to not take on any one project that represented more than one month's revenue.

What are the risks that the company executives took in the failed project?

- They took on a project that was more than they could afford to lose.
- They created a comprehensive design document that allowed others the ability to complete the work and then gave it to the customer.
- They hired new employees.
- They took on new space for their company.

Certainly not all projects are as risky as this one and this may be seen as an extreme example. Some projects are simply burdened with financial risks that could mean the difference between a highly profitable project and one that is a break-even situation. Nonetheless, a project manager must consider all the project risks and learn how to manage them to successfully complete a profitable project. In this chapter we will address those project risks and how to control them.

Study/Career Tip

Question what you read. The more you question, the more you will concentrate on the meaning, and the more you will learn. Try to determine if you agree or disagree with the information.

Critical Thinking Questions

- What types of risks to you encounter in your daily life?
- Are they big risks, or small (or perhaps a bit of both) ?
- How do you go about avoiding these risks, or at least minimizing their impact?
- Are risks okay?

Project Risk Management

Risks are present in all projects, whatever their size or complexity and whatever industry or business sector. In order to minimize project risks, project managers should adopt a formalized process known as project risk management.

Project risk management is the process of identifying, assessing, allocating, and managing all of the project risks. Through the use of expert judgment, techniques, and tools project managers are able to recognize and alleviate risk events throughout the project life cycle in a manner that is in the best interests of the project's objectives.

Project risk management is comprised of the following major processes:

- Risk Management Planning
- Risk Identification
- Qualitative Risk Analysis
- Quantitative Risk Analysis
- Risk Response Planning
- Risk Monitoring and Control

Before we take a detailed look at each of these processes, here are several project risk management terms that you should become familiar with to guide us through our review.

- **Uncertainty** is a condition that drives everything in risk management. Uncertainty is a lack of information, knowledge, or understanding of an outcome of an action, decision, or event and thus how the project will be affected either favorably or unfavorably. Uncertainty gives rise to both opportunity and risk (see below).
- Project managers continually experience uncertainty in residential integration projects. A project manager may be familiar with the installation of the deliverables offered by his company, but each home is unique—even tract homes. So, every time a new project is planned, there is an element of uncertainty and an element of risk. A project manager's ability to manage the uncertainty and to move from uncertainty to certainty, reduces the amount of risk. That is one reason why gathering information about the job site is so important.
- **Risk** measures how much uncertainty (first bullet) exists. It's directly linked to the amount of information that is known. Risks are those factors (threats, see below) that may cause a failure to meet the project's objectives. Risks may also be associated with opportunities (see below). Figure 11–1 illustrates how as information becomes known, risk declines.
- **Threats** are project risks that jeopardize the project's objectives and success. Threats to the project may be allowable if they are in balance with the benefit that may be gained by taking the risk. To illustrate, fast tracking portions of a project schedule that may be overrun is a known risk that should result in an earlier completion

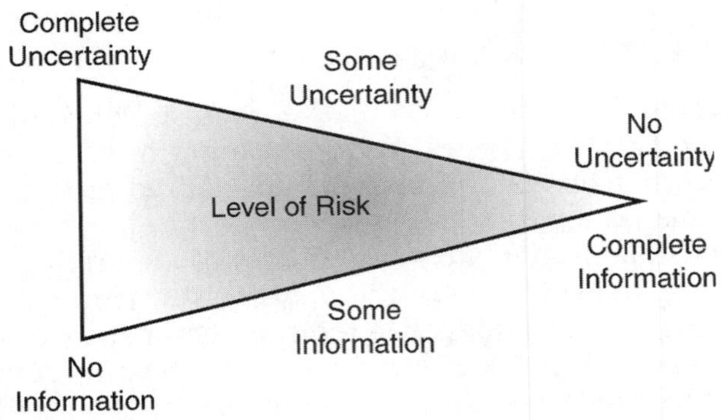

FIGURE 11–1 Risk Level

date. A project manager may wish to take this risk in order to reap the reward.

- **Opportunity** is the cumulative effect of the chances of uncertain occurrences that will affect project objectives positively. Opportunity is the opposite of risk.

Project managers must understand all the possible project risks, then gather as much information, knowledge, and understanding as possible to be able to assess the risks and minimize those risks to reach a successful and profitable completion of a project. In the world of residential integration, project management's ability to predict a particular outcome with certainty relates directly to the amount of risk the company will be taking. Studies of decision and risk analysis have verified this conclusion.

Study/Career Tip

Try to connect the information you are learning to information in other courses you may be currently taking or have taken. The more connections you make, the better you will establish the information in your long-term memory.

Critical Thinking Questions

Refer to Figure 11–1.

- If you have complete information, and thus no uncertainty, could there still be a level of risk?
- Why or why not?

Risk Management Planning

Risk management planning is identifying, analyzing, and responding to potential loss or injury during the project life cycle. By identifying what can go wrong along the way, a project manager can create a realistic **Risk Management Plan.** This plan details how these potential issues will be handled, when and if encountered. It assesses possible impact and what is to be done (and when) to avoid, remove, and control risks. It also includes the detailed processes for managing the risk. As identified earlier, these processes include risk identification, qualitative and quantitative analysis, response planning, monitoring, and control.

Risk Identification

Risk identification involves determining which risks are likely to affect the project and documenting the characteristics of each. For example, in cabling a home, the type of framing might pose a roadblock to a cabling plan. The project manager identifies the risk and records it. Risks are identified to minimize their affect on a project and to find workable solutions around them.

Identifying the risks in a project is the first step in figuring out what you're up against. It defines those things that threaten your ability to deliver what you've promised. As we said earlier, the risk is the uncertainty of not knowing exactly how things are going to turn out. There are many aspects of a project that are unpredictable, even though we try our best to recognize them and pin them down. In the residential integration industry, some of the most common areas of uncertainty are labor hours, other subcontractor schedule delays, and new technology. Problems result from these sources of uncertainty. In a residential integration project, thinking of all the potential problems is how you minimize them. The problems you don't consider can end up creating expensive headaches for you and your project team. How do you go about identifying problems? There's no magic formula. It's going to require lots of specific knowledge of the project, significant brainpower, the ability to speculate, and a fair amount of experience. Survey templates can help you gather some of the information you'll need to assess on each project.

The following are three common ways to identify project risks.

- **Brainstorming.** Brainstorming is a very common and frequency used risk identification technique. The project manager should assemble the appropriate groups of people, such as project team members and others, and conduct a meeting where all ideas are considered acceptable (within reason, of course). The project manager should set aside at least two to four hours for this brainstorming session, depending upon the size and complexity of the project. Participants should bring any and all documentation as supporting evidence.

  When brainstorming, remember to accept all ideas without negative comments to encourage the free flow of ideas. The project manager, or someone else, should write down all the ideas, then later categorize them according to risk levels and the area of knowledge as designated by *The Guide to the Project Management Body of Knowledge*. Finally, brainstorm solution ideas. Some of the best teambuilding activities are those when team members expand their own knowledge of each other and the project. Identifying potential project problems as a team is an ideal way to accomplish both at the same time. And it helps build the collective brainpower of the team.

- **Interviewing.** A project manager can interview each member of the project team, engineering team, and sales team to find out their ideas. These interviewees identify project risks based on their experience, on project information provided by the project manager, and from any other sources they feel will benefit the risk identification process.
- **Previous Projects.** A project manager should keep a file of all risks encountered on previous projects. He should also record how the risks were reduced or resolved and how the risks affected the project in terms of time, cost, and scope. That information can be used in formulating the risk management plan on projects going forward.

Study/Career Tip

Consider using brainstorming techniques when studying or reviewing materials. Form a study group to suggest what questions might be on an upcoming exam or ways to learn the material better.

Critical Thinking Questions

- When it comes to the concept of brainstorming, whether for a residential integration project or other business scenario, why do you think it's important to let all ideas (good or bad) flow?
- Isn't that a waste of precious company time and dollars? Provide examples.

Categories of Risk

The first step to understanding risks in a project is to divide the risks into categories as identified in *A Guide to the Project Management Body of Knowledge*. This helps a project manager organize how to think about these risks as well as how best to respond to them. These knowledge areas include:

- Project Scope Management Risks
- Project Cost Management Risks
- Project Time Management Risks
- Project Quality Management Risks
- Project Integration Management Risks
- Project Human Resource Management Risks
- Project Communications Management Risks
- Project Procurement Management Risks

Project Scope Management Risks. Project scope management involves initiation, scope planning, scope definition, scope verification, and scope change control. Below are a few examples of common scope management risks involved in a residential integration project.

- **Ambiguity in Scope.** A poorly written scope statement that is not clear in identifying each project deliverable can create the risk that the deliverables will not be delivered properly or fall short of customer expectations. For example, if the telephone and networking wallplate states that it includes two category 5e cables, but does not state the purpose of each cable, the customer may become confused as to the purpose of the cables, possibly resulting in poor customer satisfaction.
- **Plan and Specification Discrepancies.** Oftentimes customers provide house plans and other documentation to the sales representative to aid in the preparation of the scope statement. In some cases, there are discrepancies between various plans provided by the customer. For example, one plan might call for two category 5e cables to each telephone and networking wallplate, and another document might call for three category 5e cables. This, then, become a project risk because of inaccuracies between the two documents.
- **Extended Warranty.** The length of time the extended warranty is in effect extends the risk on the project. A 90-day warranty inherently has less risk than a 5 year warranty.
- **Special Test Equipment/Tools.** A home theater project might require the use of a special audio analyzer. Acquiring the tool, learning its proper usage, and training an employee on its use all add to the risk of the project.

Project Cost Management Risks. Cost management involves cost estimating, cost budgeting, and cost control. Below are a few of the cost management risks that may be encountered in a residential integration project.

- **Creditworthy Contractor or Customer.** A risk taken on by all residential integration companies is the ability of contractors and customers to pay within the agreed upon terms stipulated in the contract. When payments are made late, cash flow issues arise.
- **Payment Terms.** Standard payment terms should provide a positive cash flow for the residential integration company. However, a project might have payment terms that put the residential integration company in a situation where cash flow is negative during a period of the project, which puts additional risk on the project.
- **Inflation.** Some projects can take more than a year to complete. The price of electronics and other materials can go up during that time. Labor costs may also need to be adjusted.

Project Time Management Risks. Time management involves activity definition, activity sequencing, activity resource estimating, activity duration estimating, schedule development, and schedule control. Below are a few examples of time management risks frequently encountered in a residential integration project.

- **Fast Track Schedule.** A fast track schedule compresses the original schedule, permitting multiple phases of a project to exist simultaneously. This can create several risks, which include work not completed from a previous phase and thereby not allowing full efficiency of the next team.

- **Extended Project Schedule.** A project that is delayed also represents additional risk. For example, the project is getting ready to finish, however, it is learned that the guest suite will not be finished on schedule, while the rest of the home will be completed. A reasonable plan must be developed to deal with issues that arise from the delay.
- **Completion Penalties.** In some projects, a penalty might be applied to the project if it is not completed by a certain date. The risk is in not completing the work in time.
- **Special Working Hours.** Special working hours create a situation of nonstandard working hours for the installation teams; overtime charges and unnecessary down time can result.

Project Quality Management Risks. Quality management includes quality planning, quality assurance, and quality control. We can think of quality as the culmination of meeting scope, time, and cost constraints on a project. So what exactly is a quality management risk? A few risks that may impact the quality management of a project are listed below.

- **Fast Tracking a Project.** If the development timeline is compressed, it is possible to skip steps in the project, which can have a direct impact on the quality of a project. Let's say, in order to meet an accelerated project timeline, the testing of the electronics is accomplished at the customer's home rather than in-house (at the company facility). In this customer's home environment, the ability to test each feature of the system is limited and, as a result, there may be unforeseen problems.
- **Cost Reengineering.** Cost reengineering involves reducing system costs while maintaining as much of the original system deliverables as possible. For example, in an effort to reduce cost, a backup infrared repeating system for a home theater system is removed. Instead, the only point of control for the system is the touchscreen in the home theater room. The handheld remotes do not work in the theater room. They must be used in front of the equipment, which is located outside the room, and not in the line-of-sight of the screen. As a result, the touchscreen is not working and the system is unusable. The result: poor customer satisfaction.
- **Scope Creep.** This is the process of adding additional features to a project without reengineering the foundation of the system. An example of this is adding a single television wallplate to a system. Let's say the distribution panel can handle eight wallplates, and we have now added the ninth wallplate without adding a second distribution panel for the television wallplates. In order to provide signal to the added wallplate an inexpensive splitter is added to the distribution panel, providing a ninth wallplate. The wallplate

works; however, the quality of signal is greatly degraded. This might cause poor customer satisfaction due to one wallplate providing an inferior television picture to the others in the home.

Study/Career Tip

The concept of project risk management has many terms that sound similar, although their meanings can be quite different. For example, consider risk management plan, risk management planning, risk response plan, and risk response planning. There are many more examples. To make sure you can differentiate among these terms, start a personal electronic glossary by writing in what you think the definition is, then following up with the glossary in this text or from other sources.

Critical Thinking Question

There's an adage (old or otherwise) that states: "We never have time to do the job right, but we always have time to do it twice." Explain how project quality management risks play into this statement.

Project Integration Management Risks. Project integration management involves directing the project execution, monitoring and controlling project work, change order control, and closing the project. Some of the risks involved in this area are:

- **Relationship with Owner, Engineer, Contractor and Subcontractors.** This can take the form of working with one of the stakeholders on a prior project. Perhaps they were nonresponsive to issues in the past, and as a result are likely to be difficult to work with on the current project.
- **Slow Change Order Response.** If the process of generating change orders is slow, then it will be difficult to complete the work on schedule. This can be due to internal processes or due to signoff by one or more external stakeholders, such as the customer, builder, or architect.

Project Human Resource Management Risks. Project human resource management involves organizational planning, staffing, and team development. The primary risk involved with human resource management is:

- **Availability of Trained Manpower.** In some areas of the country the manpower available to perform specific tasks is limited and in some cases not available. Let's look at the example of an Image Science Foundation (ISF)-trained calibration technician. The ISF technician is trained to calibrate video displays such as plasma, DLP projectors, and other specialized video display devices. If an ISF technician is not available for hire on staff, nor available in the immediate area as a consultant, the organization will need to determine whether they will offer this calibration service or to seek out the training needed to complete the installation.

Project Communications Management Risks. Project communications management involves communications planning, information distribution, performance reporting, and administration closing. Two of the greatest risks involved with communication management are:

- **Communications with the Stakeholders.** While most people today have cell phones and e-mail accounts, there may be subcontractors, and even customers, who are not readily accessible to help in key project decisions.
- **Performance Reporting.** At the end of each phase of the project a performance report is delivered to all key stakeholders, keeping them up to date on project deliverables. If the means of communicating with the key stakeholders is not well defined, then important project decisions can be delayed. To illustrate, if the builder is to receive the performance report prior to authorizing the next scheduled payment, it is imperative the report be accurate, on time, and delivered to the appropriate person at the building contracting company.

Project Procurement Management Risks. Project procurement management involves procurement planning, solicitation of vendors and subcontractors, source selection, contract administration, and contract closure. Some of the risks involved with procurement are listed below:

- **Reliability of Outside Quotes.** When using an outside contractor to provide certain services on the project, it is imperative that the quote be accurate. For example, a subcontractor may be used to install a satellite dish. It's important to note if the installation includes cabling, and if so, to what point?
- **New Subcontractors.** New subcontractors are unfamiliar with the operational processes of the organization. The risk here is the ramp-up time needed to understand how the company operates on the job.
- **New Vendors.** New vendors can introduce a host of issues to contend with for the project team. Is the team adequately trained on the product? Does the administration staff understand how to communicate with the vendor? What are the typical shipping and back order times for the products?

Qualitative Risk Analysis

Qualitative risk analysis is the process of identifying and evaluating risk factors, present or anticipated, and determining both the probability and the impact of those identified risk factors. It is a preliminary step in project risk management and is intended to increase the probability that the project will produce the desired outcomes (project objectives) while diminishing risk factors. Some of the factors considered in the qualitative risk analysis for each system are identified in the **Risk Matrix** (see Table 11–1). Each factor in the

TABLE 11–1
Risk Matrix

Risk Category	*High Risk (H)*	*Medium Risk (M)*	*Low Risk (L)*	*N/A*
Project Scope Management	Special Test Equipment/Tools	Plan and Specification Discrepancies	Ambiguity in Scope Inflation Extended Warranty	
Project Cost Management		Payment Terms	Credit Worthy Contractor or Customer Inflation	
Project Time Management	Fast Track Schedule		Completion Penalties	Extended Project Schedule Special Working Hours
Project Quality Management	Fast Tracking a Project		Scope Creep	Cost Reengineering
Project Integration Management		Relationship with Owner, Engineer, Contractor, and Subcontractors	Slow Change Order Response	
Project Human Resource Management	Trained Manpower Available			
Project Communications Management			Communications with the Stakeholders Performance Reporting	
Project Procurement Management			Reliability of Outside Quotes	New Subcontractors New Vendors

matrix is rated according to its potential adverse impact as High (H), Medium (M), Low (L), or Not Applicable (N/A). As you can see, each of the risk categories has been assigned identified risks and the associated qualitative risk level.

Table 11–1 can be created as a template prior to starting the project and maintained during the project, updating it at each phase. The N/A column accounts for identified risks that no longer offer any level of risk or do not apply to the current situation. This particular table was created at the start of the project; however, each risk event may shift as the project moves forward.

Quantitative Risk Analysis

Quantitative risk analysis involves measuring the probability and consequences of risks and estimating their affect on the project work. ABC Integrations uses quantitative risk analysis to determine which projects they are willing to pursue. Here's a case in point.

A CASE IN POINT
Big Fish

Project One involves integrating all of the major electronics systems in a large home. The project is expected to take 18 months to complete and has three companies bidding on it. This would be the largest project taken on by ABC Integrations, but it is of interest to the executive team because it is estimated that the gross profit margin will be $200,000. It will take 40 hours of time to prepare the project scope statement and cost estimate and to present these documents to the customer. This represents approximately $3,000 in internal resources to prepare for the bid.

Because of the multiple bidders and the fact that the project is so large, ABC Integrations estimates that they have a 20 percent chance of being awarded the project. If that is the situation, the best-case profit scenario calls for a 40 percent gross profit and the worse-case scenario of 20 percent gross profit. Because of the size and length of the project, Project Manager Owen estimates that it is more likely that ABC Integrations will fall into the worst-case profit scenario.

TABLE 11–2
Project One Probability of Outcome

Situation	*Probability*	*Profit or Loss of Situation*	*Outcome (EMV)*
Not Awarded Contract	20%	–$3,000	–$600
Awarded Contract, Maximum Profit (40%)	20%	$200,000 × 40% = $80,000	$16,000
Awarded Contract, Minimal Profit (20%)	60%	$200,000 × 20% = $40,000	$24,000
		Project One's EMV	$39,400
		Project One's EMV / Month (18 months)	$ 2,188

As Table 11–2 indicates, ABC Integrations has three probable outcomes: the company will not be awarded the project, it will be awarded the project and obtain maximum profit, or it will be awarded the project with minimal profit. The sum of each of the probable outcomes must equal one (100%). The probability of the situation is multiplied by the potential profit or loss of that situation; this represents the outcome or **Expected Monetary Value (EMV)** of the situation. Each of the individual EMVs are summed to created a total EMV for the project. We can then take the EMV and divide it by the number of months this project will occupy for ABC Integrations.

Let's review a second case in point. See page 223.

Table 11–3 shows the EMV for Project Two.

Knowing the EMV for each project, Project Manager Owen can make a decision on which project to take. Or perhaps only projects with an EMV of higher than a value set by the executive team can be taken for bid. To illustrate, if the executive team set a minimum EMV/month of $2,000, then both projects will be bid by the company.

A CASE IN POINT
Little Fish

Project Two is a bread-and-butter project for ABC Integrations. It is a home theater system installation costing $18,000. While there is one other bidder on the project, the customer is a referral and Project Manager Owen estimates the probability of obtaining the project to be high at 75 percent. In addition, the project will last two months, and the cost of preparing the estimates is figured at $500. Since ABC Integrations has completed this type of project many times, the expected profit can be accurately estimated at 47 percent.

TABLE 11–3
Project Two Probability of Outcome

Situation	*Probability*	*Profit or Loss of Situation*	*Outcome (EMV)*
Not Awarded Contract	25%	–$500	–$125
Awarded Contract	75%	$18,000 × 47% = $8,460	$8,460 × 75% = $6,345
		Project Two's EMV	$ 6,220
		Project Two's EMV / Month (18 months)	$ 3,110

Study/Career Tip

Learn to recognize what is important during a class lecture. Instructors will often tell you directly or give you clues. Clues might include repetition, emphasis, writing it on the board, putting it into the slide presentation, etc. Assess your instructor's style and determine how she emphasizes important points.

Critical Thinking Question

List three factors (or more) that show the value of calculating a project's EMV.

Risk Response Planning

Risk response planning entails developing procedures to enhance opportunities and minimize threats to the project's objectives. Naturally, risk response planning is an essential part of any risk management plan. Why is it essential? Remember the old adage: Failing to plan is planning to fail. This is never truer than in project risk management. The financial health of the company depends on accurately assessing risk and minimizing it as much as possible. An installer could fall off a ladder, injuring his back, anytime he ignores safety guidelines and steps up on that last step of a stepladder. But how many installers always follow that guideline? They assess the risk.

Risk Response Plan

A **risk response plan** is a document that details all of the identified risks once project implementation has begun and how to specifically control those risks. This is not the same thing as a risk management plan that, as we have seen, does not address responses to specific risks. There are several risk response strategies that are widely accepted for use in developing a risk response plan. The project manager should select the approach that is most likely to succeed in eliminating the risk. These include:

- **Risk Avoidance.** In avoidance, a course of action that eliminates exposure to the threat is chosen. This can mean following a completely different course from what was initially planned. The space shuttle program is an excellent example of avoidance. NASA plans many flights very carefully and then scrubs them because of weather conditions. Delaying the flight of a space shuttle because of the risk is an example of risk avoidance. In the residential integration industry, an example of avoidance would be canceling a project or a portion of a project. A customer may request integrating something into the home-wide system that is not practical. Surprisingly, it could be a request to integrate the motorized cover of a swimming pool. Technically you may be able to do it, but at what risk to the entire system? That is a risk best avoided by not agreeing to it at all.
- **Risk Mitigation.** Mitigation reduces the effect of the risk. This strategy focuses on lessening the negative effects of a problem. For example, following a regular exercise program and healthy nutrition guidelines doesn't mean you'll never get sick, but it does give you a much better chance at a reasonably long and healthy life. Some may say mitigation tactics are a waste of time, money, and effort if a potential problem does not occur. In the residential integration industry, a company may have a policy of always sending two installers on a job for safety reasons. If they are the only two people at the job site and an accident occurs, the second installer can call for help. But if no one ever has an accident, which is doubtful, then it could be viewed as a waste of one installer's time and an added expense to the organization.
- **Risk Assumption.** In risk assumption, you are aware of the risk, but have decided to take no action concerning the risk. That means you are agreeing to accept the consequences or simply deal with them when and if the risk occurs. That's how most project managers treat minor risks. Assumption is also a strategy in situations where the consequences of the risk are less costly and/or less trouble than the effort required to prevent it. For example, there is a risk that other trade subcontractors, such as the electrician, will not finish their work on time, creating a slippage in the schedule. While

there is a risk of this happening, it is assumed that it will not, and the issue will be dealt with when it actually happens.

- **Risk Transference.** Transference is shifting the consequence of the risk to a third party. Insurance is a prime example of risk transference. It doesn't actually eliminate the risk; it merely makes the insurance company responsible should the risk occur. Of course, there is a cost involved—regular insurance payments. One way a project manager can transfer risk, other than insurance such as Worker's Compensation, is to spell out the responsibilities of the builder and other subcontractors before the work begins. For example, a memo to an HVAC contractor could say that he is responsible for purchasing the programmable thermostats for an integrated HVAC system. That means any problems with the thermostats are his problems even though they are tied to the integrated system.
- **Risk Prevention.** How many times have you heard "an ounce of prevention is worth a pound of cure"? Prevention is what you do to reduce the chance that a problem might occur. Prevention generally is a project manager's first line of action in handling high-risk issues, such as wearing safety glasses to prevent eye injuries or wearing hard hats on home construction sites when appropriate. The action can be simple, but ignoring it can have devastating consequences. Finding the root of a problem and identifying the causes go a long way in preventing potential problems. Corrective action to prevent these potential problems should be incorporated into the risk management plan for a project. This helps the project team remember what problems could happen if they ignore the preventive measures.

A set of standard questions is often used when formulating the risk response plan. Let's look at a specific example for ABC Integrations and answer some risk planning questions. Here is a case in point.

A CASE IN POINT
Risky Business

ABC Integrations is taking on a structured cabling project for a customer in an existing home. The home is five years old with an open basement and access to the second floor via the attic. Wallplates on the first floor are to be cabled from the basement and the wallplates on the second floor are to be cabled from the attic. The attic cables are to be brought to the basement through an existing conduit installed for this purpose.

While there are many risks involved with this project, we will look at one specific risk, the time required to cable the home. The same process is then applied to other project risks.

- **What is the specific risk?** This risk is that the cabling time could take longer than anticipated.
- **Why is it important to take or not take a risk?** If we do not take this risk, we cannot take the project.

- **How will the project team resolve the risk?** The cable team leader's expertise is used to provide a best guess of the time to cable the home. In addition, the customer is notified that additional hours would be billed at a standard rate, and the cost of any unused cabling hours would be refunded.
- **Who is responsible for implementing the risk response plan?** The cable team leader tracks the cabling time; he then notifies the project manager when and if the cabling cannot be completed within the estimated time. The project manager will notify the customer prior to starting any additional work.
- **What resources are required for the risk?** The cabling team is required to complete any additional time requirements for the cabling phase.

In addition to considering the above questions to form an initial risk response plan, the project manager will also prepare possible contingency plans and fallback plans in case they are needed. These plans are important in running an efficient operation. A project manager can plan a particular outcome, but when that outcome does not happen, she must be prepared to switch to an alternative plan or she risks being unable to complete the project. Let's take a closer look.

Contingency Plans

Contingency plans are predefined actions that the project team could take in response to an identified risk event. Going back to our cabling example, the initial plan is to cable the home within the estimated hours; the contingency plan is to provide additional hours for the cabling team to complete the cabling phase of the project, if needed, because of unforeseen site conditions.

Study/Career Tip

What are your goals for this class? Your degree? Your career? Goals are both short-term and long-term, some easily attainable, others more challenging. Keep all your academic and professional goals in front of you even if many of them won't be realized quickly.

Critical Thinking Questions

- What are some contingency plans you've made of late?
- Why did you feel you had to make them?
- Did you use those plans?
- Were they successful?

Fallback Plans

Fallback plans are developed for risks that have a high impact on the project objectives. Looking again at our cabling example, the initial plan is to cable the home as specified within the scope statement.

The fallback plan for this project instructs installers not to cable certain locations if those locations cannot accept cabling. For example, perhaps the master bedroom has a cathedral ceiling, one that does not have an attic or crawl space above it. Therefore, the wallplate cannot be cabled from the attic area. Instead, it must be cabled directly from the basement through a closet located on the first floor. If this can be accomplished, ABC Integrations is following its initial plan. If this location cannot be cabled, then a secondary location within the room may be a wall that does have access to the attic. This second location is known as the fallback plan. Let's say the television is to be located at the other end of the room. To connect the TV, a cable can be run along the carpet behind furniture to reach the television location. This plan enables installers to connect the TV to a wallplate even though that's not how they initially planned to complete the task.

Planning Meetings and Analysis

Generally, a project manager schedules planning meetings to develop the risk response plan. Those attending the meeting are the project manager, selected project team members, and possibly stakeholders, such as the general contractor, architect, and other key subcontractors on the project. The group defines the basic plans for conducting the risk response activities. Risk cost elements and schedule activities are developed and incorporated into the project budget and schedule. Risk responsibilities are assigned. General organizational templates can be used to assist in defining risk categories and definitions of terms such as levels of risk, probability by type of risk, impact by type of objectives, and the probability and the risk matrix, (see Table 11–1). The templates allow this information to be tailored to the specific project under consideration.

For small projects, a formal meeting might be unnecessary, especially when the project has been consistently successfully completed for other customers in the past. The project manager may assume total responsibility for the risk response plan in this case. However, as a rule, it is a good idea to encourage the staff to suggest possible risks and solutions for any project when an idea occurs to them.

Risk Monitoring and Control

Risk monitoring and control is the act of revising the project's scope, budget, schedule, or quality, preferably without substantive impact on the project's objectives, in order to reduce uncertainty on the project. This process continues throughout the life of the project and should be evaluated at each project team meeting. As we have seen, risks change as the project matures, new risks emerge, or anticipated risks fail to materialize. Risk monitoring and control arms the project manager, project team members, and stakeholders (as appropriate) with the information they need to make adjustments during the project life cycle to ensure that risks are being managed in the most effective and efficient way.

Summary

In Chapter 11, we have defined and discussed risk management, addressed questions a project manager considers in project risk management, and identified common risks in the PMBOK Guide areas of knowledge. We also discussed the six major processes associated with project risk management, defined contingency and fallback plans, and discussed three ways to identify risks and respond to them. We discussed quantitative risk analysis and qualitative risk analysis and provided examples.

Important points in this chapter are:

- Risk management is identifying, analyzing, and responding to potential loss or injury during the project life cycle. By anticipating and identifying what can go wrong along the way, a project manager can create a realistic risk management plan.
- The six major processes in the project risk management process are risk planning, risk identification, quantitative risk analysis, qualitative risk analysis, risk response plan, and risk monitoring and control.
- Contingency plans and fallback plans are important because they provide a ready alternative when the initial plan cannot be followed.
- Risks can be identified through brainstorming, interviewing employees, and reviewing previous projects.
- Quantitative risk analysis involves measuring the probability and consequences of risks and estimating their affect on the project work.
- Qualitative risk analysis is the process of identifying and evaluating risk factors, present or anticipated, and determining both the probability and the impact of identified risk factors.
- Five ways to respond to risk are avoidance, mitigation, assumption, transference, and prevention.
- Risk monitoring and control involves monitoring known risks, reducing future costs for those risks, and evaluating the effectiveness of the risk response.

Key Terms

Contingency plan A contingency plan is designed to deal with a particular risk event, or problem.

Expected monetary value (EMV) The product of an event's probability of occurrence and the gain or loss that will result.

Fallback plans Plans developed for risks that have a high impact on the project objectives.

Opportunity The cumulative effect of the chances of uncertain occurrences that will affect project objectives positively.

Quantitative risk analysis The process of numerically analyzing the effect on overall project objectives of identified risks.

Qualitative risk analysis The process of prioritizing risks for subsequent further analysis or action by assessing and combining higher probability of occurrence and impact.

Risk The cumulative effect of the chances of uncertain occurrences which will adversely affect project objectives. It is the degree of exposure to negative events and their probable consequences. Project risk is characterized by three factors: risk event, risk probability, and the amount at stake.

Risk management plan Documents how the risk processes will be carried out during the project.

Risk matrix The presentation of information about risks in a matrix format, enabling each risk to be presented as the cell of a matrix whose rows are usually the stages in the project life cycle and whose columns are different causes of risk. A risk matrix is useful as a checklist of different types of risk that might arise over the life of a project but it must always be supplemented by other ways of discovering risks.

Risk response plan A document detailing all identified risks, including description, cause, probability of occurring, impact(s) on objectives, proposed responses, owners, and current status.

Threats Threats are project risks that jeopardize the project's objectives and project success.

Uncertainty The possibility that events may occur which will impact the project either favorably or unfavorably. Uncertainty gives rise to both opportunity and risk.

Review Questions

1. Define project risk management.
2. What are the six major processes that comprise project risk management?
3. The term "risk" is associated with two other terms. What are they?
4. What is brainstorming in terms of risk?
5. What are some project scope management risks?
6. Why might fast tracking a project schedule be considered a risk?
7. Why would the hiring of new subcontractors be considered a project risk?
8. What is a risk matrix? Why is it useful?
9. What does expected monetary value (EMV) tell a project manager?
10. What are contingency plans used for?

Procurement Management Planning

OBJECTIVES

After studying this chapter, you should be able to:

- Define procurement management planning.
- Define outsourcing and list several reasons why companies outsource.
- Explain the two phases of outsourcing.
- List and explain the four types of contracts generally used in the residential integration industry.
- Discuss make-or-buy decisions for outsourcing and why one method would be chosen over another.
- List and explain the two planning phases of purchasing.

OUTLINE

Introduction

While residential integration companies strive to complete as much of their project work "in-house," oftentimes it makes more business sense to seek resources from outside the organization.

This chapter focuses on both planning the purchasing process to acquire goods and planning the outsourcing process to acquire services in the residential integration industry. Both aspects of procurement planning are discussed separately, although there is some overlap because the principles of procurement planning apply to both purchasing and outsourcing.

Procurement Management Planning

The **procurement management plan** is a document used to purchase goods and/or services from a source that is outside the company, commonly referred to as an *outside source*. In similar terms, it is defined as a process for establishing contractual relationships to accomplish project objectives. The term procurement is known as purchasing in many companies within the residential integration industry, and typically, the term is limited to the acquisition of goods. For example, the Purchasing Department would buy all the office supplies needed by the organization. It would buy all the materials needed for a project—the nuts and bolts, equipment, cables, and any other goods needed to complete the job.

In most residential integration companies, acquiring services is handled in a different manner. Outsourcing is the term used to define the procurement of services, an industry accepted concept we discussed in Chapter 10 in relation to time management planning.

For the purposes of this chapter, we make certain terminology assumptions. Let's take a moment to clarify these assumptions in regard to how procurement management planning terms are used in this textbook and, frequently, in the residential integration industry.

In procurement management planning, a company can be either the buyer or the seller. In this chapter, we refer to the company, or **performing organization,** as the *buyer.* That makes the company the primary stakeholder. A number of principles are outlined in this chapter that apply to the company as the buyer and primary stakeholder. Additionally, we refer to the person who procures either goods or services as the procurement manager, even though in some organizations the role of the procurement manager is carried out by a project manager, an office administrator, an engineer, or even an entire procurement or purchasing department.

Study/Career Tip

Use a To Do List. Organize your day or week. Develop the list. Then, check off each item as you complete the task. This will not only assist in personal time management and organization, but it will also assist in increasing your motivation as you see your progress.

Critical Thinking Question

- Before reading any farther in this chapter, what types of services might a residential integration company outsource?

Jot these down. You'll see several examples as you continue through these materials.

Outsource Planning

Outsourcing is a growing resource for residential integration companies, making it very important for project managers to understand how and when to outsource. To begin the process of understanding this vital tool, let's look at some of the reasons why companies use outsourcing. Companies outsource to:

- **Gain specialized skills.** Rather than training in-house personnel on specific tasks that are seldom required, a company can outsource the task to an outside contractor who specializes in that area and provides those skills only when needed. For example, a residential integration company may hire a consultant to calibrate the equalizers in a very high-end home theater, which is a project rarely taken on by that company. Even though little profit will be realized in the calibration process, it takes the quality of the installation to the next level. And that provides more opportunities to provide higher-level projects. As the number of high-level projects grows and profits from the projects increase, the company can consider hiring such a specialist fulltime as a company employee.
- **Allow a company to focus on its core business.** Let's assume that a company has decided to stick to installing a particular brand of telephone system. Then a single customer comes along who would like to have another system installed. The company can provide the infrastructure for the telephone system because it adheres to company standards, and then rely on an outside consultant to install and program the specialized telephone system. This practice provides great advantage for customers. They get what they want, *and* they receive a thorough cabling installation. The outside contractor can be invisible to the customer and can even provide service of the system after installation. A project manager can arrange, in advance, for service with the subcontractor. When the customer needs service, he contacts the integration company, who then sets up the service call with the subcontractor. If payment is required, the customer pays the integration company, which, in turn, pays the contractor.
- **Allow customer-driven solutions.** When an organization allows outsource solutions, the breadth of solutions that can be provided to the customer expands. And, it follows the philosophy of focusing on a core business. Think of how a building contractor works. That contractor doesn't actually perform the electrical, plumbing, HVAC, framing, and other services. Instead, the building contractor understands enough of each of the trades to ensure that the work is completed to meet the requirements for the home. This same philosophy can be applied to the electronic systems within the home. Specialized contractors can be brought in for the various areas in the electronic system, such as lighting, HVAC controls, security, music, home theater, computer networking, shading systems, motorization, and any other system provided by the company. The company can focus on the systems that it installs regularly and use casual workers (subcontractors) to install a wide range of systems that may be only requested infrequently.
- **Reduce both fixed and recurrent costs.** Outside contractors can quite often provide a level of efficiency only realized at their scale of economy. In other words, because they perform a single job function on an ongoing basis, they are able to provide it more cost effectively. For example, let's look at the installation of

satellite dishes. If a single contractor installs only these dishes, then his truck is completely outfitted for any number of circumstances that may arise on the job site. That contractor is ready to install a dish, using a non-penetrating mount, or a chimney mount, or ground mount, or some other specialized installation method. Relying on a contractor who can provide on-the-fly solutions to complex issues allows the residential integration company to maintain less inventory, training, and specialized skills within the organization. In effect, the company is using someone else's tools, training and specialized skills—a cost-saving practice.

Study/Career Tip

Surveying is a reading technique that brings to mind what you already know about the topic of a chapter and prepares you for learning more. To survey a chapter, read the title, introduction, headings, and the summary or conclusion. Also, examine all visuals such as pictures, tables, maps, and/or graphs and read the caption that goes with each. By surveying a chapter, you will quickly learn what the chapter is about.

Critical Thinking Questions

- Do you think residential integration firms should limit the amount of outsourcing they do on any given project?
- Why or why not?
- Is there such a thing as too much outsourcing?

Outsourcing Requirements

Establishing requirements of a service is essential to avoiding potential disputes between the company and the outside contractor. As such, it is important that residential integration companies establish a policy that addresses how outsourcing decisions will be made and who within the organization will make them.

Some pertinent questions to ask may include:

- What elements should be considered in planning an outsourcing process?
- Who establishes the requirements of the service to be outsourced?
- Who describes the details of the requirements of the service to be outsourced?
- Who decides what services to outsource?
- Who decides which contractor will be hired?
- Who is responsible for "make-or-buy" decisions?
- How and where is a list of outside contractors maintained and who maintains it?

At ABC Integrations, the requirement phase of outsourcing is the responsibility of the project manager, who controls the project budget. He is also responsible for describing what is required for a deliverable through the use of specifications, drawings, and schedule parameters. In addition, the project manager is responsible for "make-or-buy" decisions. A "make-or-buy" decision is one in which the company determines whether to manufacture internally or to buy some article or item of equipment from an external source. We will be reviewing the concept of make-or-buy in greater detail later in this chapter.

Here's a case in point.

A CASE IN POINT
It's Not My Job

A customer has requested a satellite dish be installed as part of the home theater system, which is part of the company's regular product offerings. ABC's installers know how to install the dishes, but they are currently busy with many other projects. Project Manager Owen decides to outsource the job to XYZ Installer Co. to install the satellite dish at Fred and Wilma's home.

Figure 12–1 shows an excerpt of the specification that Project Manager Owen prepared for the work to be performed. In this figure, the deliverable (the installation of the satellite) is described by stating the method of installation (Items 1 through 3) and the verification process that should be used (Item 4). Figure 12–1 also clearly indicates that Project Manager Owen has outlined subcontractor specifications in various categories, such as work completion date, price, and safety code requirements.

It should be noted that, if ABC Integrations were in a slow period, Project Manager Owen could have assigned the work internally to company employees. He could easily make the change by assigning a staff member to perform the work without changing the deliverables of that work.

Let's take a moment to review Figure 12–1 in greater detail. The bulleted list below provides an explanation of what is contained in the Satellite Contractor agreement.

- **Items 1 through 3. Details of the installation** clearly spell out how the dish is to be installed.
- **Item 4. Verification** of the installation provides actual measurement numbers that can be verified during the equipment installation. If the satellite dish does not meet the quantified measurements, the subcontractor will be responsible for returning to resolve the problem.
- **Item 5. Any particular installation pitfalls** are listed. In this case, it concerns mounting the dish on the side of the house. This is a particular "gotcha" that is common in the industry. Without specific instructions to contact the project manager, a subcontractor might screw the dish directly into the siding. That could end up creating an unhappy customer. A better method is having a carpenter provide a nice trim plate on which to mount the dish. And yes, sometimes the only way this is learned is from an unhappy customer. This type of experience is a good example of what to document in a "gotcha" file. Write down the incident and file it as it happens. These kinds of issues will happen, and a

Project Management Plan PM24
Satellite Contractor

ABC Integrations requests that you provide and install an 18"x20" DIRECTV Multi-Satellite dish antenna (Triple–LNB) for Mr. & Mrs. Smith. Please see the attached address and directions to the job site.

Please follow the instructions listed below.

1. Install the dish at the home in the area least visible from the front of the home.
2. Install four (4) RG6QD cables from the dish to a ground block mounted in the attic. ***We will provide and install the ground block.***
3. Install a #10-ground wire from the dish to the ground block located in the attic.
4. Test the installed dish for the minimum acceptable signal strength of 87, as measured at the ground block in the attic. If you cannot attain a signal at this level or higher, please contact me directly to arrange for an acceptable alternative. Measurements should pass on the following satellites: Satellite A, 101—with transponders #1 to #32, Satellite B, 119—with transponders #22 to #32, and Satellite C, 110—with converted transponders #8, #10, #12.
5. Contact me directly to arrange the installation of a mounting board if the dish is to be mounted to the siding on the home.
6. Complete all work by August 21st. We have a standing agreement of $135.00. Please forward your invoice to my attention within seven days of completing the project. Please include the customer name, site address, and date of installation. Please fax to 555-1234. Tools, test equipment, cables, dish, and other supplies are your responsibility and are included in the prearranged price of the installation.
7. All installations must meet city, county, state, and federal standards for installation of low voltage devices, aerial antennae and satellite dishes.
8. You are responsible for total cost of damages to equipment, materials, and building systems that were caused during the performance of the work.
9. All cables that enter the home must include fire-stopping and must conform to both flame and temperature ratings as required by the local building codes and with United Laboratories fire test in a configuration that is representative of field conditions. All penetrations are to be sealed for a minimum F rating of one (1) hour but not less than the fire resistance rating of the assembly being penetrated.
10. Any work needing to be done that falls outside the original scope of work should be communicated prior to beginning any work at the site. Any out-of-scope work done without prior approval will not be reimbursed. This approval must come directly from me. Call me immediately and give the details of all time and materials required to complete the installation. You will be notified quickly whether the change order will be approved. An example of a change order is a chimney or ground mount.

FIGURE 12–1 Project Management Plan PM24—Satellite Contractor

good project manager puts processes in place to ensure it's the last time it happens.

- **Item 6. Payment details** are explained. Specify not only the amount, but also the process of obtaining payment for the work.
- **Items 7 through 9. Responsibilities of the subcontractor** are outlined.
- **Item 10. Change orders** are critical to ensure there are no hidden charges in the work performed. ABC Integrations has agreed to a set amount for the installation, and as such must provide a change order request to the customer justifying any additional work that may be needed.

Study/Career Tip

There are only so many hours in a day, a week, and a term. You cannot change the number of hours, but you can decide how best to use them. To be successful in school, you must carefully manage your study time.

Here is a study time strategy: At the beginning of the term, prepare a term calendar. Update it throughout the course of the term. Be sure to record your school assignments with their due dates and scheduled tests. Also record any known school activities. Finally, record your known out-of-school activities. Check this calendar often to help keep you and your work on track.

Critical Thinking Question

There are several reasons why the Satellite Contractor agreement is so specific. List at least three with an explanation as to why they are important.

Outsourcing Requisition

The work for the contract person or team starts when the requisition is received. As we have stated, the responsibility of specifying the requirements of the contract lies generally with the project manager, and the company's procurement manager oversees the process.

During the requisition phase, the procurement manager, who handles all of the contract issues for ABC Integrations, uses one of four broad categories of contracts. The type of contract she uses depends on the degree of uncertainty the project is facing. As we discussed in Chapter 11, Risk Management Planning, when entering the contract, the objective of the buyer is to minimize risk while placing a degree of incentive for efficient performance. The objective of the seller is to minimize his risk while maximizing profit opportunity.

Different types of contracts are used for different types of purchases. The four broad categories of contracts are:

- Fixed Price
- Unit Price
- Cost Reimbursable
- Time and Materials

For the purposes of this textbook, we discuss each type of contract in terms of the way it is most likely used in the residential integration industry. This does not mean that it is the only way that all residential integration companies use each of these contract types.

Fixed Price Contracts

Fixed price contracts are the most common. The buyer incurs the least amount of risk in this situation. In this contract, the buyer and provider agree on a defined project or product and fix the price for that project or product. For example, let's say a homeowner wants a music system in the home. The residential integration company will provide a scope document that details the deliverables of the system and provide a fixed price for the project. While the price of the system is fixed, the payments can be scheduled depending on the project progress; say 20 percent of the total cost can be due prior to cabling the home, another 20 percent after cabling the home, 40 percent prior to the installation of the speakers, 30 percent prior to installation of the electronics, and a final 10 percent due upon completion of the project.

Our case study employs a fixed price contract with the homeowner as the buyer and our organization as the seller. A fixed price contract can be used due to the repeatable nature of the work. Since our case study organization installs music systems and structured cabling systems on a regular basis, the project deliverables are well known, and cost estimates can be computed easily.

The burden of management lies squarely on the seller in a fixed price contract. The great risk is that cost could outstrip the contract price, resulting in a loss for the organization. Often, small organizations become insolvent because they use a fixed price contract to take on projects that are too complex and too large for the size of the company. Here's a case in point.

A CASE IN POINT

Just Say No to Joe

ABC Integrations has done a thorough job of estimating a home theater for customers George and Laura and has set a $50K contract price. However, Joe's Theaters, a newcomer to the industry, has also prepared a proposal for George and Laura for the same equipment and priced it at $42K. ABC Integrations explains to the customers that the competitor is under priced and has left out many of the hidden materials and work that is required to complete the project correctly. Despite the ABC's explanation, George and Laura select the lower fixed cost contract.

Unfortunately, it is far easier to sell a complex project than to actually pull it off. Joe's Theaters takes the project and starts work. They take on several other theater projects and start to use the cash from the next project to finish the previous project. What Joe's Theaters is doing is not generating profit from the company's work, but instead floating from project to project on cash deposits. Such a practice eventually will catch up with the company, and it will be forced to close its doors. Joe's Theaters is no more!

Fixed price contracts should be taken very seriously and should be tracked closely for profitability.

Unit Price Contracts

In a unit price contract, the buyer agrees to pay a predetermined amount per unit of service. To illustrate, a cable supplier creates a price list of each type of cable that they sell, and your residential integration organization purchases cable based upon this price list; this is the contract.

Volume discounts can be extended in a unit price contract. In our cabling purchase example, the vendor states on the price sheet that shipping is free for purchases of more than $500. The vendor also sets a price based upon the quantity purchased—one price per reel for 1 to 10 reels of cable, and a lesser price per reel for 10 or more reels.

The lifespan of the unit price contract is not static. It changes regularly. Typically, vendors post their prices once a year. While the seller (vendor) can re-establish pricing at any time, the buyer (residential integrator) retains the right to switch vendors at any time.

Study/Career Tip

Conduct an Internet search on the phrase "types of contracts." Here you will find a host of links to Web sites that will give you more complete information on contracts. Print off noteworthy pages or otherwise catalogue the pertinent information in a journal.

Critical Thinking Question

In your daily life, think about where you see unit prices listed for products.

- How often do you purchase items that are priced based on unit? The local grocery store is a great place to start.

Cost Reimbursable Contracts

Cost reimbursable contracts involve payment to the supplier for direct and indirect costs. For example, the contract can include an hourly charge for installers on the project. There are three methods of reimbursement for costs.

In a **cost plus incentive fee,** the buyer (residential integration company) and supplier (vendor) predetermine the costs, and if the supplier meets or goes below that predetermined amount, there is a bonus given to the supplier. For example, suppose the expected cost of the project is $50,000 with a $5,000 fee to be paid to the supplier and a 15 percent bonus paid on costs below the $50,000 target. If the actual costs come in at $40,000, the buyer pays the supplier the $40,000 plus the $5,000 fee, plus an additional $1,500 bonus for coming under target costs for a total of $6,500. In this instance, both the buyer and the supplier benefit. The supplier earns extra funds and the buyer saves the difference between the $46,500 and $50,000, or a savings of $3,500.

In a **cost plus fixed fee** contract, the buyer (residential integration company) pays the supplier (vendor) for cost plus a fixed fee, commonly based on a percentage of estimated costs. This is similar to the cost plus incentive fee, however, there is no bonus given for bringing costs in under target.

In a **cost plus percentage fee** contract the buyer (homeowner) pays a percentage above and beyond the supplier's (residential integration company's) costs. For example, a residential integrator can charge the customer based upon the actual costs of the project by supplying invoices for items used on the project, such as cables, equipment, and outsourced contractors. A percentage is then added to these costs. The organization might charge an additional 20 percent beyond project costs, so if the total cost comes in at $50,000, then an additional $10,000 is paid by the buyer (homeowner). In addition, a rate for employees working on the project can be established at the outset of the contract.

Time and Materials Contracts

The time and materials contract is a hybrid of unit price and cost reimbursable contracts. An hourly rate is established for the project, and a unit price is established for the products. For example, the homeowner wants a music system, and the residential integrator provides a selling price for the cables, amplifiers, and various quality levels of speakers. In addition, the residential integration company provides an estimate of the time required to complete the installation, but does not guarantee the estimated time. Let's say the homeowner selects the equipment from the list provided by ABC Integrations, and uses the time estimate to calculate the total charges. The homeowner becomes the project manager, making all of the decisions on the project. This type of contract is suitable for small projects that last a day or two.

Study/Career Tip

Working out or playing sports relieves stress. You should have incentives after your schoolwork is done: go out, talk with a friend, read for pleasure, have fun. Reward yourself for having a successful study day.

Critical Thinking Question

- Before you read the following material, which do you think are the two most common types of contracts used in the residential integration industry? Why?

Comparing the Types of Contracts

Each type of contract shares the risk between the seller and the buyer. Figure 12–2 shows the continuum of risk associated with each type of contract. A fixed price contract places all of the risk on the seller, while the time and materials contract places all of the risk on the side of the buyer. The cost plus contracts share the risk between the seller and buyer. The two most common contract types used in the residential integration industry are the fixed price and the time and materials contracts.

Fixed price contracts are typical for small- to medium-sized projects where all of the requirements are well known. An example of this is a music system for the customer's new home that has been installed by ABC Integrations many times. The time and materials contract often is used for the very small and the very large contracts. In a very small contract, such as delivering and installing a television, the total project value does not justify a formal cost estimate process. At the very large and

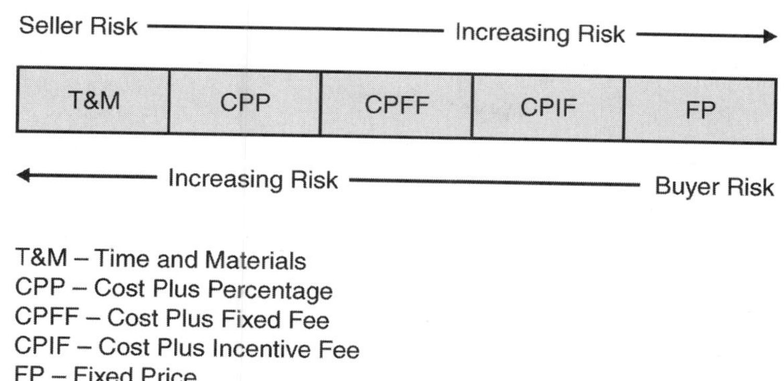

FIGURE 12–2 Buyer and Seller Risk with Various Contract Types

complex end of the spectrum, the project deliverables are unknown, and in that case, it is impossible to assign a fixed cost to the project.

Make-or-Buy Decisions for Outsourcing

A commonly used tool in procurement planning is the **make-or-buy analysis.** A make-or-buy analysis involves comparing internal costs of delivering a service to the cost of outsourcing the service. Here's a case in point.

A CASE IN POINT

Team Effort

ABC Integrations needs to assign a cable team for a whole house integration system. First, the project manager considers three different possible teams: Team A is completely outsourced, Team B has two internal installers, and Team C is comprised of two internal installers and two outsourced installers. Team A has no startup costs associated with it, but the team charges $45 per hour per person of installation time. Team B has $28,000 in startup costs, which consist of the truck purchase and required tools and training for team members. Team C has $34,000 in startup costs, and it is assumed that new startup costs are required approximately every seven years. This is the expected life span of the tools and the truck for the team. It is also assumed that each direct employee can produce 1,400 billable hours per year.

Figure 12–3 summarizes the costs of each team, with the billable hours in thousands along the horizontal and the total costs in thousands of dollars along the vertical. Team A has no fixed cost associated with it, and as such is a great choice if the organization bills less than 2,500 hours per year. However, as the company grows past the 2,500 hours per year threshold, then Team B is used until it reaches 2,800 hours per year, which is the maximum that a two-person team can provide. With the addition of two outside contractors and more startup costs, the team can grow to 5,600 hours per year.

Team C represents a blend of internal and external sourcing of labor. Selecting the right team for ABC Integrations depends upon the sales projections for the coming year. The executive team, in conjunction with sales, sets the sales projections for the coming year. If they determine that 3,000 to 4,000 labor hours of cabling is anticipated for the organization, then the project manager chooses Team C.

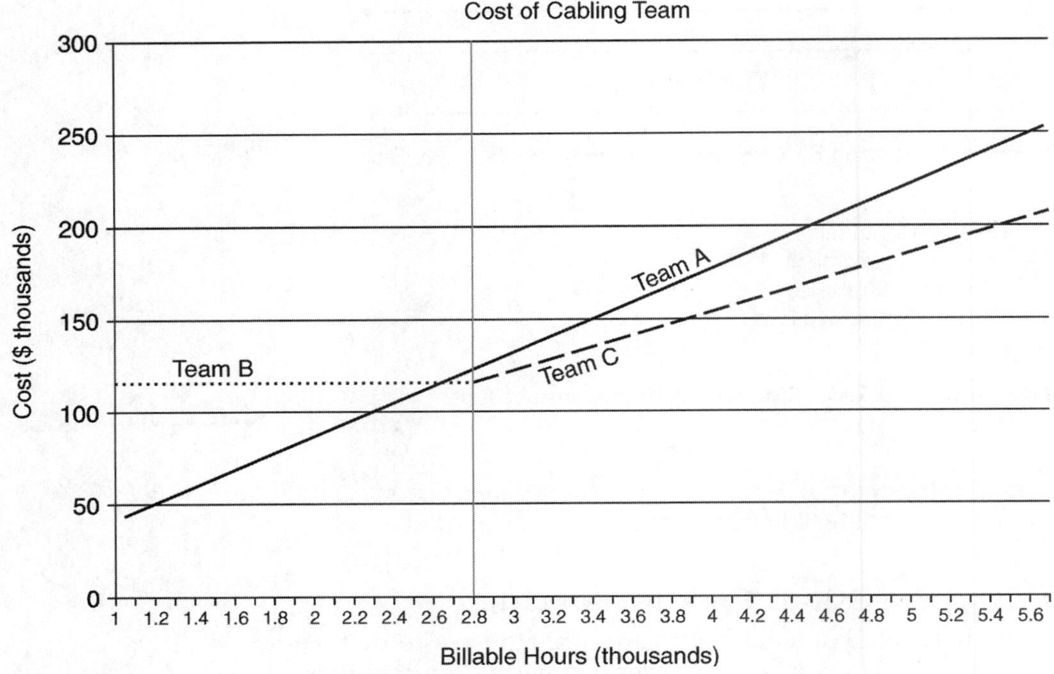

FIGURE 12–3 Cost of Cabling Team

Other Decisions in the Make-or-Buy Analysis

Money isn't the only consideration in a make-or-buy analysis; other factors go into the decision process as well. Some of those factors include:

- **Quality of work** is related directly to the experience level of the person doing the work. That means when the work is performed infrequently, outsourcing may be a good alternative. Conversely, if the quality of work cannot be obtained outside the organization, it may need to stay in-house.
- **Availability of the workforce** determines the workload that can be accomplished. By having access to subcontractors, the workload can be expanded to meet project demands. For example, the cabling team can be comprised of two fulltime installers, and up to two more subcontractors can be added to the team to expand its capability. This practice is a good solution when the burden of four fulltime installers is more than the organization can take on.
- **Frequency of work** can be a determining factor in deciding whether to make or buy. If a satellite dish installation is required only once per month, it may be advantageous to have that work performed by someone outside of the organization. If, on the other hand, a satellite dish is installed several times per day, then a dedicated employee may be better suited to the task.

Study/Career Tip

Do you have a resume? The two most commonly used and accepted resume formats are the chronological and the functional. Frequently, elements of both are combined. A chronological resume is most widely used and preferred by recruiters and interviewers. It is good for someone with a consistent work history. A functional resume focuses attention on your accomplishments and is often used more successfully if you are trying to change careers or industries or are trying to start a career after school. If you have a current resume, great! If not, it is never too early to create one.

Critical Thinking Question

- What types of make-or-buy decisions do you make in your daily life?

Product Purchases and Acquisitions Planning

The purchasing process is reactive in nature. A company needs goods for a company product, which triggers the procurement manager or Purchasing Department employee to begin the process of obtaining those goods. Depending on the size of a company, there are many different levels of procurement administration. It can be as simple as following up on a low-cost product to ensure delivery is on time, or it can be a lengthy and detailed purchasing cycle for complex products. Those products might include the installation of a home theater system or other whole-house systems. Developing a plan for purchasing products can be divided into two phases:

- Purchasing Requirements
- Purchasing Requisition

Let's look at each area in detail.

Purchasing Requirements

Requirements are all the physical items and materials that are needed to complete a successful project. In a small project, the requirements could be a plasma TV, mounting brackets, interconnects, and various miscellaneous supplies. In a large project, the requirements could fill pages. At ABC Integrations, the procurement manager identifies the goods required by a company product. Here's an example. ABC Integrations' procurement manager has identified a project requirement as the audio interconnects used to connect the music system preamplifier with the amplifier. She then specifies the exact interconnect and note whether an equivalent quality product is an acceptable substitute. For complex products, the procurement manager can ask the engineering department to prepare the actual specifications for a product and retain responsibility for the final purchasing decision. Once all the requirements are identified and the purchasing decision has been made, procurement moves to the next step.

The following are some of the elements to consider in planning the requirements phase.

- Who identifies the requirements?
- Who decides if a substitute product may be acceptable?
- Who provides the product specifications?
- Who makes the final purchasing decision?

Purchasing Requisition

Requisition is the step that identifies the source for the goods or product requirements. Where will they be purchased? Is one supplier or vendor a better source than another? To illustrate, here's a case in point.

A CASE IN POINT
Checks and Balances

At ABC Integrations, Procurement Manager Pam Putnam acts as a check and balance for products that may be requested by Project Manager Owen Ginizer. Procurement Manager Pam may believe the specifications for a particular piece of goods are overly restrictive, thus limiting the number of firms that can bid on the work. Or, Project Manager Owen may specify a sole source, when other sources offer equal quality for a reduced cost. Consider audio connections. Procurement Manager Pam seeks pricing and product information from dozens of cable manufacturers. She becomes an expert in how to get the best goods for the least price. She presents several options for purchase to Engineer Maggie Pi and to Project Manager Owen, who make the final selection based on technology requirements, the availability in the marketplace, quality, and cost.

A procurement manager also must make and maintain contacts with suitable vendors and suppliers in the industry. Why? Vendors are close to their products and may know advance information about new products. They also can provide technical support in the use and installation of current and new electronic and non-electronic products. They also are likely to notify the procurement manager when special deals are available for regularly purchased products.

Relationships with vendors can be established by consistently using a particular vendor's products. Other areas that can be used to develop sources are catalogs, trade journals, trade conferences, and written vendor reviews. Vendor and supplier sales personnel are an excellent source of information concerning possible vendors.

The requisition of products takes the form of the Bill of Materials (BOM) generated as part of the cost estimate. Table 12–1 shows a BOM created as part of the cost estimate. In some cases, exact make and model are specified. The table shows a DVD player from Gamma Electronics with the model DIR7.2 specified. Note that the cables have no manufacturer specified, yet the BOM provides a budget unit price and a brief description of the required item. The procurement manager

TABLE12–1
Bill of Materials (BOM) for a Home Theater System

Materials				
Quantity	*Category*	*Description*	*Unit*	*Price*
1	Display	Alpha Screens: P50X10—50″ Plasma Monitor, 1366768	$10,999.00	$10,999.00
1	DVD	Gamma Electronics: DPS-7.3—DVD Player, RS232	$ 799.00	$ 799.00
1	Processor	Gamma Electronics: DIR 7.2—Surround Receiver, 7100, RS232	$ 1,399.00	$ 1,399.00
1	Controller	Iota Controls: RFX6000—RF Extender for Pronto Remote, Black	$ 149.00	$ 149.00
1	Controller	Iota Controls: TSU7000—Pronto Remote, Color, Black	$ 999.00	$ 999.00
3	Speakers	Tau Acoustics: CLR7.5—Bookshelf speaker, Black	$ 150.00	$ 450.00
1	Speaker	Tau Acoustics: SR585—8″ In-Ceiling Speaker, White	$ 450.00	$ 450.00
1	Speaker	Tau Acoustics: SB10F—Subwoofer Amplifier, Black	$ 600.00	$ 600.00
4	Equipment Rack	Mu Pacific: EVT-1 Vent Panel, 1 Space, EA	$ 10.99	$ 43.96
1	Equipment Rack	Mu Pacific: RK-16 16 Space Black Laminate KD Rack, Black, EA	$ 144.00	$ 144.00
1	Equipment Rack	Mu Pacific: RKW Casters for RK Racks, Set of 4	$ 44.50	$ 44.50
2	Equipment Rack	Mu Pacific: RSH-4S Custom Rack Shelf (See Rack Drawings), EA	$ 279.90	$ 559.80
1	Equipment Rack	Phi Power: M4310—8 Outlets (4 Switched), Plus front panel convenience outlet	$ 249.00	$ 249.00
1	Cables and Connectors	Upsilon Cables: CCV-10M—10 Meter Component Video Interconnect	$ 199.99	$ 199.99
1	Cables and Connectors	Cables: 2 Meter Component Video Interconnect	$ 69.99	$ 69.99
1	Cables and Connectors	Cables : 2 Meter Digital Coax Interconnect	$ 34.99	$ 34.99
2	Cables and Connectors	Zeta IR: 283M—Blink-IR Mouse Emitter	$ 13.99	$ 27.98
5	Cables and Connectors	Upsilon Cables: Double Banana Plug	$ 3.50	$ 17.50
250	Cables and Connectors	Upsilon Cables: CL412—14 Gauge Speaker Cable	$ 0.80	$ 200.00
1	Miscellaneous	Alpha Screens: PST-2000—Standard Plasma Wallmount Bracket	$ 199.99	$ 199.99
1	Miscellaneous	Tau Acoustics: NCBR—New Construction Bracket	$ 39.99	$ 39.99
			Total Equipment	$17,676.69
			Sales Tax	$ 883.83

uses this requisition information to solicit vendors for final pricing and delivery terms.

The following are some of the elements to consider in planning a requisition process.

- Who initiates the purchasing process?
- How will the goods be purchased? Credit card, purchase order, contract arrangement, receivables and payables system?

- What determines an acceptable supplier or vendor?
- Who makes the final decision on where the item will be purchased?
- Who will track the purchase and how will it be tracked?
- What purchasing records will be kept and how will they be stored?
- How will received items be recorded?
- How will received items be distributed to the appropriate party?

Summary

In Chapter 12, we discussed procurement planning for outsourcing and purchasing and considered requirements and requisition elements. We also discussed several reasons why companies outsource. We reviewed the various categories included in establishing requirements when outsourcing a service, and discussed four types of contracts as they are used in outsourcing.

Important points in this chapter are:

- Procurement planning is developing a process for purchasing goods and/or services from a source that is outside the company.
- Outsource planning can be divided into establishing requirements and requisition.
- Companies outsource to gain specialized skills, to allow a company to focus on its core business, to allow customer-driven solutions, and to reduce both fixed and recurrent costs.
- Residential integration companies use four basic contracts in outsourcing: fixed pricing, unit pricing, cost reimbursable, and time and materials contracts. The fixed price and time and materials contracts are the two types most commonly used in the residential integration industry.
- Make-or-buy decisions depend on the availability of labor resources within the company, quality of work, and frequency of work.
- Purchasing planning is divided into establishing requirements and requisition.

Key Terms

Cost plus fixed fee A type of contract in which the buyer reimburses the seller for the seller's allowable costs (allowable costs are defined by the contract), plus a fixed amount of profit (fee).

Cost plus incentive fee A type of contract in which the buyer reimburses the seller for the seller's allowable costs (allowable costs are defined in the contract), and the seller earns its profit if it meets defined performance criteria.

Cost plus percentage fee Provides reimbursement of allowable costs of services performed plus an agreed upon percentage of the estimated costs as profit.

Make-or-buy analysis A decision process in which it is determined whether to manufacture internally or buy from external sources some component, article, or item of equipment.

Performing organization The enterprise whose employees are most directly involved in doing the work of the project.

Procurement management plan A type of procurement document used to request proposals from prospective sellers of products or services. In some application areas, it may have a narrow or more specific meaning.

Review Questions

1. What does the term procurement refer to in regard to the residential integration industry?
2. What are some reasons why residential integration companies outsource?
3. What is a "make-or-buy" decision?
4. What is the most common type of contract used in the residential integration industry? How does the contract work?
5. Cost reimbursable contracts have three methods of reimbursement for costs. What are they?
6. When are fixed price contracts typically used?
7. When are time and materials contracts generally used?
8. Aside from money, what other decisions may go into a make-or-buy analysis?
9. What are the two phases for purchasing products?

Project Communications Planning

OBJECTIVES

After studying this chapter, you should be able to:

- Define and discuss communications planning.
- Identify project communications requirements.
- Examine communications planning policies.
- Discuss determining communications methods.
- Discuss the issues concerning communicating with outside contractors.
- Describe the value of running effective meetings and the project manager's role in those meetings.
- Discuss the Sales-to-Operations Turnover Meeting and its purpose.
- List and discuss the documents that could be included at the Sales-to-Operations Turnover Meeting.
- Review the role of the sales representative in the post-turnover meeting environment.

OUTLINE

Introduction

Once a customer has signed a contract, extensive communications and planning is necessary to move from Point A—selling a project to the customer—to Point B—actually determining how to accomplish the project. In addition, it is also necessary to plan how the sales representative is going to let the project manager and the project team know what the customer wants and expects during the course of the installation. Additionally, the project team needs to know what the details of the project are to reach a finished project, be it a home theater, a house-wide music system, or an integrated lighting control system. In order to accomplish these goals, it is vital that effective communications be a part of the overall project strategy.

To that end, this chapter examines the importance of effective communications planning, the steps it takes to provide such communications, the parties involved, and the methodology for turning over the project from sales to operations.

Study/Career Tip

When in an interview, be spontaneous, but also be well prepared. You need to present yourself as authentic and professional, yet real. Try to engage in true conversation with your interviewer without showing your possible nervousness. Consider conducting several trial runs with another person as practice. It's the same as anticipating the questions you'll be asked on a final exam.

Critical Thinking Question

Jot down as many ways as you can think of to communicate. Compare your list with what you'll discover in this chapter.

- Which method(s) do you think are the most widely used in the residential integration industry?

Communications Planning

In project management terms, **communications planning** is described as the determination of the information and communications needs of the project stakeholders, specifically, who needs what information, when they will need it, and how it will be given to them. It is an overall planning approach that guides all communications activities throughout the course of the project.

As we have learned, project stakeholders are those individuals or groups who are actively involved in the project, and whose interests may be positively or negatively affected as a result of the project execution or the project completion. Project stakeholders may include:

- The customer
- The sales representative
- The executive team
- The architect
- The builder
- The project manager
- Project team members
- Outside contractors.

While all of these entities are in need of project communications information, generally speaking in the residential integration industry the onus of responsibility for continued, effective communications is on the project manager. Let's take a closer look.

The Project Manager

Top-notch project managers are not only good people managers, they also have other traits in common; they have excellent communications skills, as well as organizational and planning skills. They develop long- and short-range plans in completing a project. And to tackle a large, complex project, they treat each phase of the project as a separate component with a distinct beginning and ending. They understand that each phase of a project consists of planning, implementing, and closing stages. In this way, they are able to begin a residential integration project in small manageable pieces.

One facet of a project manager's skill set is communications planning. As we have learned, the project manager determines the information and communication needs of the customer, the builder, the architects, other subcontractors, and his own project team. The project cannot be completed successfully if the project team does not know what the customer wants. The project team also needs to be aware of the builder's construction schedule and plans, the architect's design plans, and other subcontractors' schedules. How that information is distributed to everyone involved with the project varies from company to company. As we discussed in Chapter 5, Information Gathering, large companies may use a company intranet that allows authorized persons to access the information they need. A smaller company may use a main computer and e-mail to communicate. A smaller company may still use paper files, duplicate important information for each team member, and then use e-mail to communicate with those outside the company. However the information is distributed, all companies need to share project information with the project stakeholders. Identifying their informational needs and determining how to meet those needs is an important factor in project success.

Generally, communications planning is accomplished very early in the project life cycle. At ABC Integrations, the executive team develops communications policies that are in effect before the sales representative even approaches a customer. The executive team prepares templates for use in information gathering, sets a policy for a turnover meeting to pass the torch from the sales representative to the project management team, and sets up specific protocols for naming computer files and releasing current versions of project information documents. It is then the responsibility of the project manager to review the project information regularly throughout the project and to update documentation as the information changes to make sure it reflects the current status of the project. E-mails can be sent to the appropriate people to let them know an updated document has been released.

To illustrate, a customer may decide to add an additional television to the integrated system halfway through the project. The project manager and the engineer will then need to review the installation documents to determine if the television can be added without an additional cable or satellite connection. That information needs to be conveyed to the installers once that decision is reached so the additional television can be added without decreasing the quality of the signal that the customer expects. Of course, that communication is only one piece of information needed for the installation to be successful.

Communications planning involves several internal steps. These include:

- Identifying project communications requirements
- Developing communications planning policies
- Determining communications methods
- Working with outside contractors
- Running an effective meeting
- Leveraging project manager communications skills
- Orchestrating the sales to operations turnover.

Let's review each in greater detail.

Study/Career Tip

No one likes taking tests. But there are ways you can help yourself do better on them by staying relaxed and confident. Remind yourself that you are well prepared and are going to do well. If you find yourself anxious, take several slow, deep breaths to relax. It really does help. And most importantly, don't talk about the test to other students just before taking it; anxiety is contagious!

Critical Thinking Question

- What are some ways that lack of communication can backfire on a project?

Identifying Project Communications Requirements

Project communications requirements represent all of the informational needs of the project stakeholders. Those requirements are defined through the identification of what types of information are compulsory and in which format the information needs to be disseminated. An analysis of the value of the informational needs helps to pinpoint what must be communicated in order to achieve project success, or where a lack of communications will contribute to project failure.

Project communications requirements involve the following groups:

1. **Project Team.** The project team has many communications needs. Its needs can be filled, in part, through the scope statement, project schedules, and change orders.
2. **Project Stakeholders.** The architect, builder, and related trades have different communications needs. Their needs can be met

through e-mails, fax, phone calls, and detailed memos to explain their responsibilities on the project.

3. **Customer.** As the primary stakeholder and sponsor, the customer's communication needs are very important. Once the sales person turns a project over to the project manager, it becomes the project manager's responsibility to keep the customer informed of the status of the project. Regular communication through e-mail, phone, and face-to-face meetings helps to meet the customer's expectations.

Project communications depend on several factors. These include:

- **Type of company organization and its organizational chart.** As we discussed in Chapter 3, The Project Team, the type of organization determines the amount of authority given to the project manager. This may include the job responsibility of being the "communications manager" in addition to other job accountabilities.
- **Logistics.** This refers to how each of project team performs work on the project, and at what locations. In the case of ABC Integrations, Engineer Maggie Pi works primarily out of the office on programming, documentation, and other engineering matters, performing many of her duties on the computer. That means much of the communication to and from her can be handled electronically. Communications such as a company intranet, used to share internal information, would work well for Maggie. Joe Installer spends the overwhelming majority of his time at job sites, and does not use a computer for day-to-day work. That means paper-based work orders and project reports are critical in communicating with Joe.
- **External informational needs.** Stakeholders outside the organization dictate the type of communication to and from them. For example, the electrical contractor may request all communications be handled via fax, while the architect may request e-mail.

Study/Career Tip

Active, effective listening is a habit, as well as the basis for effective communication. Active listening focuses on who you are listening to in order to understand what is being said. If you are actively listening, you should be able to repeat back in your own words what was said to the speaker's satisfaction. Bring this skill to the classroom.

Critical Thinking Question

- Why do you think that active listening is so important for a project manager who is running a residential integration project?

Developing Communications Planning Policies

As we have seen with other project planning processes, residential integration companies develop a set of policies about how project communication is accomplished. These policies support the company mission

Excerpt from Communications Plan PMXX

The purpose of the Sales-to-Operations Turnover Meeting is to provide a smooth transition from sales to operations prior to the beginning of work at the customer site. This meeting will help minimize cost slippage on the project by improving internal communications and further defining the scope statement. This meeting will ensure that our customer's expectations are fully communicated to the project management team.

The Process

Step 1. The Sales Representative is responsible for calling the meeting.

Step 2. The Sales Representative is responsible for bringing the following documents to the meeting. Each document must follow the file naming conventions specified in the "Project Defining Management Policy" document.

- PM02 Site Survey
- PM03 Customer Survey
- PM04 Structured Cabling Survey
- PM05 Scope Statement
- PM06 Structured Cabling WBS
- Signed Contract

Step 3. The Sales Representative must move the above files to the appropriate directory, specified in the "Project Defining Management Policy" document. In addition, a printed copy must be provided to the Project Manager *at least* one day prior to the meeting.

Step 4. The project must be reviewed by the sales representative and the cost engineer prior to the meeting to ensure the accuracy and completeness of the project data.

Step 5. The meeting is to be scheduled as soon as possible after the awarding of the job. *No equipment is to be ordered prior to the meeting.*

FIGURE 13–1 Excerpt from Communications Plan PMXX

statement and are a reflection of how the executive team intends to run the business. ABC Integrations' mission statement says . . .

> *"ABC Integrations provides high-quality turnkey systems for entertainment, comfort, and convenience in the modern digital home."*

From this statement we can determine that the communications planning process will be based on turnkey systems. That means the same system is installed repeatedly. Therefore, the company can create a project communications plan from a set of well-defined templates, which means that project communications details always remain the same from project to project. And *that* means communications among project stakeholders is held to a minimum.

Once the policies are established by the executive team, the sales and project teams use these policies to determine operations details. Let's look at ABC Integrations policies and planning process. Figure 13–1 depicts the Sales-to-Operations Turnover policy and the step-by-step

	Color and Purpose				
	Blue—Administrative	*Red—Sales*	*Green—Project Management*	*Yellow—Installation*	*Brown—Engineering*
Project Scope Statement accepted versions	X	X	X	X	X
Copy of signed contract and checks	X				
Project memos		X	X		X
Bill of materials (BOM) by phase with quantities			X	X	X
BOM with pricing	X	X			
Cabinetry, theater, and other drawings			X	X	X
Lighting control documents			X		X

FIGURE 13–2 File Folders

process key players must follow in order to comply with that policy. We will be reviewing additional details about the sales-to-operations turnover procedures later in this chapter.

Another effective method of executing a communications planning policy is to catalogue project information through the use a project folder system. In the case of ABC Integrations, the administration person prepares multiple folders in various colors for each installation team, including engineering, installation, project management, sales, and administration (see Figure 13–2). A copy of the final project scope statement and other appropriate information goes inside each folder. Note that not all stakeholders are in need of all of the documentation, only those communications tools that directly contribute to the project success or, lacking that information, cause project failure.

Study/Career Tips

Here are some tips for conducting Internet research:

- Narrow your topic and its description; pull out key words and categories.
- Use a reliable search engine. Ones that offer a directory of topics can be especially helpful.
- Find the best combination of key words to locate information you need and enter these in the search engine.
- Review the number of options returned. If there are too many links, add more keywords. If there are too few options, delete some keywords or substitute other key words.

Critical Thinking Question

- What might be another way that a residential integration company could organize and disseminate the right documentation to the right project team members without using traditional file folders?

Determining Communications Methods

A variety of communications vehicles are used in the residential integration industry. What communications method is utilized depends on who is sending the communication and who is receiving the communication. The following questions are helpful in determining what form of communication is appropriate in any given instance.

- **What is the urgency of the communication?** Does the outcome of the project depend on it? Has the customer requested the information? Does a project team member need the information to move the project forward?
- **What is the availability of the form of communication?** Fax, e-mail, paper, company intranet, Web site, verbal, cell phone, landline, face-to-face, walkie-talkie, message in a bottle, hieroglyphic obelisk? Obviously, certain forms of communications will be more available and more practical than others. Who receives and who sends the communication will determine the form it will take.
- **Who is receiving the communication?** Project team member, builder, architect, subcontractor, customer, administrative personnel, or upper management?
- **What is the physical environment of the project?** Laptop computers need a hot spot for Internet access. Muddy, wet construction sites are not suitable for computers. Cell phones calls can be dropped in dead areas.

While there are many forms of communications available (and more technology is becoming available every day), it is best that all stakeholders use the best method that helps to expedite project work within the limitations of their situation.

Working with Outside Contractors

Outside contractors can push a project manager's communications skills to the limit. Outside contractors are naturally independent, and conflict can arise over who is running the project. Yet the residential integration firm has already assigned a project manager to the overall project. So who is in charge of what? Carefully defined project memos that outline the company's responsibilities and the contractor's responsibilities before the project begins can help reduce and sometimes eliminate disagreements before work begins. The memo can outline the groundwork for job responsibilities, and a phone call, e-mail, or face-to-face interview to follow up can ease possible authority tensions.

A project manager also must realize that outside contractors have a number of ongoing projects and may not have much time to spend discussing one project in particular—the very project that is nearest and dearest to the project manager's business heart. Fitting a meeting into their schedule and an outside contractor's schedule can be a challenge. Scheduling meetings, or even telephone appointments, with builders is especially difficult. What's important to the project manager of a residential integration company is not necessarily important to a builder or other outside contractor. One way to assert authority is to remind the contractor in a tactful way that the customer hired the company for the project because he wants an *integrated* system in his house.

Developing some flexibility in your schedule is also important. In discussing the project schedule, there are certain tasks that must be completed by outside contractors before the low-voltage work can be done. However, other tasks can be moved forward regardless of an outside contractor. When delays by outside contractors threaten to delay project completion in a timely manner, a project manager should be prepared to add additional personnel to the installation team to move that part of the project forward faster than originally planned. Let's look at several ways a project manager can handle contractor delays.

- Fast track your schedule around the contractor, especially a builder.
- Set an early deadline for subcontractors you hire. Make sure there is extra time available before the critical part of your project needs to be finished.
- Schedule any work that does not depend on the completion of a previous phase (whether performed by outside contractors or company teams) as early as possible.
- Be prepared to complete the work with a substitute contractor, if possible. In a pinch, another contractor may be available to step in and complete the work. There might even be someone on your project team capable of completing the work.
- Accept that some delays are beyond your control. Frequently, you will be unable to work around the contractor, and you will have to accept the delay. When the delay affects the planned completion date for the project, let the customer know as soon as possible and explain the reason for the delay.

As we have demonstrated, outside contractors are important to most project managers in the residential integration industry. And not all interaction with outside contractors is negative. Far from it. However, it is important to remember that a project manager does not have the same control over subcontractors that he has over members of his own project team, which is bound by company polices.

Study/Career Tip

Searching for a job? Consider this tactic. Leverage your uniqueness in your cover letter, resume, and during an interview. Employers generally prefer to hire those with a strong identity about who they are and what they can offer the company. Take the time to discover what unique skills and background you bring to the company's table.

Critical Thinking Question

Without reading further, write down several ways in which you feel a meeting could be more effective, and thus productive in the long run. Be specific, using examples from meetings you've attended that you considered a waste of time.

Running an Effective Meeeting

How often do you hear that meetings are a waste of time; they're dull, boring, and ineffective, and hardly ever result in decisions? "Nothing ever gets done" is a frequent comment we hear.

While we have all probably sat through what we consider to be a bad meeting, a short, well-run gathering can change that opinion and improve communications among project stakeholders. But effective meetings don't just happen. There needs to be preparation on the part of the person (generally the project manager) who will be scheduling and calling the meeting. The following are time-honored suggestions for running a successful and pertinent meeting.

Scope and Time

The first step is to control the scope and time of the meeting. To accomplish this, be prepared to:

- **Ask only the people absolutely essential** to the agenda of that particular meeting to attend. Meetings with a large number of people are difficult to manage, and it is harder to stay on topic and to get things accomplished.
- **Write an agenda** that includes the meeting goals. The agenda should reflect what you hope to achieve. What action do you want taken, and who is the responsible party? Distribute the agenda well in advance of the scheduled meeting and ask participants to come prepared to work and to bring appropriate documentation along as needed.
- **State meeting intentions up front.** Make sure meeting participants—team members, department managers, executives, or contractors—know your intention to run an effective meeting. Move through the agenda with a firm manner; state and record the decisions and actions to be taken. Assign who is to carry out those decisions and actions. And set a deadline for them. It is harder to hold people accountable for their actions if there is no audit trail of what was decided.
- **Set a time limit.** Set a time limit not only for the meeting, but also for each item on the agenda. It is not necessary for each person at the meeting to give an opinion, especially if they are repeating a thought already expressed. Encourage meeting participants to offer additional or different thoughts only.

Tools of the Trade

There are different types of meetings for different purposes. To illustrate, a quick five-minute meeting can be held at the beginning of each week to run through assignments for clarification purposes.

Other meetings can be useful for resolving issues at various levels. Determine what types of meeting you need (daily, weekly, at the beginning or end of a project phase) that will accommodate the communications needs of the stakeholders.

Several aids to communication can help get your message across:

- **Written agenda.** A properly written agenda can be a powerful communication tool. One way to design an agenda is to set up a table on a landscape format that states the agenda in the first column, allows space to record discussion points in the second column, provides a third column for actions, a fourth for the individual assigned to the task, and the fifth a date for completing the task. The project manager should require that the action column always be filled in, even if it's only to say a decision on the item will be reached at the next meeting. Also, the agenda item should focus on the solution to a problem rather than simply state the problem. For example, instead of listing "Project Delays" as an agenda item, call it "Resolving Project Delays." The first way the agenda item is written may lead to a general discussion of project delays and never finding a solution to the issue.
- **Flowchart.** People who have problems understanding complex network diagrams can relate better to a top-to-bottom flowchart. It can isolate a period of time and a limited number of tasks or phases.
- **Gantt chart.** A Gantt chart presented at a meeting can be useful in visually showing a schedule problem you might be facing. However, it often is not very useful for project control.
- **Network diagram.** The most effective communications tool for a team meeting is the network diagram. It makes it easier to explain a problem you expect to occur.

Leveraging Project Manager Communications Skills

Communicating effectively is a skill that takes time and lots of practice. Feedback from others is also necessary for a project manager to learn the extent to which her communications skills are deemed valuable. Some of the attributes of a proficient communicator are the ability to:

- Know when to talk and when to be quiet.
- Provide constructive feedback.
- Write notes and memos clearly and concisely.
- Write technical material and reports that are clearly understandable.
- Understand and respond to nonverbal communications from others.
- Encourage open communication of ideas and thoughts.
- Think and speak "in the moment" successfully.
- Determine whether others understand a message or communication.
- Use language and tone suitable to listeners.
- Correct others diplomatically.
- Prepare a presentation and deliver it to large groups of people.

- Articulate effectively when speaking with subordinates and support personnel.
- Show self-confidence.
- Negotiate successfully.
- Listen carefully to the other person in a conversation.
- Articulate effectively in face-to-face and phone conversations with the company's management personnel.
- Articulate effectively in face-to-face or phone conversations with project team members and other staff personnel.

Obviously, not all project managers display this complete level of communications skills. Nonetheless, having a well-rounded repertoire of these competencies goes a long way in sending the right set of messages to all parties involved in the project.

Study/Career Tip

Are you lacking motivation to study? Most distractions come from within you. If you're having trouble concentrating, try to figure out what's bothering you and take actions to correct it.

Critical Thinking Questions

- How many of the communications skills listed above do you possess?
- Which ones are you lacking?
- How can you overcome these deficiencies so that you can become a very proficient communicator?

Orchestrating the Sales-to-Operations Turnover

An essential communications tool in the residential integration industry is the meeting for the Sales-to-Operations Turnover. Its purpose is to create a smooth transition from the sales team to operations, which is the division of an organization that carries out the major planning and operating functions. In the residential integration industry, operations consist of project managers, installation teams, and engineers. In a small company, a single person may be responsible for operations planning. The important point is that operations plans must be determined before the physical project can be started.

Once the customer signs the contract for a specific project, a member of the sales team requests and schedules a meeting for the purpose of turning over the project to the operations group: the **operations turnover.** Improving communications between sales and the operations personnel reduces cost slippage, a primary purpose of the meeting. Another purpose for the meeting is to ensure that the sales team has an opportunity to communicate and explain the customer's expectations, especially those areas of the scope statement (see document PM05 in Appendix A) that are outside the typical efforts of the company. The project operations team uses that information to plan how to proceed with the project.

In a retrofit installation, the sales team provides the details of the site conditions. Let's say the first floor is to be gutted, exposing the studs; and the second floor has its walls untouched. This information provides the basis for determining what the project team needs to do. Using the site information, the project operations team can plan how to install the cables and position speakers on the first floor, and determine that a different method will be needed for the installation on the second floor. Without such information, the installation team will arrive at the site unprepared for two different types of installation methods. That not only causes confusion, but also delays the work process. And that will cost the company money.

Project plans that are established at the Sales-to-Operations Turnover Meeting are stored in computer files for easy access by all team members. The plans are frequently given to the installation team in paper form to carry to the job.

Let's take a closer look at the logistics of organizing and preparing for this all-important project meeting.

Planning the Sales-to-Operations Turnover Meeting

To be sure, planning the Sales-to-Operations Turnover Meeting is more than just putting on the coffee and making sure the doughnuts are fresh. Each company develops its Sales-to-Operations Turnover Meeting as best suits that particular organization. The ultimate outcome of the turnover is the shift of responsibility for the project from the defining stage in sales to the operations phase, which is project management. Creating and writing a formal, methodical plan helps eliminate communications breakdown between sales and project management and provides for a smooth transition between the two business groups. Additionally, it assures that customer expectations are understood and can be met. And it provides a formal outlet to discuss possible issues that might occur, as well as an opportunity to discuss resolutions to those potential problems. But how is this meeting planned and organized? See the case in point on page 262.

Plans for carrying out the Sales-to-Operations Turnover Meeting probably vary as much as the number of residential integration organizations that exist. The specifics of the ABC Integrations plan are intended to act as a guide in establishing a company's own procedures. The important point to remember is that a formal Sales-to-Operations Turnover handoff is critical in the smooth transition from sales to project management.

Study/Career Tip

Capture your understanding of your class material in an active way. Make up examples, create mnemonics, write out summary notes, identify key words, highlight in your textbook, add marginal notes. Use the approaches that work best for you in learning and retaining the information.

A CASE IN POINT

Step-by-Step

Step 1. Sales Representative Sal Moore understands he is responsible for calling the Sales-to-Operations Turnover Meeting. Up until this point in the project, Sal has been the lead manager. He has ensured that the customer's (Fred and Wilma Smith) requirements are accurately captured and documented in a series of information gathering forms and finally in the scope statement document. This meeting represents a shift of control over the project from sales to operations (project management.)

Step 2. Sal brings all necessary project documents to the meeting. These documents include the information gathering forms, the scope statement document, the WBS documents, and the cost estimate. In addition, ABC Integrations' policy on Sales-to-Operations Turnover calls for a specific manner in which these files are named and stored within the company's file system.

Step 3. This step underlines the structure in Step 2. Sales Representative Sal has a printed copy of all documentation described in the previous step available for Project Manager Owen Ginizer a day prior to the meeting. This allows Owen time to review all of the information for accuracy, and to bring questions and concerns to the meeting. (It is helpful to have all parties do their homework prior to this meeting so that a meaningful dialogue can occur, rather than just a simple show-and-tell of the project information).

Step 4. Prior to the meeting, Sales Representative Sal and Cost Engineer Maggie Pi review the project to ensure all details have been accounted for in the project scope statement and cost estimate. In addition, they need to verify that the contact information, the site address, and the required project management forms have been completed. Just as with Project Manager Owen, the sales representative and cost engineer must do their homework to contribute meaningful dialogue at the meeting. Items like missing contact information, such as the general contractor's fax number or the electrician's cellular telephone number, are often caught during the review process.

Step 5. Sales Representative Sal schedules the meeting as soon as possible after customers Fred and Wilma sign the contract. Scheduling the meeting quickly lets everyone concerned know the project is a priority and has a high level of importance for the company. It also helps underscore ABC's policy that the project cannot just happen without the turnover meeting. ABC's policy is always hold a formal turnover to operations; never skip it. However, in certain instances, the project can be fast tracked.

Step 6. ABC Integrations does not order any equipment prior to the turnover meeting. Project Manager Owen has not yet reviewed and approved the project. This means that any materials ordered may have been done erroneously. And that eats into company profit. Again, this highlights the need for a methodical approach to a residential integration project.

Critical Thinking Question

- What types of mishaps will very likely occur if a Sales-to-Operations Turnover Meeting is skipped or otherwise abbreviated?

Other Documents That May Be Used in the Turnover

Many other documents may be used during the turnover from sales to operations. Which documents are used depends on the complexity of the project and the size and structure of the organization. Some of these documents are described below.

LOW-VOLTAGE CABLING PLAN

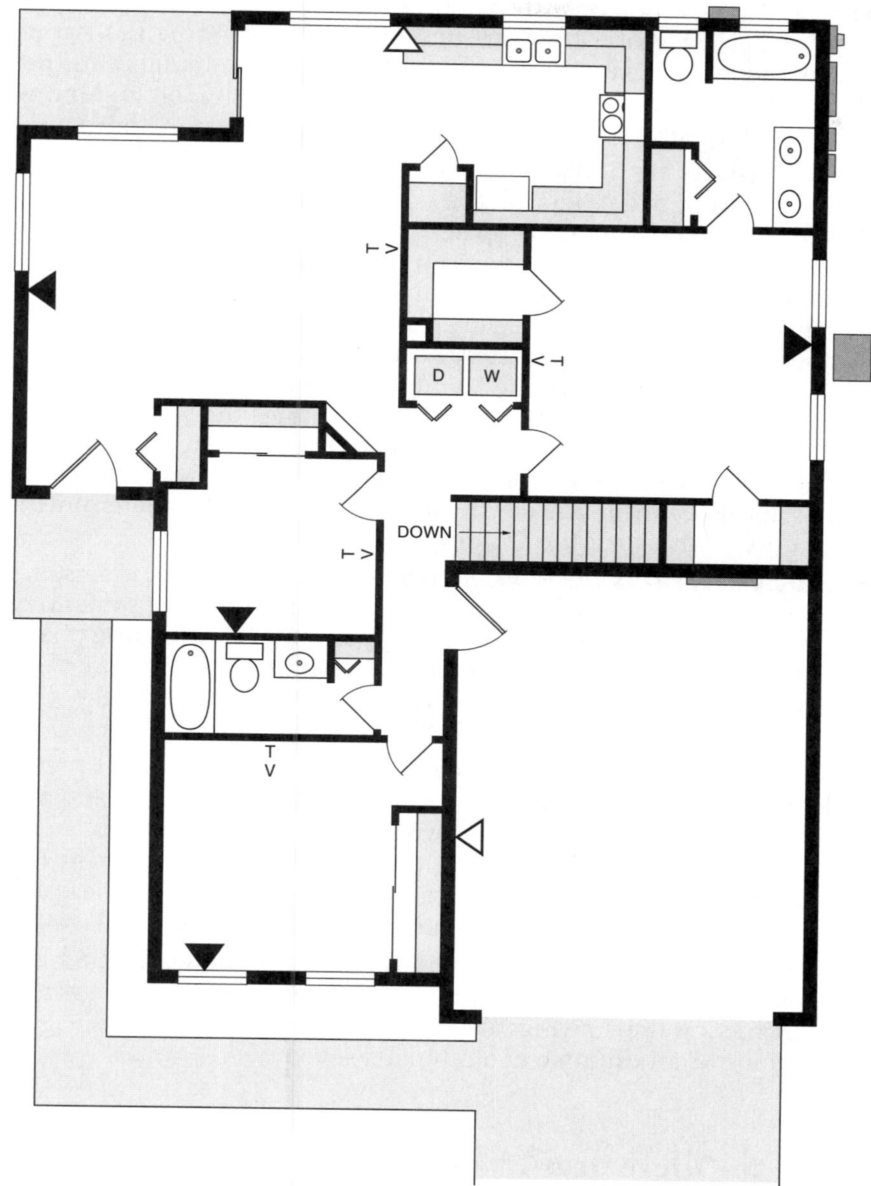

FIGURE 13–3 Low-Voltage Cabling Plan

- **Cabling documentation** is provided when the home has been cabled previously as part of another project, or as earlier project completed by another residential integration company. Figure 13–3 shows a typical cabling layout on a floor plan of the home.
- **Installation documentation** is provided when any prior installations are in the home. If this project is an addition to an existing

system, the installation documentation of the prior project is critical in determining project deliverables and objectives of the new project. These documents include **as-built documents** from the prior project. An as-built document can be the same as a scope statement. However, it has been updated at the completion of the project to reflect the final system details. In addition, programming worksheets, such as a list of preset radio stations, are included.

- **Floor plans** are provided in some new construction projects. The floor plans provide a view of the home from the ceiling to the floor. The plans may include electrical, HVAC, plumbing, and low-voltage systems in the home. In addition, other floor plans may include a furnishings layout, which aids in the placement of equipment, wallplates, and speakers. Figure 13–4 shows an electrical plan of a home, and Figure 13–5 shows the same plan with furniture placement. Both plans show the layout of the home, but the electrical plan shows where lighting and power outlets are located, while the furniture plan shows the location of the furniture in the home. The furniture plan can be used to determine locations of televisions and speakers, while the electrical plan can be used to determine if an outlet exists near the TV location.
- **Block diagrams** are provided to help the project team understand how the system is to be connected. These diagrams are sometimes generated during the cost-estimating portion of the defining phase. The diagrams, as well as any supporting documents in the cost estimating process, help the team understand the project. Figure 13–6 shows an example of a block diagram.
- **Touchscreen layouts** are provided to help with customer preferences. Typically, they are generated during the programming and build phase of the project. However, there are cases where, as part of the sales presentation, the touchscreens are laid out sooner. The layouts are intended to help program the system, but they can be useful sales tools. Figure 13–7 shows an example of a touchscreen layout.
- **Cabinetry drawings** are provided in retrofit projects because the cabinetry is designed as the project is defined. Typically, with new construction the drawings are completed later in the process. Figure 13–8 shows an example of a cabinetry layout.

Sales-to-Operation Turnover Meeting Preparations

As we learned in "A Case in Point, Step-by-Step," in preparing for the Sales-to-Operations Turnover Meeting, the sales representative distributes copies of the project documents to the project manager *at least* one day prior to the turnover meeting. The project manager reviews each document for the following items:

- **Policies.** The policies for the defining phase are included in the scope management plan. The project manager reviews the project documents to ensure they meet those policies.

ELECTRICAL PLAN

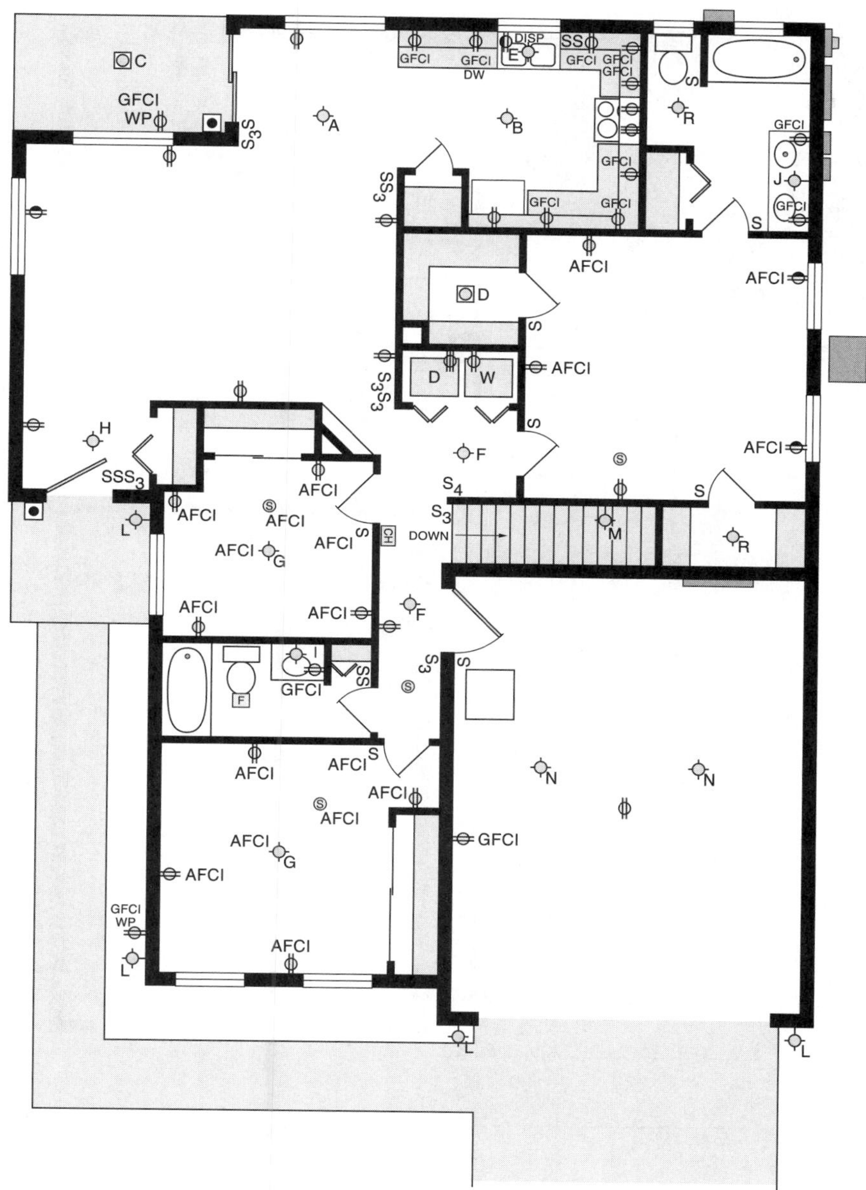

FIGURE 13–4 Electrical Plan

- **Gross Profitability.** Gross profitability is the profit of the project after direct labor and material expenses. Gross profitability is calculated as part of the quantitative risk analysis step in risk management planning (Chapter 11.) The project manager makes sure the gross profit meets the standards set out in the scope management plan. The best way to determine if a project is profitable is to take the gross profit

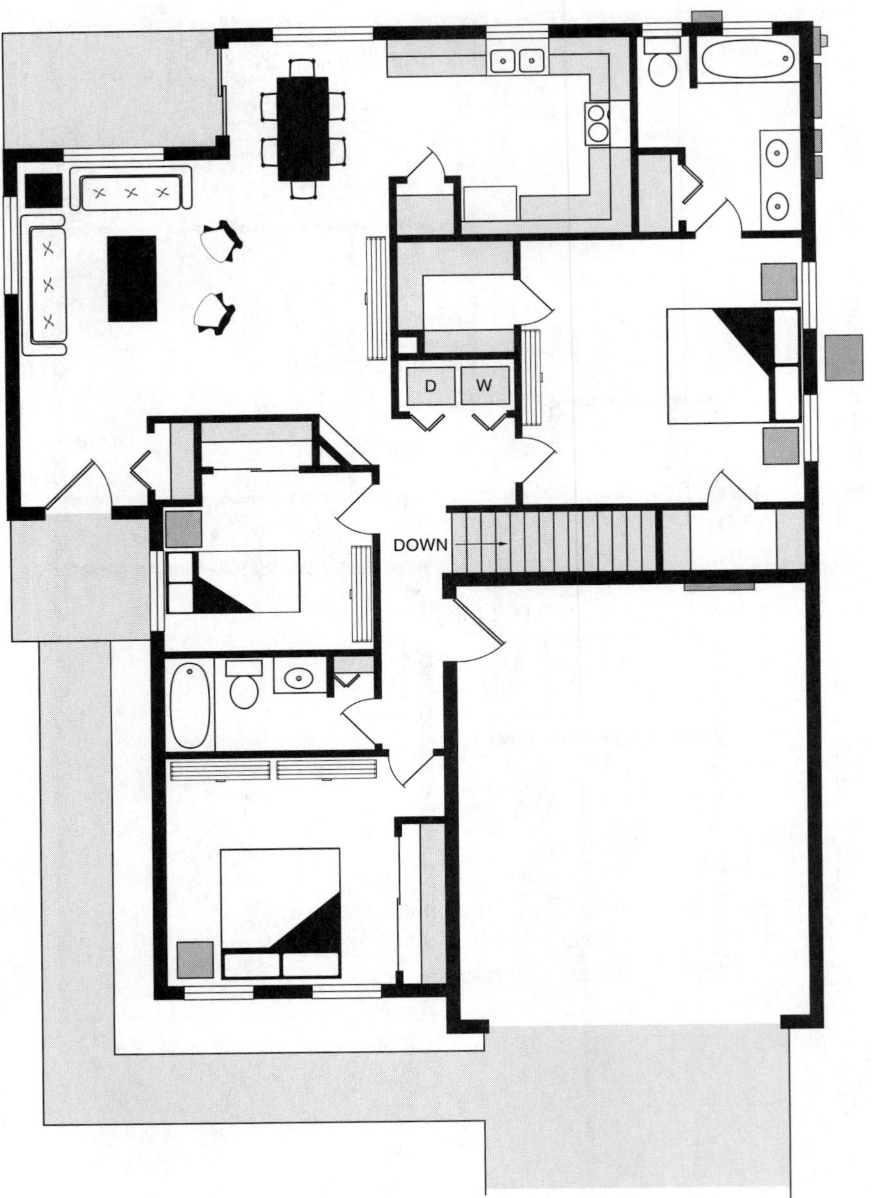

FIGURE 13–5 Furniture Plan

and divide it by the total project hours. This mathematical process yields a profit-per-hour amount, which can be compared to other projects of various sizes. Let's say, for example, one project yields $50,000 in gross profit, and requires 500 labor hours, or $100 per labor hour of gross profit. And let's say another job yields $10,000 in gross profit and requires 50 hours of labor, or $200 per labor hour of profit.

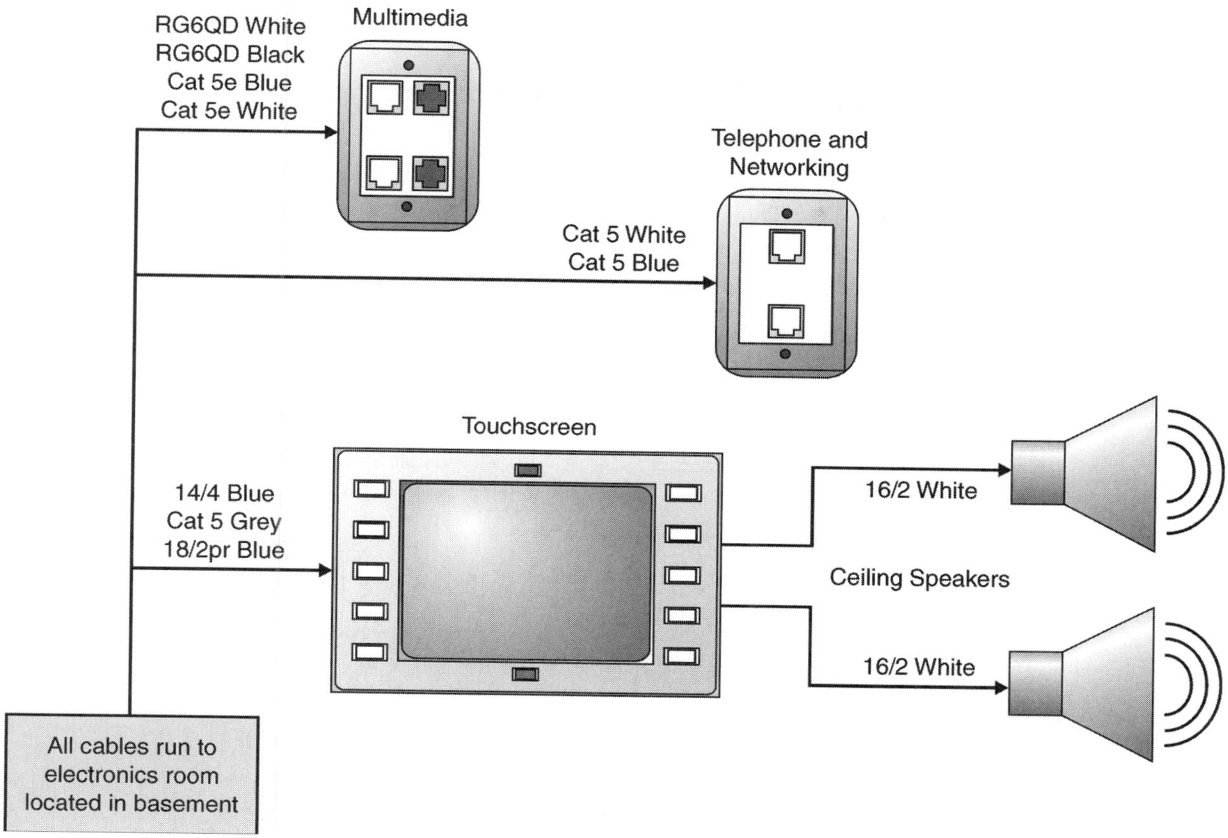

RG6QD White
RG6QD Black
Cat 5e Blue
Cat 5e White
Multimedia
Telephone and
Networking
Cat 5 White
Cat 5 Blue
Touchscreen
14/4 Blue
Cat 5 Grey
18/2pr Blue
16/2 White
Ceiling Speakers
16/2 White
All cables run to
electronics room
located in basement

FIGURE 13–6 Block Diagram

Comfort (HVAC)
Living Room
Home
Current Scene
Change room ...
72
70
Heat
Home
Overnight
Current Temperature
78
Cool
Away
Store ...

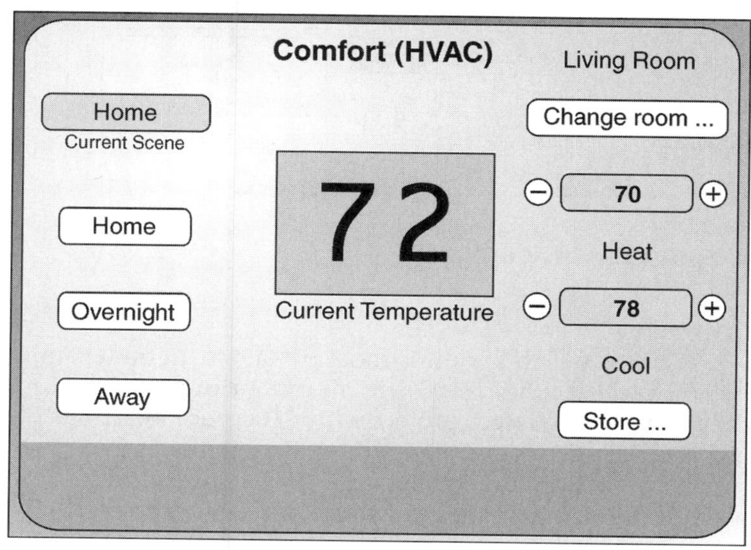

FIGURE 13–7 Touchscreen Layout

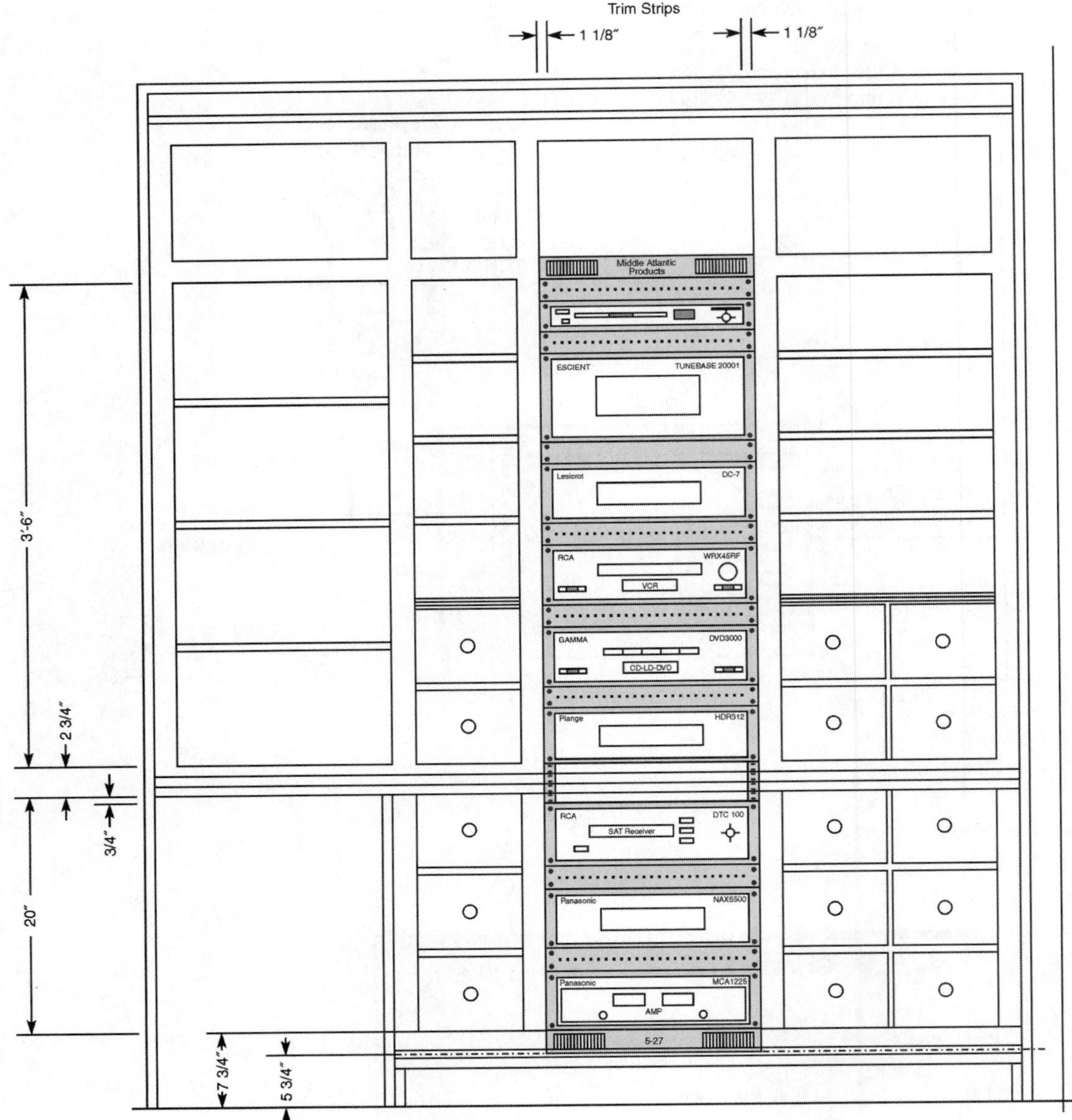

FIGURE 13–8 Cabinetry Drawings

It is easy to see which project is more profitable. In our example, the smaller project yielding $200 per labor hour of gross profit is more profitable than the larger project with $100 per labor hour of gross profit. Smaller projects can be more profitable than larger ones. The company owner or comptroller could set a lower limit for the per-labor-hour profit based upon the financial plan of the organization.

- **Project Workload.** The project manager should check the workload of the project against the current project team schedule. It is important to determine how soon the cabling needs to be started and how many labor hours are required for that phase. Is the current project team capable of carrying out the workload in a timely manner? Studying this information helps the project manager figure out whether hiring subcontractors is necessary to complete the work. The project manager should bring up any issues that may arise with the schedule during the turnover meeting. Depending on the company policies and the severity of the scheduling issues, the project manager also may make the company owner or director of operations aware of these issues. Will the planned project push the company ahead of sales projections expressed in the business plan? That issue is critical to the success of the company. Too many sales can be as deadly as too few sales.

- **Unusual Site Conditions and Permit Requirements.** Unusual site conditions include any situation that is out of the ordinary. It could be an uncommon circumstance concerning the house construction. It could be a homeowner's special request, such as not making noise in the house during a retrofit job while the baby sleeps, or having to leave the home when the owners cannot be present. Another unusual circumstance could be requiring additional labor hours or requiring a change in the normal routine of the installation. That information helps the project manager plan the work appropriately.

  The project manager also needs to make sure of which permits are required so that they can be obtained before the work begins. Every municipality has its own regulations on the cost and type of permit required for low-voltage systems. Some have a flat charge for the permit; others require a percentage of the project. In the case of a percentage, the project may need to be split into what work is done prior to the Certificate of Occupancy (CO) and what is done afterward, thus minimizing the permit costs.

A CASE IN POINT

Business Forecast = Stormy Weather

ABC Integrations' financial forecast for the year was set at $1.2 million in sales revenue. This is the measure of what is to be delivered by the company. The company wants to grow, so the dollar amount for contracts to be closed is set at $1.4 million. That provides a continued growth of sales revenue for the following year.

The project manager builds a team that is capable of delivering $100,000 in products and services per month. Now the trouble begins. During the last quarter of the previous year, the sales team only closed $150,000 in new contracts, less than half of what is required, and as a result the average monthly deliverables during the first quarter dropped to $70,000 per month, resulting in much less work than the team could handle. Now, during the first quarter, the close rate catches up to $450,000 in new projects, thus making up for lost time in closed sales. However, much of the work closed in Q1 is delivered in Q2, resulting in a strained workload. Overtime charges and employee burnout result, following a quarter with an underutilized installation team. The result is that, even though ABC Integrations has hit its numbers, the uneven sales caused a very unprofitable situation due to both under sales and over sales.

Study/Career Tip

When taking notes during a class lecture, model them after how textbooks are designed. (See the Table of Contents of this text as an example). Include a title or main heading for each daily lecture and any subheadings where appropriate. A course outline may be helpful in this regard as well. When it comes time to review specific topics and subtopics it will be much easier to find this information in your notes.

Critical Thinking Questions

- What do you think the sales representative's role will be once a residential integration project physically gets underway?
- Does he simply move on to the next potential sale or will he continue to interact with the customer?
- What might be the advantage in doing that?

The Sales Representative's Role after the Turnover

Once the project is turned over to the project manager, the sales representative moves from lead position to a supporting position for the project and conducts continued communication with the customer. Throughout the project life cycle, the sales representative continues to be responsible for making sure that customer expectations are met. The sales representative interacts regularly with the customer and gathers project information when needed. In some companies, the project manager also interacts with the customer during the project life cycle, while in others all communication with the customer is funneled through the sales representative. A good sales representative knows that each time he meets with the customer it is a potential opportunity to sell. So when it is time to determine wallplate colors, for example, the sales representative should use that occasion to up-sell the project, which could include better quality equipment such as speakers and amplifiers. In addition, she may review options that were previously not purchased to see if it is now a good time to add them to the project.

Summary

In Chapter 13, we discussed communications planning, how the project manager needs to be an effective communicator, the importance of communications in the Sales-to-Operations Turnover Meeting and the logistics surrounding that meeting. We also discussed the issues involved in using outside contractors.

The important points in this chapter are:

- Project communications planning is an essential element to the completion of a successful project. Detailed communications planning—knowing and communicating each step along the way to a successful project—has a positive effect on any project.
- A successful project manager is an effective communicator with abilities needed to

communicate successfully with others—project team members, stakeholders, and customers.

- The type and form of the information and its value defines the communication requirements. The Sales-to-Operations Turnover Meeting provides a smooth transition from sales to project management. Operations plans must be determined before the physical project can be started.
- Documents provided to the project manager prior to the Sales-to-Operations Turnover Meeting include the information gathering forms, the scope statement document, the WBS documents, and the cost estimate. A number of other documents can also be presented. These could include cabling documents, installation documents, floor plans, block diagrams, touchscreen layouts, and cabinetry drawings.
- A project manager checks all the documents he receives at the turnover meeting, keeping in mind company policies, gross profitability, project workload, and unusual site conditions and permit requirements.
- A challenge for project managers is communicating with outside contractors, who are independent. It's important they understand their specific responsibilities. A project manager must plan his schedule to allow for subcontractor delays.

Key Terms

As-built documentation Drawings and diagrams that provide an accurate representation of how the product or facility is actually built.

Block diagrams Simple drawings of how cables in an integrated system are to be connected.

Cabling documentation The documentation that explains any prior cabling that has been competed in a home.

Communications planning Determining the information and communications needs of the project stakeholder—who needs what information, when they will need it, and how it will be given to them.

Floor plan A view of the layout of the floor structure from an overhead perspective is called a floor plan. The floor plan contains information such as location and size of windows and doors, room size, location of exterior and partition walls, location of electrical fixtures, outlets and switches and plumbing fixture locations and other items. Each floor has its own floor plan.

Installation documentation Documents that show any prior installations in the home.

Operations turnover The process of shifting the responsibility for a project from the defining phase (sales) to the operations phases (project management).

Review Questions

1. What is a definition of communications planning?
2. Why is it so important to identify project communications requirements?
3. Name a common communications planning policy.
4. Is there a set rule for what type of communication is used between project parties?
5. What is a way in which both the project manager and the outside contractor understand their roles within a project to avoid misunderstandings?
6. What are some things a project manager can do to manage the scope and time of a meeting?
7. What are some tools project managers can use to get their message across at a meeting?
8. What is the main purpose of the Sales-to-Operations Turnover Meeting?
9. List three other documents that may be used in the Sales-to-Operations Turnover Meeting.
10. What is the role of the sales representative with the customer after the turnover meeting has occurred?

Quality Management Planning

OBJECTIVES

After studying this chapter, you should be able to:

- List and explain several characteristics of quality.
- Define and discuss quality management planning.
- Define and explain cost-benefit analysis.
- Define and explain benchmarking.
- Discuss additional quality management planning tools.

OUTLINE

Introduction

What is quality? We say we know it when we see it, but how do we define it or describe it? It's based on our perception rather than on a set of facts. It's a somewhat nebulous term. Unclear. Indistinct. Indefinite. We can't say quality is a product that lasts a specific length of time. We can't say quality is a product that costs a certain amount of money. We can't say quality is a product that encompasses a certain range of materials.

So, what is quality? Quality is the value or worth of a product to the person who buys the product. What is the product worth to you? Is the product worth spending your money? Quality involves the general standard of a product, standards that are often developed over years in an industry. In the movie industry, for example, Lucas Films (of *Star Wars* fame) developed THX certification to indicate quality in audio and video products.

In this chapter, we will examine the characteristics of quality, quality management planning, and key quality management tools widely accepted throughout the residential integration industry.

Study/Career Tip

There are several pioneers in the quality management movement in the world, among them Deming, Juran, and Crosby. Visit **www.deming.org; www.philipcrosby.com;** and **www.juran.com** for more information on these individuals and their institutes.

Critical Thinking Question

Think about something that you recently purchased that you thought was of "good quality."

- What made you think that of the item?

Characteristics of Quality

In order to develop a quality management plan, it's necessary to understand the characteristics needed to define quality. Some of the characteristics of quality that apply specifically to the residential integration industry are:

- **Dependability.** The mantra of dependability is the ability to be trusted to deliver what you have promised. Does your company provide dependable deliverables? Is the installation of complex cabling systems dependable? A project manager can assure product dependability by delivering products from reputable vendors, suppliers, and manufacturers. A successful project manager quickly learns which products in the industry are trustworthy—those with a reputation for long-standing dependability. Part of a quality management plan is following the company's plan in maintaining a file of vendors, suppliers, and manufacturers who have met the company's expectations on past projects.
- **Reliability.** The near twin of dependability is reliability. The two go hand-in-hand. A reliable product or service is one that is able to act in a way that is expected or required. Are your installers and technicians reliable and well trained for the job they are expected to perform?
- **Service.** A successful project manager must provide quality service. A way to provide quality service is to develop a project team's abilities by encouraging team members to become certified in the various levels of the industry. Organizations, such as the Custom Electronic Design and Installation Association (CEDIA), and the Consumer Electronics Association (CEA), hold certification courses in installation, design, and project management throughout the year. Conferences are places to pursue increased knowledge in the industry. Formal education courses are still another avenue of learning. Experience and apprentice programs also result in competent project teams. Planning a process for project members to become more proficient, and thus more reliable, in their jobs enhances the quality of each project that is completed.
- **Ease of Use.** Designing integrated systems that are easy to use is a hallmark of the industry. No matter how well a system is designed,

there is also the element of customer training that needs to be addressed as part of a project quality plan. Programmable thermostats keep a home comfortable throughout the year. The initial programming of the thermostat is provided by company's field technician. But adjustments to the system long after installation are the realm of the customer. Providing adequate training and well-written instructions to the customer and additional telephone support enhances the level of quality of the system. Sometimes the quality of the product remains hidden when no one knows how to use it. This is true for the various integrated systems installed in the home.

- **Convenience.** Another characteristic of quality in integrated systems is convenience. Does the system's convenience meet the customer's need? Can the customer touch one button when he leaves the house to turn all systems to an "away mode"? Can the customer touch a different button to turn on the systems when he returns home? Are there manual overrides to the system in case of system failure?
- **Respect for the Customer.** A successful project manager not only respects her staff and peers, she encourages the project team to show respect and courtesy to all the stakeholders. These are not tangible characteristics, yet they provide a perception of quality that can override minor flaws or provide the opportunity to correct them when they occur.
- **Respect for the Customer's Property.** A recent trend by many service people is to wear paper booties when entering a customer's home. It's a sign that they respect and value their customer's property. Again, it encourages a perception of quality. A technician who takes care to wear paper booties or clean sneakers on a new carpet is perceived as more likely to perform his job with as much care.

Study/Career Tip

Do you mark up your textbook as you read through it and subsequently study from it? If so, you will have an understanding of the main ideas of the text reading, and you have a marked text that will help you prepare for exams. The markings are yours, your choices; that is why it is useful to buy unmarked texts. Other people's decisions will not help you.

Critical Thinking Question

The term quality is often associated with excellence.

- What other synonyms can you list that imply quality?

Quality Management Planning

Quality management planning is a series of processes that are required to ensure that the project will satisfy the needs for which it was undertaken. Quality planning is a key process in project management planning and is greatly dependent on the triple constraints of scope, time, and cost. Changes in any of the constraints affect the quality of the project. To be sure, if the scope of a product is broader than the project team can handle confidently, quality goes down. If the scope of the project is well within the project team's level of confidence, quality goes up. If the time is accelerated beyond the project team's capabilities, quality becomes doubtful. If enough time is allowed to complete a project, the opportunity for quality increases. And if the cost of electronics purchased is too low, the residential integrator is unlikely to be providing a quality deliverable. Purchasing high quality products to provide high quality deliverables costs a reasonable amount of money.

Each residential integration company is responsible for developing activities to determine if its policies result in meeting the customer's needs, wants, and requirements. Without a systematic quality management plan, the quality of the systems installed in the home is left to chance. It is also doubtful that the customer's expectations, the primary focus of quality management, will be met. Customers who view a project's outcome as unsatisfactory can be expected to look elsewhere for future services. Here's a case in point.

A CASE IN POINT
Quality is Job #1

At ABC Integrations, a company policy requires all electronics to be tested thoroughly at the company's facility before they are delivered to the home. At the end of the cable installation phase, the project manager checks each wallplate connection to make sure it is correctly installed. At the end of each phase, the project manager verifies the working condition of the system, keypads, and touchscreens. And finally, the sales representative and the project manager check each installed system to make sure it is working before the project is turned over to the customer. The quality management plan also has a policy that requires the sales representative to continually monitor if the customer's expectations are being met.

A quality management plan also should include a policy to continually improve project processes. A successful project manager maintains a "lessons learned" file that she gains from previous projects. In addition, she distributes a "best practices" document to the project installation teams and requires the team leaders to review the documents with their teams on a regular basis. Let's review these quality management techniques in greater detail.

- **Lessons Learned.** This is an excellent way to get the entire project team involved in continually improving the project management process. Every time something goes wrong on the project, a project team member fills out a lesson-learned form and submits it to the project manager, who, in turn, reviews each of these forms on a regular basis with the executive team to make changes to company policies. For example, an installer may write about a particular brand of speakers that is not easily installed in retrofit installation situations. While the speakers are suitable for new construction, the organization can use the information from the installer to not provide that brand of speakers in retrofit projects in the future.

- **Best Practices.** Best practices are used to ensure a consistent process from project to project. For example, ABC Integrations installs cables for first floor wallplates from below, bringing the cables up from the basement. Since most basements are left unfinished, the wallplates can be updated in the future with new cables. As with lessons learned, best practices should be documented and discussed among project team members to ensure quality and consistency across all projects.

Study/Career Tip

There are many organizations that support the quality movement. Some of these include: The American Society for Quality (**www.asq.org**), the National Committee for Quality Assurance (**www.ncqa.org**), and The Baldrige National Quality Program (**www.quality.nist.gov**). Take a few moments to check out these organizations.

Critical Thinking Question

Cataloguing and sharing best practices is always a good idea, no matter what the industry.

- What best practices have you adopted when it comes to studying?

Write them down and review them occasionally. You might be able to fine-tune them to your advantage.

Quality Management Planning Tools

In addition to a quality management plan, organizations can and should utilize other widely accepted methods for measuring quality. These include:

- Cost-Benefit Analysis
- Benchmarking
- Other Quality Management Planning Tools

Let's review each tool.

Cost-Benefit Analysis

A **cost-benefit analysis** is a tool widely used to decide whether or not to institute a new course of action in completing a company project. As its name implies, the project manager looks at the benefits of the new process and the cost of implementing it. If the costs outweigh the benefits, then it should not be implemented. But if the benefits outweigh the cost, then the project manager can proceed with putting it into effect.

To illustrate, purchasing company cell phones for all project team members could save time and therefore money, because it will allow the team to contact the project manager to answer questions and allow the project manager to contact the team as needed. For instance, when a project is being fast tracked, it may be necessary to contact the team immediately to send them to a different work site. The project manager needs to decide if the cost of paying for cell phones for each team member outweighs the benefit, and then potentially choose the option that only the team leaders should carry company paid cell phones.

The cost-benefit of an activity need not be immediate. It can also be realized over a long period of time. For example, the cost-benefit analysis could include a payback time. A company could decide to purchase an installer truck and outfit it with a standard set of installation tools, cabling equipment, and test equipment. A cost-benefit analysis determines how long the same truck needs to be used to determine how it benefits the company the most. The intent is to save money over the long run. Most companies set a specific time period for a decision to purchase an item in order for it to repay itself.

In its simplest form, cost-benefit analysis is carried out using only financial costs and financial benefits. It does not usually include nonfinancial benefits. In our truck example, it would not include the convenience to the installers, possible advertising on the side of the truck, added storage space for tools and equipment, or time saved loading an installer's own vehicle with company items. A more sophisticated cost-benefit analysis could try to put numeric figures on the intangible items to provide a more accurate analysis. For example, what would the value be of reducing the stress of forgetting a piece of test equipment or other needed tool and having to return to the company facility to get it? Stress is one factor, and the time it would take to return and get the equipment is another factor. In doing a cost-benefit analysis, a project manager considers both tangible and intangible benefits. Here's a case in point.

A CASE IN POINT

Analyze This

ABC Integrations regularly installs Cat 5e cables. While these cables can be tested for continuity, meaning all cables have been connected properly, a specialized tester is required to ensure the cables and connectors use the full bandwidth potential. This tester is known as a Cat 5e Cable Analyzer. It tests the cables across the full bandwidth. The cost for this tester is $6,000. The question is: Should ABC Integrations purchase the analyzer?

ABC Integrations can pay an outside contractor $35 per wallplate to perform the test, while internal time cost is estimated to be $10. If ABC Integrations is expecting a payback on the analyzer, it will need to install 240 wallplates before making any profit on this piece of equipment. Based upon the expected sales projections for wallplates, Project Manager Owen Ginizer can make a decision on whether to purchase the analyzer or hire an outside contractor to perform the work.

Study/Career Tips

Do you study in a group? If not, here are some benefits you might want to consider. If you want to establish an effective study group, choose two other students who share your desire to succeed (three is the magic number for a group). Set a specific time and place to gather, and set a topic to discuss. Lay out ground rules. Expect everyone to have completed an initial study of the material and to bring questions to the group. Compare class lecture notes across the group. Focus the discussion on two points: the unanswered questions about the material from the lecture and the reading and the predicted questions for the next test.

Critical Thinking Question

Although you have probably never used the terminology in this context, we all do a cost-benefit analysis on purchased (or perhaps not) items now and again. We certainly don't give buying gasoline a second thought, but what about the purchase of a new car? Weighing the pros and cons of such a decision can be a decidedly conscious choice.

Key Points of Cost-Benefit Analysis

Cost-benefit analysis is used to:

- Determine change within the company.
- Work out the cost of the change.
- Work out the benefit to the company.
- Determine how long the payback period is.
- Determine the value of tangible financial facts.
- Determine the value of intangible, nonfinancial facts, as appropriate.

Benchmarking

Benchmarking helps a company develop standards of quality. It compares current company processes and procedures with those of other projects to generate ideas for improvement and performance evaluation.

How to Benchmark

Benchmarking is a widely used project management tool most companies use to stay competitive. Here are some of the ways to use benchmarking to improve quality.

- **Pricing.** Identify your own benchmarking partners and find out from them what is achievable, and then, whether you can achieve a similar level of performance. For example, ABC Integrations sells each of their plasma televisions at the same price as the local retail store.
- **Surveys.** Market surveys can tell you what is being offered in the industry, and you can compare that information with what your company offers. Surveys of your customers by an independent

survey firm can be very helpful in letting you know where to improve. An independent survey company may also do an industry-wide survey that your company can use to compare with its own practices and products. However, be mindful that there is a cost associated with using an outside survey firm.

- **Process.** Think in terms of benchmarking a process (a small group of tasks or activities), rather than an entire system (a group of many processes). It is easier and more manageable that way. To illustrate, the time required to install a pair of speakers can be measured from project to project and categorized by the type of project (retrofit and new construction), as well as the type of speaker (ceiling, wall, and bookshelf). This information can be used as a benchmark as to the amount of time required to install speakers on future projects.
- **Alignment.** Choose a benchmark topic that is aligned with the overall strategy and goals of the company. A lead team at the strategic level needs to oversee the benchmarking project and make sure that it is in line with what is happening in the business as a whole.
- **Tangible Topic.** Pick a topic that is tangible and easy to measure. The benchmarking team could select speaker installation as its topic rather than employee communications. The broader topic is difficult to measure and too large to stay focused on.
- **Service.** Remember to maintain a quality standard for service delivery.
- **Customer Satisfaction.** The primary benchmark is customer satisfaction. Without it, you lose the next project with that customer.

Study/Career Tip

Don't be afraid to revise your study schedule. If your current study schedule isn't working for you, revise it. The reason you have a schedule is to help you develop good study habits. Once you have acquired good study habits, it will become easier for you to build and maintain a schedule each semester.

Critical Thinking Questions

- What types of things do you benchmark in your life?
- Might they include your abilities at a particular sport, the number of questions you get right on *Jeopardy,* your driving record?

The concept of benchmarking isn't just for the business world, we use it in our daily lives as well.

Other Quality Management Tools

Project managers can draw on many quality tools and techniques. These are not necessarily limited to the residential integration industry. Some of these tools are listed below:

- **Checklists.** Every process developed within the organization can support a checklist, which provides each team member with clear directions to completing each step in the process.

- **Brainstorming.** These are opportunities for everyone in the organization to pitch in and share ideas that inevitably spawn other ideas.
- **Industry Certifications.** Hiring and retaining employees who hold certifications from organizations such as CEDIA and CEA.
- **Industry Best Practices Standards.** Know what are considered to be best practices in the residential integration industry and adopt those standards. In fact, this textbook is based on the *A Guide to the Project Management Body of Knowledge* (PMBOK Guide) best practice standards formulated by the project management organization, Project Management Institute (**PMI; www.pmi.org**).

Summary

In Chapter 14, we defined quality and quality management planning, specifically within the context of characteristics that apply to the residential integration industry. We discussed cost-benefit analysis and benchmarking. We also discussed checklists, brainstorming, industry certifications, and industry best practices standards as additional quality management tools.

The important points in this chapter are:

- Quality is an opinion or perception, rather than something based on fact.
- There are many characteristics of quality. Some that apply to residential integration are dependability, reliability, ease of use, convenience, and respect for the customer and his property.
- Cost-benefit analysis compares the costs and benefits of change against whether or not to implement the change.
- Benchmarking is used to set a standard to improve quality in various ways, including pricing, surveys, process, alignment, tangible topics, service, and customer satisfaction.
- Additional tools that can be used in quality management are checklists, brainstorming, industry certifications, and industry best practices standards.

Key Terms

Benchmarking Benchmarking is used to create a standard for quality improvement.

Cost-benefit analysis Cost-benefit analysis compares the costs of benefits of a change to determine whether to implement the change.

Quality management planning A subset of project management that includes the processes required to ensure that the project will satisfy the needs for which it was undertaken.

Review Questions

1. What are some of the characteristics of quality as they specifically relate to the residential integration industry?
2. Define quality management planning.
3. Why is implementing a "lessons learned" practice important?
4. What is the purpose of best practices?
5. How does a cost-benefit analysis work?
6. What is benchmarking?
7. What are some other quality management planning tools that a company can utilize?

Cost Management Planning

OBJECTIVES

After studying this chapter, you should be able to:

- Define cost management planning.
- Explain the purpose of cost budgeting.
- List five ways to view a bill-of-materials.
- Explain why different ways of presenting a bill of materials are necessary, how they are used, and by whom.
- Explain the job costing report.

OUTLINE

Introduction

As we learned in earlier chapters, probably the greatest single hurdle a residential integration company has to overcome is that of committing to a project with insufficient resources: namely money. The trick to maintaining sufficient funding and resources is to accurately cost estimate a project. That is easier said than done, and many companies learn the hard way.

The initial cost estimate is done early in the project life cycle. Once the project gets underway, meaning that goods and services have been ordered, the costs that were once estimates are now realities. In this chapter we will see how residential integration firms track those costs using a variety of tools and methodologies.

Cost Management Planning

The **cost management plan** is a critical part of the overall project management plan. One of the primary functions of the project manager is to make sure the project produces a profit while maintaining quality, excellence, and customer satisfaction. Customers pay thousands of dollars for most residential integration projects, and they have a right to expect high-quality installation and deliverables when the project is finished. The company also has a right to expect a profit at the end of the day if they have done their job correctly. One significant way to manage that profitability is through the use of **cost budgeting.**

Study/Career Tip

Job search Web sites such as **monster.com** and **careerbuilders.com** (just to name two) have career advice centers as part of their offerings. If you are in the market for a new job or are seeking your first job, consider visiting these or other career sites. The information is free and may be just the ticket to helping you land that coveted job.

Critical Thinking Questions

- Do you maintain a financial budget?
- If so, do you compare the budgeted amount for a certain good or service against what was actually spent?

Cost Budgeting

Cost management planning consists of setting up the processes needed to track the company's project figures for materials and labor. Cost budgeting is the process in which the estimated figures are compared with the actual figures, which are recorded at the end of the project. These comparative figures are then used to improve the model for the next project. For example, the cost of installing the integrated cabling system may be raised or even lowered based on the information derived from implementing the cost budgeting plan.

So how does the project manager begin the process of cost budgeting? The first step is to add together all the individual estimated costs for the work packages that are listed in the Work Breakdown Structure (WBS). He adds together the amount of cable, brackets, and wallplates that were estimated to be used, along with the total number of estimated hours needed to install each work package.

The sum of the individual pieces of each work package in a project creates a total **cost baseline** that can be used to measure the performance of the project. The project manager then compares the cost baseline with the totals of each similar item in the work packages actually used for the entire system. The comparison shows the **cost differential** between the cost estimate and the actual cost as a credit or a debit.

The cost budgeting process measures the accuracy of the project manager's estimates for materials and labor. To illustrate, the project manager might have estimated that 1,875 feet of Cat5e cable will be needed to complete a project. However, the project may have actually needed 2,000 feet of cable. The work order bill of materials reflects that difference. In analyzing the information provided by the cost budgeting process, the project manager can decide whether or not to investigate the cause of the cost differential.

Study/Career Tip

Consider applying for an internship, either during the school year (as your course workload allows) or during the summer. While most internships are non-paying, what you gain in experience will position you well when you look to enter the job market. For example, hiring managers are more likely to consider someone who has already interned in the field than someone who has not. It shows you are committed to that field as a career (at least at this point in your life).

Critical Thinking Questions

As you read on, you will see that a bill of materials (BOM) can be viewed in a variety of ways.

- Why do you think this is necessary?
- Shouldn't there just be a "one size fits all" document?

Bill of Materials

As we learned in Chapter 8, Cost Estimating, the output of the cost estimating process is a bill of materials (BOM), which details the quantity and costs for each resource required during the installation and completion of the project. Again, this includes the materials and labor required. In the residential integration industry, a bill of materials includes a list of equipment required for the project and any miscellaneous materials such as brackets, interconnecting cables, and connectors. It also includes installation services, such as cabling, equipment installation, project management, and training time.

The bill of materials can take many forms depending on the intended purpose and audience. There are five ways to view the bill of materials. Each of these forms is created by using the cost estimating template (see PM07 in Appendix A). These different views include the:

- Cost Estimate Customer View
- Cost Estimate Internal View
- Purchase Requirements
- Work Order
- Sales Invoice

While our case study utilizes a spreadsheet for these calculations, it should be noted that a wide variety of specialized software exists that can be used to create cost estimates and the many forms of the bill of materials. Let's take a closer look at the different ways a bill of materials may be viewed.

Cost Estimate Customer View

The customer view of the bill of materials presents the cost estimate as logical product units that the customer can easily understand. If the

Quantity	*Description*	*Amount*
1	Hanging Telephone Wallplate	126.58
7	Telephone and Networking Wallplate	1,333.22
7	Multimedia Wallplate	2,034.76
2	Surround Sound Wallplate	759.64
3	Music and Video Output Wallplate	1,281.81
6	Speaker Cabling	485.82
3	Speaker Wallplate	518.43
1	System Telephone and Home Networking Media Panel	475.88
1	Cable Television Panel	564.18
1	High Definition Digital Satellite Panel	906.86
1	Off-Air High Definition Television (HDTV) and FM Reception Panel	450.84
	Total Products	4,016.77
	Total Services	4,921.25
	Sub Total	8,938.02
	Sales Tax	200.84
	Grand Total	9,138.86

FIGURE 15–1 Cost Estimate Customer View

contract is a fixed price contract, this represents the actual cost to the primary stakeholder. This is true regardless of the actual cost to the company. At ABC Integrations, the customer estimate separates the various wallplates into individual products and a price for a single installed unit is presented. Figure 15–1 shows ABC Integrations' list of wallplates and the installed price for each, as well as the total price of the system. In this way, the customer can understand how adding and subtracting individual components will affect the overall system price. Adding a second hanging telephone wallplate, for example, adds an additional $126.58 to the project price. It should be noted that specific parts information (individual brackets, cable lengths, labor hours, and other details) is not provided in the customer view of the cost estimate. Those facts are for the company's internal use only. This keeps the complex details of how the work package is created from the customer, thus minimizing the possibility of this key information ending up in a competitor's hands.

Cost Estimate Internal View

While the customer view shows an overall price for each unit or each wallplate, an internal view of the cost estimate can be created to show the individual products and services associated with the each wallplate. This combination of product and installation data is presented in the internal view of the work packages. This view is available to the executive team and the project manager. In some companies, the sales personnel also have access to this view. The installation team, both installers and technicians, generally does not have access to this view. Determining who has access to the various views of the bills of materials should be part of the scope management plan.

Figure 15–2 shows the cost estimate of all of the products and services as a subgroup of the work package. We see the hanging telephone wallplate includes 110 feet of CableScope Cat5e cable as well as 40 minutes of installation time to install the cables, shown as "Installation, Cabling Phase, Mins." This view is helpful in understanding how the price is derived and how each of the work packages is built. We can think of a work package as the secret recipe for the system.

Study/Career Tip

Starting your first "real" job? The chances of your success will probably have little or nothing to do with your academic prowess or technical skills. More likely, it will hinge on your ability to relate to your officemates and to fit into the organizational culture. Earn the respect of your new colleagues by being a person who is willing to learn, listen, and contribute.

Critical Thinking Questions

- How can the cost estimate internal view, or work packages, be further broken down?
- Why would this be helpful?

Purchase Requirements

As we learned in Chapter 12, Procurement Management Planning, a purchase originates with a purchase order to the vendor. Figure 15–3 shows the purchase requirements created for ABC Integrations' vendor CableScope, a supplier of cables. It includes all of the material requirements for the cabling phase. It is created from summing the quantities of each product inside of each work package, such as the telephone and networking wallplate shown in Figure 15–2. This purchase order includes the cable requirements of the entire system. The procurement manager or procurement department uses the information in this view to determine what quantities of the material should be purchased in order to get the best price. It also helps the procurement manager determine the number of reels of cable that may need to be purchased or whether partially used reels may be available from the company's in-house supplies.

Work Order

The work order is a view of the bill of materials designed for the installation team. In Figure 15–4, we see a work order for the cabling team for our sample project at ABC Integrations. In this view, we see the same list of equipment shown in the Purchase Requirements view (Figure 15–3). However, the pricing information has been removed. In addition, a column has been added to allow the installation team to record the actual quantity of cable used while performing the

Quantity	*Description*			*Amount*
1	Hanging Telephone Wallplate			126.58
	110	CableScope: CT5ebl Cat5e in box, Blue, Ft	7.70	
	1	Connection: Snap-in Hanging Wallplate, White, Ft	3.65	
	2	Connection: 8-Conductor 5e AJ45 Jack, White, Ft	13.36	
	1	Boxes: New Construction Plaster Ring, 1 gang, Ft	1.67	
	40	Installation, Cabling Phase, Mins	40.00	
	20	Installation, Termination Phase	20.00	
	10	Installation, Move-In Phase	10.00	
	10	Engineering, Block Diagrams	15.00	
	12	Project Management	15.00	
7	Telephone and Networking Wallplate			1,333.22
	110	CableScope: CT5eBL Cat5e in box, Blue, Ft	7.70	
	110	CableScope: CT5eWH Cat5e in box, White, Ft	7.70	
	1	Connection: Decora Trim Plate, 1 gang, White, Ft	1.98	
	1	Connection: 2-Port Snap-in Decora Insert, White, Ft	3.24	
	3	Connection: 8-Conductor 5e RJ45 Jack, White, Ft	20.04	
	1	Connection: 8-Conductor 5e RJ45 Jack, White, Ft	6.68	
	1	Boxes: New Construction Plaster Ring, 1 gang, Ft	1.87	
	50	Installation, Cabling Phase, Mins	50.00	
	35	Installation, Termination Phase	35.00	
	20	Installation, Move-In Phase	20.00	
	10	Engineering, Block Diagrams	15.00	
	17	Project Management	21.25	
7	Multimedia Wallplate			2,034.76
	110	CableScope: CT5eBL Cat5e in box, Blue, Ft	7.70	
	110	CableScope: CT5eWH Cat5e in box, White, Ft	7.70	
	110	CableScope: RG6QD-WH RG6-QD Coaxial Cable, White Ft	23.98	
	110	CableScope: RH6QD-BK RG6-QD Coaxial Cable, Black. Ft	23.98	
	1	Connection: 4-Port Snap-in Decora Insert, White, Ft	3.24	
	1	Connection: 2-Port Snap-in Decora Insert, White, Ft	3.24	
	3	Connection: 8-Conductor 5e RJ45 Jack, White, Ft	20.04	
	1	Connection: 8-Conductor 5e RJ45 Jack, White, Ft	6.68	
	2	Connection: F-F Snap-in Module, White, Ft	8.50	
	1	Boxes: New Construction Plaster Ring, 1 gang, Ft	1.87	
	65	Installation, Cabling Phase, Mins	65.00	
	50	Installation, Termination Phase	50.00	
	25	Installation, Move-In Phase	25.00	
	10	Engineering, Block Diagrams	15.00	
	23	Project Management	28.75	
2	Surround Sound Wallplate			759.64
	110	CableScope: CT5eBL Cat5e in box, Blue, Ft	7.70	
	110	CableScope: CT5eWH Cat5e in box, White, Ft	7.70	
	110	CableScope: CT5eGY Cat5e in box, Grey, Ft	7.70	
	110	CableScope: CT5ePK Cat5e in box, Pink, Ft	7.70	
	210	CableScope: RG6QD-WH RG6-QD Coaxial Cable, White Ft	45.78	
	210	CableScope: RH6QD-BK RG6-QD Coaxial Cable, Black. Ft	45.78	
	2	Connection: 4-Port Snap-in Decora Insert, White, Ft	6.48	
	2	Connection: 2-Port Snap-in Decora Insert, White, Ft	6.48	
	3	Connection: 8-Conductor 5e RJ45 Jack, White, Ft	20.04	
	1	Connection: 8-Conductor 5e RJ45 Jack, White, Ft	6.68	
	4	Connection: F-F Snap-in Module, White, Ft	17.00	
	2	Boxes: New Construction Plaster Ring, 2 gang, Ft	5.78	
	80	Installation, Cabling Phase, Mins	80.00	
	60	Installation, Termination Phase	60.00	
	10	Installation, Move-In Phase	10.00	
	10	Engineering, Block Diagrams	15.00	
	24	Project Management	30.00	

FIGURE 15–2 Cost Estimate Internal View

PURCHASE REQUIREMENTS

ABC Integrations
123 Main Street
Anytown, PA 01234
(800) 555-1212

Proposal Date: 12/28/2005
Client: Smith, Fred & Wilma
123 Little Street

Description	*QNTY*	*RATE*	*AMOUNT*
CableScope: CT5eBL Cat5e in box, Blue, Ft, Part No: 1001	1871	0.04	74.84
CableScope: CT5eGY Cat5e in box, Grey, Ft, Part No: 1003	220	0.04	8.80
CableScope: CT5ePK Cat5e in box, Pink, Ft, Part No: 1004	550	0.04	22.00
CableScope: CT5eWH Cat5e in box, White, Ft, Part No: 1002	1760	0.04	70.40
CableScope: RG6QD-WH RG6-QD Coaxial Cable, White, Ft, Part No: 1005	1850	0.11	203.50
CableScope: RG6QD-BK RG6-QD Coaxial Cable, Black, Ft, Part No: 1006	1190	0.11	130.90
CableScope: S14-4-BL 14 Gauge, 4 Conductor Speaker Cable, Blue, Ft, Part No: 1007	990	0.23	227.70
CableScope: S16-2-BL 16 Gauge, 2 Conductor Speaker Cable, White, Ft, Part No: 1008	450	0.11	49.50
		TOTAL	787.64

FIGURE 15–3 Purchase Requirements for Cabling Phase

installation. Once the cabling phase has been completed and the actual and **differential quantity** (bid or estimate minus actual) column has been filled in, this report can be returned to the project manager, along with a list of outstanding issues that will be used to create a performance report of the phase. At ABC Integrations, the cabling team lead fills in actual quantities of cables as the phase progresses. As we determined earlier, these performance reports are valuable tools for both the project manager and the executive team.

Study/Career Tip

Do you proofread your work before handing it in? In proofreading, you can take nothing for granted. Word processing spell/grammar checkers are a great help, but they will not catch everything. The only way to really double check your work is to read it aloud. It may sound silly, but by hearing what you've written you often catch mistakes such as an omitted or repeated word that you have not seen.

Critical Thinking Question

- What do the project manager and the executive team do with the performance reports that are handed in from the field?

Work Order

ABC Integrations
123 Main Street
Anytown, PA 01234
(800) 555-1212

House Size: 4,000
Project: Structured Wiring System
Proposal Date: December 28, 2005

Client: Smith, Fred & Wilma
123 Little Street
Anytown, PA 12345

PRODUCTS	*BID QNTY*	*ACTUAL QNTY*	*BID ACTUAL*
CableScope: CT5eBL Cat5e in box, Blue, Ft, Part No: 1001	1871		
CableScope: CT5eGY Cat5e in box, Grey, Ft, Part No: 1003	220		
CableScope: CT5ePK Cat5e in box, Pink, Ft, Part No: 1004	550		
CableScope: CT5eWH Cat5e in box, White, Ft, Part No: 1002	1760		
CableScope: RG6QD-WH RG6-QD Coaxial Cable, White, Ft, Part No: 1005	1850		
CableScope: RG6QD-BK RG6-QD Coaxial Cable, Black, Ft, Part No: 1006	1190		
CableScope: S14-4-BL 14 Gauge, 4 Conductor Speaker Cable, Blue, Ft, Part No: 1007	990		
CableScope: S16-2-BL 16 Gauge, 2 Conductor Speaker Cable, White, Ft, Part No: 1008	450		
SERVICES	*BID HOURS*	*ACTUAL HOURS*	*BID ACTUAL*
Installation, Cabling Phase, Mins, Part No: 0101	20		

FIGURE 15–4 Work Order for Cabling Phase

The differential quantity information can be fed back into the cost estimating process to ensure the models are more accurately created for future projects. If the actual quantities are significantly different from the bid quantities, the project manager can investigate the cause. Perhaps additional wallplates have been installed that were not included in the scope statement.

Figure 15–5 shows a table included in the scope statement for this project. This table was developed with the customer in mind, but becomes a useful tool for the cabling team in addition to the work order. Figure 15–5 shows the quantities and type of wallplates to be installed, while Figure 15–4 shows the cables that are required for the project. The installation team can look at Figure 15–4 and see that a hanging telephone and multimedia wallplates are planned for the kitchen. The team can then look at Figure 15–5 and see what equipment they need to install those wallplates and how long it is estimated the installation will take.

Sales Invoice

As we discussed in Chapter 4, Understanding Cost Accounting, sales revenue is the money received for goods and services provided. As you recall, the two methods for recording revenue are cash-basis and

Your Structured Cabling Selections

The table below shows each wallplate in your home. We will field locate each wallplate in the rooms shown below with a "cabling" sticker. This will give you an opportunity to approve all locations before the home is cabled.

Location	*Hanging Telephone*	*Telephone and Networking*	*Multimedia*	*Surround Sound*	*Music Input/ Output*	*Speaker Cabling or Wallplate*
Kitchen	1		1			1
Breakfast Room						
Study		1		1	1	
Living Room		1	1			1
Dining Room						1
Family Room		1		1	1	
Deck						1
Garage						
Master Bedroom		1	1		1	1
Master Bath			1			1
Bedroom 2		1	1			
Bedroom 3		1	1			
Bedroom 4		1	1			
Totals	1	7	7	2	3	6

Distribution Panal	*Capacity*
Telephone and Networking	16 Networking 16 Telephone
Television	8 Wallplates
High Definition Satellite	8 Wallplates
Off-Air HDTV and FM Reception	Included

FIGURE 15–5 Work Order for Cabling Phase—from Scope Statement

accrual-basis. In a cash-basis organization, the revenue is recorded when money is received for goods and services. In an accrual-based system, revenue is recorded when goods and services are delivered to the customer. Most residential integration companies use an accrual accounting system. That's because progress payments often are received prior to delivering and even ordering equipment. Our case study company ABC Integrations is running an accrual-based accounting system.

The sales invoice, shown in Figure 15–6, is yet another view of the bill of materials. This view is given to the bookkeeper to register the delivery

SALES INVOICE

ABC Integrations
123 Main Street
Anytown, PA 01234
(800) 555-1212

Proposal Date: 12/28/2005
Client: Smith, Fred & Wilma
123 Little Street
Anytown, PA 12345

Description	*QNTY*	*RATE*	*AMOUNT*
CableScope: CT5eBL Cat5e in box, Blue, Ft, Part No: 1001	1871	0.07	130.97
CableScope: CT5eGY Cat5e in box, Grey, Ft, Part No: 1003	220	0.07	15.40
CableScope: CT5ePK Cat5e in box, Pink, Ft, Part No: 1004	550	0.07	38.50
CableScope: CT5eWH Cat5e in box, White, Ft, Part No: 1002	1760	0.07	123.20
CableScope: RG6QD-WH RG6-QD Coaxial Cable, White, Ft, Part No: 1005	1850	0.22	403.30
CableScope: RG6QD-BK RG6-QD Coaxial Cable, Black, Ft, Part No: 1006	1190	0.22	259.42
CableScope: S14-4-BL 14 Gauge, 4 Conductor Speaker Cable, Blue, Ft, Part No: 1007	990	0.46	455.40
CableScope: S16-2-BL 16 Gauge, 2 Conductor Speaker Cable, White, Ft, Part No: 1008	450	0.22	99.00
Installation, Cabling Phase, Mins, Part No. 0101	1170	1.00	1,170.00
		TOTAL	2,695.19

FIGURE 15–6 Sales Invoice for Cabling Phase

of the products and services. In addition to the sales invoice, the project manager provides the bookkeeper with a copy of the filled out work order to account for over- or under-used parts. For ABC Integrations, using an accrual-basis accounting method, the "differential quantity" column is entered into the invoice with a zero sales price. This removes or adds the appropriate quantities from inventory without adjusting the overall sales price of the contract. Remember, the contract is a fixed-price type contract, which means the sales invoicing is fixed, regardless of actual quantities of products and services used to complete the project.

Study/Career Tip

When do you review your lecture notes? The best time to do that is right after class. It's an excellent way to add notations that you did not have time to write during class time. It also is a way to clarify what you've written. If you still do not understand a concept, seek out the help of your instructor or revisit your textbook for meaning.

Critical Thinking Questions

- Aside from the company bookkeeper, who else in the organization might use a sales invoice?
- Why?

ABC Integrations
123 Main Street
Anytown, PA 01234
(800) 555-1212

Job Costing Report

Client: Smith, Fred & Wilma
123 Little Street
Anytown, PA 12345

System: Structured Wiring System
House Size: 4,000
Proposal Date: 12/28/2005
Sales Tax Rate: 0.05
Sales Tax Amount: $200.84

Products						
Phase	*Cost*	*Sell*	*List*	*Gross Profit*	*Margin*	*Discount*
0	$ 5.78	$ 11.56	$ 12.72	$ 5.78	50%	9%
1	762.60	1,525.19	1,677.71	762.60	50%	9%
2	744.47	1,488.93	1,637.82	744.47	50%	9%
3	495.55	991.09	1,090.20	495.55	50%	9%
4	–	–	–	–	0%	0%
Totals	*2,008.39*	*4,016.77*	*4,418.45*	*2,008.39*	*50%*	*9%*
Services						
Phase	*Cost*	*Sell*		*Gross Profit*	*Margin*	
0	$ –	$ 1,276.25		$ 1,276.25	100%	
1	–	1,170.00		1,170.00	100%	
2	135.00	2,070.00		1,935.00	93%	
3	–	405.00		405.00	100%	
4	–	–		–	0%	
Totals	*135.00*	*4,921.25*		*4,786.25*	*97%*	
Products and Services						
Phase	*Cost*	*Sell*	*List*	*Gross Profit*	*Margin*	*Discount*
0	$ 5.78	$ 1,287.81	$ 1,288.97	$ 1,282.03	100%	0%
1	762.60	2,695.19	2,847.71	1,932.60	72%	5%
2	879.47	3,558.93	3,707.82	2,679.47	75%	4%
3	495.55	1,396.09	1,495.20	900.55	65%	7%
4	–	–	–	–	0%	0%
Totals	*2,143.39*	*8,938.02*	*9,339.70*	*6,794.64*	*76%*	*4%*

Payments				
Phase	*Terms*	*Amount*	*Percentage*	*Expected Balance**
0	Due upon acceptance	$ 2,741.66	30%	
1	Upon completion of wiring	$ 2,741.66	30%	$ 2,466.17
2	30 days prior to installation of head-end panels	$ 2,741.66	30%	$ 1,648.90
3	30 days prior to installation of speakers	–	0%	$ 252.81
4	30 days prior to installation of electronics	–	0%	$ 252.81
5	Due upon substantial completion	$ 913.89	10%	
	Total Payments	$ 9,138.86		

* Expected Balance is sum of payments less expected invoicing by phase end.

FIGURE 15–7 Job Costing Report

Job Costing Report

The **job costing report** is a view of the cost estimate that summarizes the cost of products and services for the project. Figure 15–7 shows a job costing report for our sample cost estimate, shown in Figure 15–1. This report includes cost, sell, list, gross profit, margin, and discount for each phase of the project.

Summary

In Chapter 15, we discussed cost management planning, the process and purposes of cost budgeting, and five ways to view the bill of materials. We also reviewed the job costing report. Important points in this chapter are:

- Cost management planning involves cost budgeting and the generation of bills-of-materials.
- The bill-of-materials can be prepared in various ways. The different views are used for different purposes.
- The customer view of the bill-of-materials shows the total price of individual work packages needed to complete a project.
- The internal view of the bill of materials includes a detailed list of all parts within a work package needed to complete the package and its installation.
- The purchase requirements view of the bill-of-materials shows the quantities of material needed to complete a project. This becomes the purchase order that is forwarded to the vendor. The quantities are based on the quantity that is to be purchased in order to get the best price.
- The work order view of the bill-of-materials shows the actual amounts of materials used and hours worked to complete a project. It also shows the cost differential between the bid amount and the actual amount.

Key Terms

Cost baseline A time-phased budget used to measure and monitor cost performance.

Cost budgeting The process of aggregating the estimated costs of individual activities or work packages to establish the cost baseline.

Cost differential The difference between the estimated or bid cost and the actual cost. It can either be a credit or a debit.

Cost management plan The document that sets out the format and establishes the activities and criteria for planning, structuring, and controlling the project costs. A cost management plan can be formal or informal, highly detailed or broadly framed, based on the requirements of the project stakeholders. The cost management plan is contained in, or is a subsidiary plan of, the project management plan.

Differential quantity The difference between the baseline and actual quantities.

Job costing report A report that summarizes the costs of the project by phase and is broken out by products and services.

Review Questions

1. What is cost budgeting?
2. Explain a cost differential.
3. What does the cost estimate customer view detail for the customer?
4. Who uses the cost estimate internal view document? What is its purpose?
5. Who uses the purchase order, and for what purpose?
6. The installation team uses the work order. What happens to it when the cabling phase is completed?
7. Who uses the sales invoice?
8. What is a job costing report?

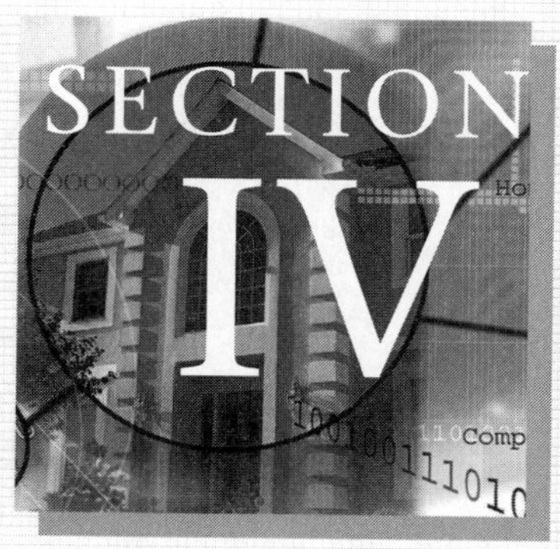

Executing, Monitoring/ Controlling, and Closing the Project

Executing the Project

During the execution portion of the project, the project manager follows the processes created to carry out the work defined in the project management plan to achieve the project's requirements defined in the project scope statement. In this section, we develop ABC Integrations' processes shown in Figure S4–1. They are to perform quality assurance, to acquire the project team, and to develop the project team, which will then be responsible for information distribution, requesting seller responses, and selecting sellers.

Perform Quality Assurance

Performing quality assurance is the process created and used to apply the planned, systematic quality activities to ensure that the project employs all processes needed to meet the requirements of the project.

Acquire the Project Team

Acquiring the project team is the process created and used to obtain the human resources needed to complete the project.

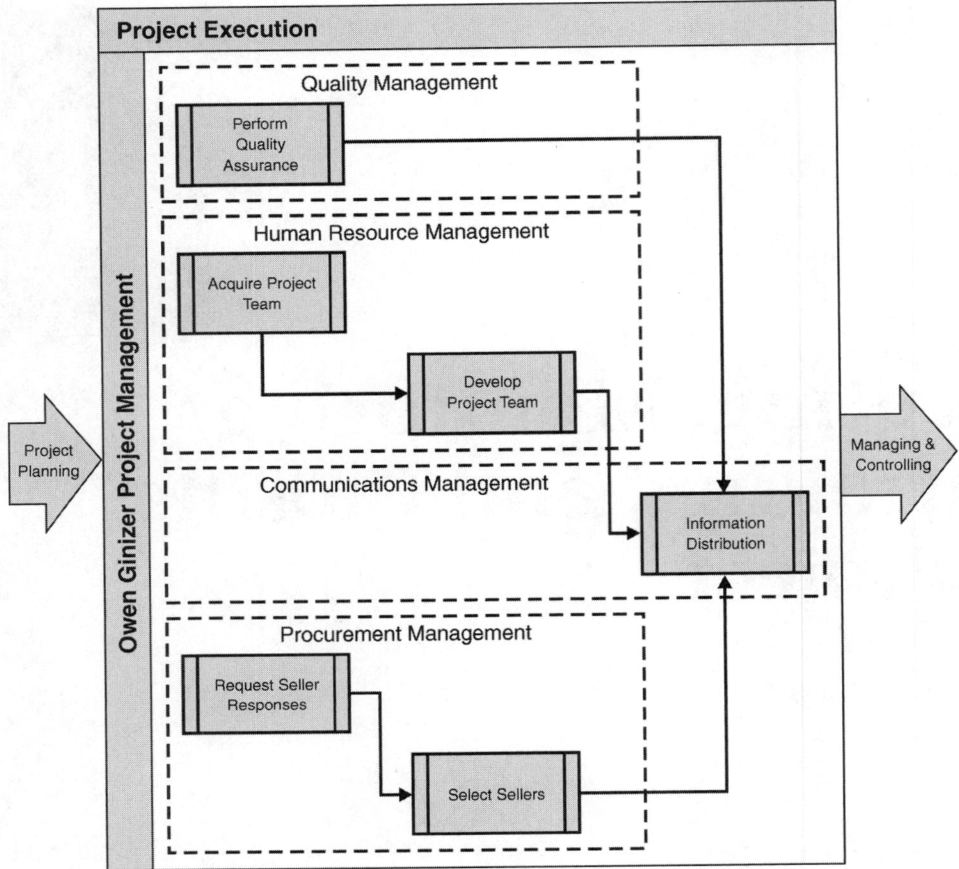

FIGURE S4–1 Project Execution

Develop the Project Team

Development of the project team is the process created and used to improve the competencies and interaction of team members to enhance project performance.

Request Seller Responses

Requesting seller responses is the process created and used to obtain information, quotations, bids, offers, and/or proposals, as appropriate during a project.

Select Sellers

Selecting sellers is the process created and used to review offers by choosing from among potential sellers and negotiating a written contract with the chosen seller.

Information Distribution

Information distribution is the process created and used to make information needed for the project readily accessible to project stakeholders within an appropriate length of time.

Monitoring and Controlling the Project

During the monitoring and controlling portion of the project, processes are created and used to initiate, plan, execute, and close a project to meet the performance objectives defined in the project management plan. In this section, we develop ABC Integrations' processes shown in Figure S4–2. They are scope verification, scope control, schedule control, cost control, quality control, project team management, performance reporting, stakeholders management, risk monitoring and control, and contract administration.

Scope Verification

Scope verification is the process created and used to formalize acceptance of the completed project deliverables. The project manager carries out the verification process.

Scope Control

Scope control is the process created and used to control changes to the project scope, including the processes for change orders, variance analysis, re-planning, and configuration management.

Schedule Control

Schedule control is the process created and used to control changes to the project schedule. It involves schedule management, a schedule baseline, performance reports, and approved change requests.

Cost Control

Cost control is the process created and used to control costs so that the project can be completed within the approved budget. Cost control influences the factors that create cost variances and control changes to the project budget.

Perform Quality Control

Quality control is the process created and used to monitor specific project results to determine whether they comply with relevant quality standards and to identify ways to eliminate causes of unsatisfactory performance.

Manage Project Team

Managing the project team is the process created and used to track project team member performance, providing feedback, resolving issues, and coordinating changes to enhance project performance.

Performance Reporting

Performance reporting is the process created and used to collect and distribute project performance information, which includes status reporting, progress measurement, and forecasting.

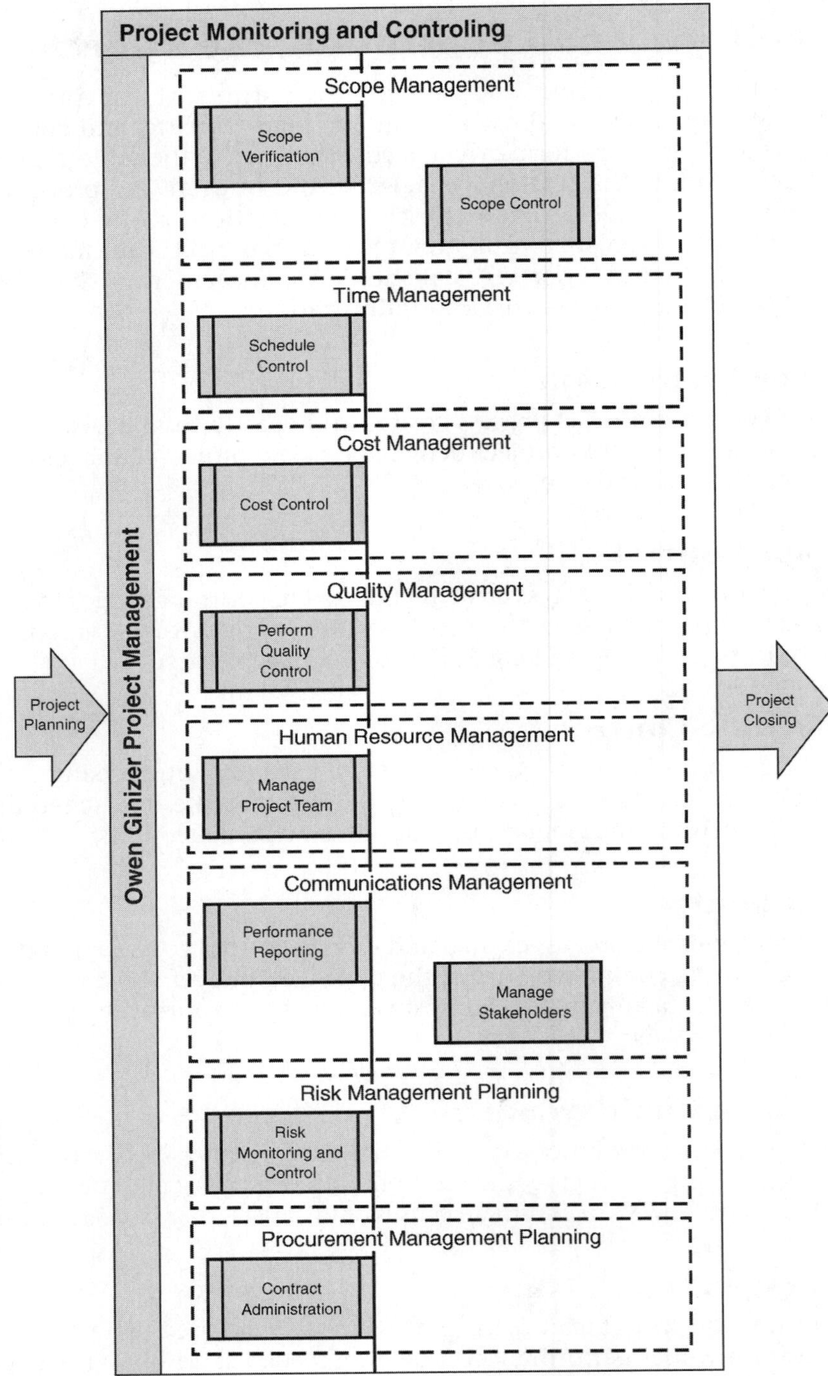

FIGURE S4–2 Project Monitoring and Controlling

Manage Stakeholders

Managing stakeholders is the process created and used to manage communications and interactions with project stakeholders—builder, architect, subcontractors, customer, and project team. It provides a way to meet their requirements and resolve issues that might arise with stakeholders.

Risk Monitoring and Controlling

Risk monitoring and control is the process created and used to track and identify risks, monitor residual risks, identify new risks, execute risk response plans, and evaluate success throughout the project life cycle.

Contract Administration

Contract administration is the process created and used to manage the contract and the relationship between the buyer and the seller. It involves reviewing and documenting the performance of the seller to establish corrective measures and provide a basis for future relationships with the seller, managing changes related to the contract, and managing the contractual relationship with the buyer of the project, usually the homeowner.

Closing the Project

During the closing portion of the project, the contract is completed and formally closed. In this section, we develop ABC Integrations' process shown in Figure S4–3. It is contract closure.

Contract Closure

Contract closure is the process created and used to complete and settle each contract, including the resolution of any open items, and the process used to close each contract. At ABC Integrations, the contract is closed by the project manager.

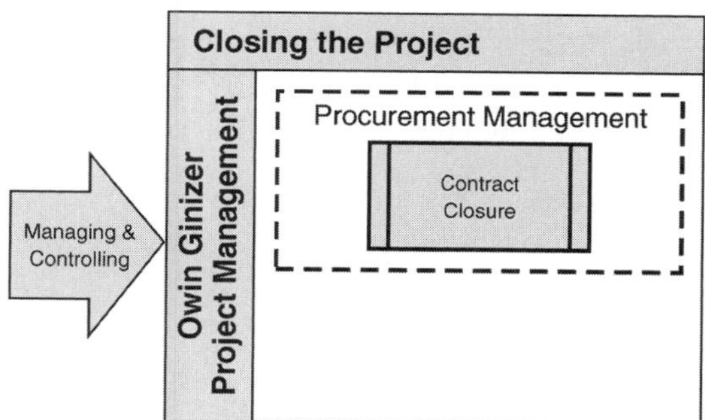

FIGURE S4–3 Closing the Project

Project Execution

OBJECTIVES

After studying this chapter, you should be able to:

- Define quality assurance.
- Explain the purpose of quality assurance.
- Define continuous improvement and explain why companies use it.
- Discuss five factors a project manager needs to remember in acquiring a project team.
- Discuss formal and informal factors involved in developing a project team.
- Discuss ways to provide a company environment that allows motivation and inspiration to occur.
- Develop a plan for information distribution.
- Explain source and seller solicitation and fulfillment.
- Discuss contract closeout.

OUTLINE

Introduction

Perform Quality Assurance

Acquire the Project Team

Develop the Project Team

Information Distribution

Source and Seller Solicitation and Selection

Contract Closeout

Introduction

Once all of the project planning phase is completed, it's time to put those plans into action. The project plans should be now be executed and the systems ready to be installed as described in the scope statement. Finally, all that hard work in planning the project will pay off! But there are a few more steps that have to happen in order to reach the project's objectives and goals.

The following steps from various project management processes must be performed in order to implement the project. These include:

- Perform quality assurance (project quality management)
- Acquire the project team (project human resources management)
- Develop the project team (human resources management)
- Information distribution (project communications management)
- Solicitation and source selection (project procurement management)
- Contract closeout (project procurement management).

In this chapter we will review each step more thoroughly.

Study/Career Tip

There is much advice about a proper study area, such as good lighting, proper ventilation, and a large enough work space to spread out work. Here are some tips on what *not* to have in your study area.

- A distracting view of other activities you may want to be involved in
- A telephone, including a friend or roommate's cell phone
- A blaring stereo (although soft background music can be an aid. Some students swear by classical music.)
- A 27-inch (or larger!) color TV
- A roommate or friend who wants to talk a lot or gets easily distracted
- A refrigerator stocked to the max that keeps calling your name.

Temptations abound. Do your best to complete your studying in an atmosphere where the only enticement is an "A" on that next paper or exam.

Critical Thinking Question

- What other terms do you know that are associated with quality and quality assurance?

Perform Quality Assurance

The project manager uses planned, systematic quality activities throughout the project installation to ensure all the processes needed to meet the project objectives are accomplished. This process is called **quality assurance.** The sales representative can be responsible for quality assurance in small companies, and a separate team can be responsible for quality assurance in large companies. Perhaps a project team that is not involved in the current project can take responsibility for providing quality assurance if the company is not large enough to hire a team devoted specifically to those tasks.

Continuous Improvement

To remain competitive, companies strive for **continuous improvement** in their products and services. A continuous improvement process provides a routine that is repeated regularly to improve the quality and efficiency of all company and project processes (and to reduce waste within the company).

For example, at ABC Integrations, in-house technicians build the equipment racks that will hold all the electronics. As a quality assurance activity, another in-house technician tests all the equipment in the rack to make sure it is in working order before it is disassembled and packed up for delivery to the customer's home. By continually using this quality assurance process, technicians figured out a way to organize the electronics in the rack in a more efficient way by always putting them in the same order and marking each piece when it is prepared for delivery. A simple improvement, to be sure, but one that helped the company to save time in its installations. Those designated to provide quality assurance also test the various system designs to confirm the designs used by the project team are efficient. Tests are carried out to determine how well a design fits the need.

Acquire the Project Team

As we know, much of a company's success depends on the project team. Acquiring a successful project team is the result of a systematic process defined in the scope management plan. A project manager should pay particular attention to hiring the members of the project team. In some companies, the project manager has complete responsibility for hiring the team members, while in others he has less authority. Regardless, a project manager's input in the process is an important factor in most residential integration companies. Companies should consider the following when beginning the hiring process:

- **Type of Organization.** How is your company organized? Is it a functional, projectized, weak-matrix, or strong-matrix organization?

- **Job Descriptions and Responsibilities.** Have detailed job descriptions and responsibilities been prepared? Does the applicant meet the requirements of the position?
- **Availability.** Is the person applying for the job available to take the position? What is the market availability for the position? Have many people applied for the position? Are the skills needed for a position hard to find? Can training the appropriate person provide the skills needed for a position?
- **Ability.** Does the applicant possess the abilities needed for the position?
- **Experience.** How much experience should an applicant have? Is the applicant's experience appropriate for the position? Does ten years of experience as an installer qualify an applicant for a lead position? Or, must the applicant have experience in the lead position to qualify?
- **Interest.** Does the person applying for the job seem interested in the position? Is he excited about the position? Is the person applying for the job interested in areas that complement the job he's applying for?
- **Costs.** Are the salaries within the capability of the organization to pay? What would it cost to train an unskilled applicant? What is the cost of available courses an applicant may need? What is the cost of creating an apprenticeship?

Not all of these items must necessarily be considered when forming a new project team. For example, identifying the type of organization and the use of job descriptions and responsibilities might have been identified and communicated to the hiring manager when previous project teams were assembled. The key is to recognize all the factors that should go into bringing a team together and to focus on the ones that have the most immediate impact.

Study/Career Tip

Job fairs are a great way to shop for new job opportunities, to network, and to make personal contacts with recruiters at various organizations. The interview lines may be daunting, but don't fail to maximize this opportunity. Talk to every company that fits your experience and ambitions. If you meet with 20 recruiters, at the end of the day you will know 20 people by name. That sure beats sending a blind resume to "Personnel Director."

Critical Thinking Questions

- Why is identifying the type of organization (functional, projectized, weak-matrix, or strong-matrix) important when hiring employees?
- What impact does the type of organization have on the hiring process?

Develop the Project Team

Developing a project team should be a continuous improvement endeavor. While each team member arrives with a certain skill set, getting the team to work in a cohesive fashion improves the overall competencies and interaction of team members, thus enhancing project performance. This includes:

- Advancing worker skills in order to improve effectiveness, efficiency, and the ability to complete project tasks
- Improving feelings of trust and cohesiveness among team members in order to raise productivity.

Training the Project Team

One way to develop a project team is to offer employee training. Training includes all formal and informal learning experiences that are intended to improve the abilities of the project team. Examples of formal training are:

- **Industry Training.** Seminars and workshops offered by industry trade organizations, such as the Custom Electronic Design and Installation Association (CEDIA), the National Sound Contractors Association (NSCA), and the Computing Technology Industry Association (CompTIA)
- **Manufacturer Training.** Seminars, workshops, and Webinars routinely offered by manufacturers
- **Vendor Training.** Seminars, workshops, and Webinars offered by vendors
- **Organizational Training.** Internal seminars or workshops offered by your company
- **Academic Training.** Courses offered at local technical schools, continuing education schools, colleges, or universities.

Examples of informal training include:

- **On-the-Job Training.** Learning from other project team members during the course of project installations
- **Mentoring.** Learning from project team leaders through example and special attention
- **Company Standards.** Learning from company best practices, lessons learned, and anything else documented on behalf of the project team to improve performance
- **Vendor and Manufacturer Instructions.** Documented instructions provided by suppliers for their products
- **Technical Publications.** Trade magazines, technical journals, and books
- **Internet.** Informal research/surfing to get industry information.

Motivating the Project Team

Motivation is a powerful tool used to raise the energy level of a team that results in meeting or exceeding the standard for job performance. A successful project manager has the ability to motivate his team members to perform at their best. The ultimate result is high quality workmanship. Whether motivation is an external force or an internal force, self-motivating force is a question that is debated among experts. Many say that motivation comes from within, and a supervisor or boss cannot really motivate those who work for her. However, it is possible to create an environment conducive to motivating and inspiring team members. Motivating an individual is about seeing a need within that person and providing a way to satisfy that need, which encourages them to perform to the best of their abilities. A positive work environment that allows motivation to occur can be fostered in the following ways:

- *Value the team and its work.* As project manager, it is your responsibility to explain each team member's role and how each role contributes to project objectives and goals. Make a point of recognizing responsibilities that may normally go unnoticed with the entire team.
- *Express confidence.* As a project manager, let your team know you have confidence in the scope of their knowledge, abilities, and work ethic. Show them you trust their ability by not re-checking and micromanaging every step they take. Quality assurance is intended to improve process, not stifle work performance. Expect your team members to perform well. Assign goals and objectives that stretch their abilities and allow them to grow. Instill trust by providing freedom and decision-making authority appropriate to the position.
- *Recognize good performance.* A program designed to honor "The Installer of the Month," or a similar program can show appreciation to team members even though it may be laughed and joked about among team members. Mention of a job well done at a project team meeting is a less formal way of recognizing individual or team efforts. Let team members know when they go beyond a company standard. Let the team know what you consider is a high standard of performance. Send a memo to upper management when your team exceeds expectations.
- *Lead by example.* As a project manager, set and adhere to a high standard of work. If a project team sees you accept mediocre work in your position, they'll follow suit. Be a supportive manager who keeps the team's interests in mind when discussing issues with upper management. Exhibit honesty and integrity at all times.

Information Distribution

The process of information distribution is how information is moved from the originator to those who need it and then tracking the path it takes. There are many levels of information distribution, depending on the need. For example, a simple e-mail may be satisfactory to let

Study/Career Tip

Have you considered the benefits of having a mentor? A mentor can:

- Improve your self-confidence and feelings of self worth
- Increase your motivation
- Move you to seek out new opportunities
- Raise the achievement bar
- Build on your communications skills.

Mentors are overwhelmingly volunteers. See if your school has a formalized mentoring program. If not, seek out an individual whom you feel will provide you with the support you need. Who knows, some day you may be able to return the favor to someone else.

Critical Thinking Question

- What are other ways a project manager can motivate his project team?
- Think in terms of company-wide recognition or other internal reward system.

one team member know another will be 20 minutes late. A complex documentation system is necessary to present and store various versions of a scope document. Memos to contractors can be e-mailed and followed up with phone calls.

The distribution process explains what level is appropriate for each need, and it explains the process at that level. A project manager may need to use detailed documents to report the progress of project. She may need to specifically explain the responsibilities of each team member or the responsibilities of a subcontractor. The information documents may state who is in charge of each phase of the project, the scope of each person's responsibilities, the materials they will need, and the desired outcome of the project.

In Chapter 13, Project Communications Planning, we presented ABC Integrations' communications plan using colored file folders that are distributed to each team member for each project. To track these documents, before being distributed, each is stamped either incoming or outgoing, dated, and the original document is filed at the office according to the date. When information is faxed to someone outside the company, the fax is first printed on a brightly colored paper that is used only for faxes. The original documents also are stamped, dated, and filed at the office according to date. Photocopies are placed in the colored file folders and given to each team member. In this way a team member can use the documents as a reference, and if lost, the team member can request a copy from the office administrator, who can retrieve the original using an approximate date that the document was issued.

In the case of outside contractors, memos are tracked to help avoid potential conflicts. Originals of memos are dated and filed according to date and copied to the builder, architect, and sometimes even the customer. That way the subcontractor and other stakeholders are fully aware of the subcontractor's responsibilities in a project, as well as the company's responsibilities.

A company that uses a computerized system to send memos and other documents also has a record of when a memo was written, and sent, and to whom it was sent. The original version and subsequent versions are kept to provide a history of each project.

Source and Seller Solicitation and Selection

During the solicitation step of the purchasing process, the procurement manager contacts the potential sellers that have been identified in the procurement planning phase. She must make and maintain contacts with suitable vendors and subcontractors in the industry. The procurement manager is responsible for fulfilling identified requirements from qualified sources. This is accomplished by acquiring sources such as catalogs, trade journals, trade conference literature, and vendor reviews. As in the planning aspect of procurement, sales personnel for vendors are an excellent source of information concerning possible vendors and subcontractors.

There are four major concerns that need to be addressed during the solicitation process—contract origination, source qualifications, contract fulfillment, and contract negotiation.

Study/Career Tip

Did you know?—The big months for hiring are January and February, and late September and October. If you are looking for a job, make contact right at the start of these hiring cycles in order to have the best chance of being hired.

Critical Thinking Question

- What makes a Request for Quotation, a Request for Proposal, and an Invitation to Bid a bilateral (two-sided) contract?

Contract Origination

In the purchase of large quantities of supplies, contracts are a common way to acquire the items. These contracts can be originated in two fundamental ways—**unilaterally** or **bilaterally.** When made unilaterally, a **purchase order** is provided to the vendor. When made bilaterally, one of three documents is sent for solicitation: a **Request for Quotation (RFQ)**, a **Request for Proposal (RFP)**, or an **Invitation to Bid (ITB).** Let's briefly review each type of contract.

ABC Integrations
123 Main Street
Anytown, PA 01234
(800) 555-1212

Purchase Order

DATE December 30, 2003
PO# 101

Purchase From
XYZ Distributors

FOR: Smith Project

DESCRIPTION	QNTY	RATE	AMOUNT
Xfactor Cables: 1480AB4 Audio Interconnect, 1 meter each	25	$ 9.64	$ 241.00
Xfactor Cables: 1480AB4 Audio Interconnect, 2 meter each	15	$ 12.68	190.20
Xfactor Cables: 1480AB4 Audio Interconnect, 3 meter each	10	$ 19.68	196.80
Xfactor Cables: 1480AB4 Audio Interconnect, 5 meter each	10	$ 22.36	223.60
		TOTAL	$ 851.60

Please contact Owen Ginizer x103 with any delays in shipping.
Expected arrival in seven business days from purchase order date.
Ship common ground freight.

THANK YOU!

FIGURE 16–1 Purchase Order

The Purchase Order

A purchase order, also known as a P.O., is a very common document used in a variety of industries. Generally, companies set policies for purchase orders, such as who can authorize one and up to what dollar amount. Purchasing requirements that exceed the cost threshold require the signature of someone higher up in management.

Figure 16–1 depicts a typical purchase order. Multiple copies of the purchase order are created and distributed as shown in Table 16–1. The purchase order is a legal document and serves as a contract between the buyer (the residential integration company) and seller (product

TABLE 16–1
Purchase Order Distribution

Copy Number	*Who*	*Purpose*
1	Receiving	As a check against incoming goods
2	Purchasing	To determine outstanding orders
3	Project Management	To show outstanding orders

vendor or dealer). It's important to treat a purchase order carefully because it is a contract. A purchase order should include quantity, model number, description, unit purchase price, delivery instructions, and payment terms.

Request for Quotation (RFQ)

A Request for Quotation (RFQ) is a bilateral contract in which the buyer (residential integration firm) provides a formal invitation to provide a quotation on specific goods and/or services. This is most common in low-cost supplies and materials. To illustrate, the audio interconnect could be handled in this manner prior to the purchase order. Figure 16–2 shows the same products as in Figure 16–1. However, the cost information has been removed to create a request for quotation. The quote could then be sent to a distributor, or several distributors for pricing information. In our example, the audio interconnects are available at several distribution centers, and our procurement manager is seeking the best possible price for the product.

Request for Proposal (RFP)

Another version of a bilateral contract is a request for proposal (RFP). An RFP, typically, is chosen when the product and service are for a high dollar value. When soliciting proposals for the residential integration systems in the home, builders and homeowners commonly use the RFP method. The service is not usually standardized. How each company goes about providing a system in the home is unique. The purpose of the RFP is to solicit unique solutions to the homeowner's special issues, entertainment, and technology needs. Figure 16–3 shows a request for proposal. Note that it is significantly more detailed in the information it requires the vendor to provide.

Invitation to Bid (ITB)

The third method in which a contract is bilaterally developed is by releasing an invitation to bid (ITB). The project is well-defined and the buyer is requesting a contract price for the product or service. In our satellite installation example, the contract could be released as an

ABC Integrations
123 Main Street
Anytown, PA 01234
(800) 555-1212

Request for Quotation

DATE December 30, 2003
RFQ# 101

Purchase From
XYZ Distributors

DESCRIPTION	*QNTY*	*RATE*	*AMOUNT*
Xfactor Cables: 1480AB4 Audio Interconnect, 1 meter each	25		
Xfactor Cables: 1480AB4 Audio Interconnect, 2 meter each	15		
Xfactor Cables: 1480AB4 Audio Interconnect, 3 meter each	10		
Xfactor Cables: 1480AB4 Audio Interconnect, 5 meter each	10		
		TOTAL	$ –

Please contact Owen Ginizer x103 with any delays in shipping.
Expected arrival in seven business days from purchase order date.
Ship common ground freight.

THANK YOU!

FIGURE 16–2 Request for Quotation

invitation to bid. Every detail of the work is clearly defined with only the contract price remaining to be set. The residential integrator could send the letter to several local satellite installers to obtain the best possible price. Figure 16–4 depicts an ITB.

Source Qualification

The procurement manager is responsible for maintaining a list of qualified sources for the purchase of goods. The list is continually updated as new vendors are available, or vendors go out of business, simply

CONFIDENTIAL ***Johnson Architects***

Request for Proposal

To: Owen Ginizer
CC: Fred Smith
From: Jonathan D. Sign
Date: 4/19/2006
Client: Fred Smith
Site: 132 Hubbard Way, Anytown, PA
Re: Home Theater System

Johnson Architects, a residential design firm in Vero Beach Florida, is seeking a consultant or consultants to assist in its initial design and installation of a home theater system for the Smith Residence. See the associated "Agreement for Services" which would typically follow this proposal, assuming the client finds a consultant that he or she likes and enters into an agreement with them.

Situation
The home theater is located in the family room area, see attached floor plans. The room is a high light area of the home with an open feel. The goal of the theater is to allow the family to enjoy the latest television viewing options to include satellite, HDTV, gaming, and Internet connection.

How to submit a proposal
Interested people should submit the following, no later than May 1, 2006, to Jonathan D. Sign. If there are questions, call Jonathan at 772-999-9999.

1. A proposal describing your qualifications (or the qualifications of the team of consultants) and how the tasks described above would be carried out

2. Afirm estimate of fees to be charged, and an estimate of expenses that would be incurred

3. Names, phone numbers, and contact people of three similar projects who have been your clients during the last18 months, whom we can call on as references.

FIGURE 16–3 Request for Proposal

change locations, or no longer offer a certain product. Data are also maintained on all companies that have been selected previously for purchases. Let's look at some of the information that should be kept for each supplier.

- **Product Information.** A list of products offered by the supplier, including price, shipping terms, and bulk quantity purchase discounts. This information is typically a confidential price sheet provided by the manufacturer's representative. In the event the products are purchased from a distributor, the pricing may be kept online at a distributor Web site in a special dealer login area.

CONFIDENTIAL

Johnson Architects

Invitation to Bid

To: Owen Ginizer
CC: Fred Smith
From: Jonathan D. Sign
Date: 4/19/2006
Client: Fred Smith
Site: 132 Hubbard Way, Anytown, PA
Re: Home Theater System

Johnson Architects, a residential design firm in Vero Beach Florida, is seeking a consultant or consultants to provide and install the following list of equipment at the Smith Residence.

Situation
The game room requires the installation of a Plasmatron GY54 54" plasma television. You are to provide a double swing arm mount. The plasma will connect to a cable television jack and AC outlet already provided at that location.

How to submit a proposal
Interested people should submit the bid, no later than May 1, 2006, to Jonathan D. Sign. If there are questions, call Jonathan at 772-999-9999.

1. A firm estimate of fees to be charged.

2. Names, phone numbers, and contact people of three similar projects who have been your clients during the last 18 months, whom we can call on as references.

FIGURE 16–4 Invitation to Bid

- **Delivery History.** The time from shipment to arrival, which provides a forward look at the product's required lead time. In addition, the method of shipment and the condition of the goods on arrival are tracked. We noted in the origination section that a copy of each purchase order is kept in receiving to annotate arrival dates of the goods. Once the purchase orders have been delivered in full, these records should be kept on file. That way the procurement manager can reference the information on occasion to gain an understanding of the manufacturer's ability to deliver product in a timely fashion.
- **Returned Materials.** Each product that has been returned to the supplier must be tracked, as well as the reason for return, such as failure due to site conditions or an out-of-the-box failure. For example, dimmer switches can sometimes be broken by an electrical contractor or a project installer. A vendor may or may not replace or repair the

broken switch. Sometimes a group of dimmers is broken before the box they come in is opened. In addition, how the supplier handles the returned materials should be noted. Are the returned goods cheerfully accepted or not? Ongoing data on product returns are critical to the cost-estimating engineer. This information is used to determine if certain parts need to be substituted in future designs. If, for example, a keypad from a particular manufacturer has a high rate of failure, then the cost estimator will need to determine if there is an internal issue with how that keypad is installed, or if another source of keypads needs to be found.

- **Supplier Response.** When a residential integration company needs further information, how the supplier responds to the request should be noted. This added service is valuable to integrators in the installation of their products. And even though most residential integration companies are small in comparison to large box-moving, retail chains, the integrators may need more service than the larger companies. That's one reason why it's important for integrators to maintain a positive relationship with their suppliers and vendors. Ever try to get something fixed or replaced at a moment's notice? You're more likely to get a response from someone who knows you, rather than with a company just selected from the Yellow Pages. Here's a case in point.

A CASE IN POINT

To the Rescue

ABC Integrations installed a theater system for customers Fred and Wilma. Project Manager Owen Ginizer thinks the project is ready to present to them tomorrow. However, there is a snag. During the final system checks, the projector no longer turns on. Project Manager Owen knows ABC Integrations must work quickly with the vendor of the projector to either resolve the problem in the field or to obtain a replacement in an expedient manner. Fortunately, the procurement manager knows the projector vendor well, and the vendor responds in time with the information needed to repair the system. Quick supplier response is frequently critical to the success of the residential integration company.

- **New Product Vendors.** New product vendors are constantly entering the marketplace. That means new sources can be tapped to fulfill project requirements. As a result, procurement managers usually maintain a source qualification list to make it easier to select new suppliers. Vendor sales personnel are an excellent resource for information concerning potential suppliers because they are usually well informed about the capabilities of products. In addition, trade shows, trade journals, supplier catalogs, and trade registers are used to research new vendors.
- **Review Process.** When bringing new products into a company's product offerings, the staff must be careful to dig below the surface in the reviewing process. Every piece of equipment must be reviewed for integration purposes prior to accepting it as a qualified material for an integration project.

Study/Career Tip

If you are struggling in class because you don't understand the course material, look beyond just re-reading the text. There's another way for you to learn. Talk to your instructor, academic advisor, or classmate; join a study group; or even consider working with a tutor. Look for solutions that will work for you.

Critical Thinking Questions

- Why is the source qualification review process so critical?
- What are some possible consequences if that review process does not occur?

Seller Qualification

As with source qualification, the procurement manager, with the input from the project manager, is responsible for maintaining a list of qualified subcontractors. Finding a qualified subcontractor requires constant attention to the local marketplace. For example, in finding a satellite subcontractor, a project manager can look in the local telephone book and interview candidates as he would for a full-time hire. Possible candidates can be found at the job site doing other work, from local distributors, and through referrals from sales representatives. It is necessary to treat the newly selected subcontractor the same as a new employee, with a trial period and close monitoring of the work output. Just because the job has been outsourced does not mean it does not need management or quality control.

In receiving the bids from the subcontractors, the procurement manager reviews several subcontractors who are capable of performing the installation. When each of the subcontractors is capable of meeting the quality criteria, the final selection might be based upon the site location and lead-time requirements of the project. Perhaps one contractor will be used for projects to the west and another for the eastern side of the organization's territory, while yet another, with a higher price, might be chosen for quick turn-around situations.

Source Contract Fulfillment

The final phase in purchasing is the fulfillment phase. In this phase, the order is placed with the vendor and the shipment is tracked, accepted, and logged into the company accounting system. The order is sent to a supplier via a purchase order, shown in Figure 16–1, and copies are kept as shown in Table 16–1. The purchase orders are kept as a means of tracking what items have been received and what items are still on order. The project manager needs to know if parts will not be available in time for installation in order to decide whether to postpone the installation or find substitute parts.

Seller Contract Negotiation

Negotiation over the final contract terms is accomplished as the last step in the contract process. In larger contracts, such as the contract awarded to the residential integration company by the homeowner, the negotiation could involve a single person or an entire team. The builder, architect, homeowner, and even an owner representative can be involved with the negotiation process. The residential integration company could make the sales representative, engineer, and project manager, as well as the company owner, available to help in the negotiation of the contract. The negotiation meeting is a dynamic situation that brings the buyer and seller together in hopes of reaching a set of common objectives.

Negotiation can be divided into five stages: protocol, probing, bargaining, closure, and agreement. The protocol stage introduces the project stakeholders, thus setting the stage for the negotiations to follow. The probing stage starts the search process. The residential integration company may inquire about key needs of the system, such as "Will you require broadband Internet distributed throughout the home?" The buyer might ask a set of probing questions, such as "Will I be able to operate both my wireless and wired network devices?"

The bargaining stage happens when concessions are made. The buyer (homeowner) might decide to have fewer pairs of speakers to lower the overall price of the contract, and the seller (residential integrator) might agree to meet timeline restrictions. The closure stage sums up each of the two positions and final concessions are made. These concessions are smaller adjustments to the contract than those made in the bargaining phase, such as wallplate color, or final positions of speakers. Finally, the contract is agreed upon and accepted by both parties.

Study/Career Tip

Each semester, time yourself reading a chapter in each one of your textbooks. The times may vary greatly given the number of pages and the complexity of the material. Track how many pages an hour you can read and make a note of it. Once you have an accurate count for each textbook, it will be easier to plan out your reading and study time on a course-by-course basis.

Critical Thinking Questions

The art of contract negotiation has five stages: protocol, probing, bargaining, closure, and agreement.

- Which do you think is the most important?
- Least important?
- Why?

Contract Closeout

The completion of the negotiation marks the start of the award phase of the contract process. The formal contract between the two parties is written based upon the agreement reached in the negotiation phase.

Both parties sign the final agreement. The contract is comprised of a number of clauses expressing the agreement between the buyer (homeowner) and seller (residential integrator). Some of those clauses include:

- **Scope Statement.** This document includes responsibilities of both parties and details the project objectives.
- **Guarantee.** This section answers the following questions: How long will the products be covered under the guarantee? Is on-site repair included? Is regularly scheduled maintenance included?
- **Price.** This section presents the contract value and explains how change orders are to be handled throughout the project life cycle.
- **Insurance.** This concerns items such as worker's compensation and employee bonding.
- **Inspections.** This section spells out when and how the work is to be inspected and who is responsible for the cost of the inspections.
- **Termination.** In the event the contract is terminated during the project life cycle, this section covers how the process works. For example, the seller (residential integrator) could allow the buyer (homeowner) to cancel at any time; however, the buyer is responsible for any special order items, plus all charges for time spent on the project.

Summary

In Chapter 16, we defined quality assurance and continuous improvement. We also discussed factors important to acquiring and developing a project team. We explained how to promote an environment conducive to motivation. We provided a plan for information distribution. We discussed source and seller solicitation and fulfillment for purchasing. In purchasing, we discussed three major concerns in the solicitation phase: the type of information that should be kept on suppliers, contracts as they are used in purchasing, and the difference between unilateral and bilateral contracts.

Important points in this chapter are:

- Quality assurance is the process of planned, methodical quality activities performed throughout the project life cycle.
- The purpose of quality assurance is to ensure all the processes needed to meet the project objectives are performed during the project installation.
- Continuous improvement is a process used to help companies stay competitive in the marketplace. It also promotes efficiency and reduces waste within the company.
- Continuous improvement is a process that provides a routine that is repeated regularly to improve the quality and efficiency of all company and project processes.
- A number of factors are important in acquiring a project team. They are the type of organization, job descriptions and responsibilities, availability, experience, interest, applicant abilities, and cost.
- Formal methods of employee training include industry training, manufacturer training, vendor training, organizational training, and academic training.
- Informal methods of employee training include on-the-job training, mentoring, company standards, vendor and manufacturer written instructions, trade and technical magazines and journals.
- Ways a project manager can promote an environment open to motivation and inspiration are to value the team and its abilities, to

express confidence in the team, to recognize good performance, and to lead by example.

- The process of information distribution is how information is moved from the originator to those who need the information and how the information is tracked.
- Four major concerns that need to be addressed during the seller solicitation process are origination, qualification, fulfillment, and negotiation.
- Information that should be kept for each supplier includes product information, delivery history, returned materials, and supplier response.
- Clauses commonly added to contracts include the scope statement, a price, insurance, inspections, and termination of the contract.

Key Terms

Bilaterally A contract is originated bilaterally when the buyer (residential integrator) requests a quotation (RFQ) for the product, a written proposal (RFP), or an invitation to bid (ITB).

Continuous improvement The process of continually striving to improve. It is a by-product of a quality assurance plan.

Invitation to Bid (ITB) A document that requests suppliers to submit a bid in order to secure work or material.

Purchase order A contract between the buyer and seller to purchase goods and services.

Quality assurance A proactive method to ensure quality. It creates quality policies and procedures that ensure that project standards are correctly and verifiably met during the project.

Request for Proposal (RFP) A document generated when a contract is originated bilaterally. The contract is developed by releasing an RFP for a high dollar contract. The RFP method commonly is used by builders and homeowners when soliciting proposals for custom residential integration systems in the home.

Request for Quotation (RFQ) A document that lists products with no pricing that is sent to suppliers asking them to fill in the prices. Most commonly used in low-cost supplies and materials.

Unilaterally A contract is originated unilaterally when a purchase order is provided to the vendor. Multiple copies of the purchase order are created.

Review Questions

1. What is continuous improvement?
2. What are some things to consider when building a project team?
3. What is one way of developing a project team?
4. What are some ways a project manager can motivate the project team?
5. How does information distribution work?
6. What is a unilateral contract? Provide an example.
7. What is a bilateral contract? Provide several examples.
8. What is source qualification?
9. What clauses in a contract express the agreement between the buyer and the seller?

Monitoring and Controlling

OBJECTIVES

After studying this chapter, you should be able to:

- Define and discuss scope verification.
- List each section of the scope statement and explain how to verify it.
- Discuss scope control.
- List the items used and updated in scope control.
- Discuss how project changes are controlled.
- Discuss schedule control and its purpose.
- Describe a comparison schedule bar chart.
- Discuss cost control and the issues it involves.
- Discuss quality control in residential integration.
- Discuss the factors involved in performance review.
- Explain risk management monitoring and controlling.
- Describe contract administration.

OUTLINE

Introduction

Scope Verification

Scope Control

Schedule Control

Cost Control

Quality Control

Performance Review

Risk Monitoring and Control

Contract Administration

Introduction

Monitoring and controlling a project happens simultaneously with all the processes that occur throughout the project life cycle. Some are so natural we may not even realize they are part of a control or monitoring process. For instance, the customer reviews the scope document (customer proposal) before signing it, thus authorizing the start of the project. Or, the project manager reviews the project scope statement before he shows up for the project Sales-to-Operations Turnover Meeting. A second check of a memo sent to a subcontractor, or a second look at a verified cost estimate are other instances. All these efforts monitor and control the processes and the outcome of the project. The processes can be informal or formal mechanisms set in place to assure a positive project outcome.

In this chapter, we will discuss the many ways in which a project is monitored and controlled throughout the project life cycle.

Scope Verification

A primary monitoring and control process is formally verifying the project scope statement. Once the project is accepted, the next step is to verify it, as discussed in Chapter 9, The Scope Management Plan. As we learned, the project scope statement clearly states each aspect of the work to be accomplished in language that is easy for a customer to understand. The sales team has spent a great deal of time on the scope document, interviewing the customer to ensure all his needs and requests are met. Now it is time to verify the customer's requirements.

The project manager carries out the verification process. It is the project manager's first opportunity to tie the reality of the home under construction to the needs of the customer (as specified in the scope statement). The verification process can be simple or complex. The selection of which documents are appropriate for the project depends on the size of the overall job. For small projects, such as installing a plasma TV, it's likely that little documentation is needed. However, in large complex projects for the whole house, floor plans, riser diagrams, and **cable schedules** are needed in order for the project team to understand the project objectives.

Study/Career Tip

Instead of simply relying on your textbook to study, consider additional readings and alternate sources of information. This new perspective can help to create a richer understanding of the content, interact with additional facts about the material, and help you to practice and familiarize yourself with new vocabulary and concepts.

Critical Thinking Question

- If the sales representative placed stickers in the appropriate locations in the home to be cabled when the scope statement was being developed, why go through that exercise again?

Another part of the verification process is for the project manager to "sticker" the home, as we determined in Chapter 9. Once the home has been "stickered" and the customer has approved the final locations, the scope statement is considered verified. Then a change order is created to reflect the differences between the "stickered" home and the original scope statement and is submitted to the customer for approval.

The cable stickers make it easy for the installation team to come in and run the cabling. Theoretically, the installation team needs zero documentation beyond the customer's address. The team can come into the site with the standard tools and standard cabling, and install the cabling with little to no discussion with the sales representative or project manager. All they need to do is follow the stickers.

However, every now and again there is a reason to discuss a home that has some peculiarities to it, or tasks may need to be done outside of the normal standard. This could include something like a satellite dish that must be cabled to a crawl space, rather than to an open attic. The project manager needs to be mindful of these out-of-the-ordinary situations and plan labor and materials costs accordingly.

Study/Career Tip

Hiring managers and recruiters are busy people. One job opening may attract hundreds or even thousands of resumes. To help yours stand out in the crowd, include a strong career summary (career objective) statement. The goal of this section is to present a hard-hitting introductory declaration filled with your most sought-after skills, abilities, accomplishments, and attributes.

Critical Thinking Question

- Why do you think the process of scope control is so important?

Scope Control

Customers do change their minds. They may want to add another television to the integrated system, or change the number of wallplates. During the life cycle of a lengthy project, electronic products can become unavailable, or a better product can come on the market.

Project scope control is the process that influences factors that create project scope changes. That means when the customer asks an installer to add a wallplate to the other side of the room, the installer is responsible for following the change order process. It also means controlling the effect of those changes. Scope control means that every change goes through the company's change order process whether it is a request from the builder, customer, or an improvement proposed by the project manager.

Project scope control also manages the changes when they occur. Scope creep is the result of changes that are not controlled. Let's say a customer asks an installer to move a wallplate after the verification process. In some cases, it might only be a five-minute process—moving it up six inches. But in other cases it can involve much more, especially if the cables have to be moved to a new location. The purpose of the control process is to keep the project on track and to charge an appropriate amount for the change requests.

The following items are used in the scope control process and are subsequently updated on an as-needed basis:

- **Approved Change Requests.** All changes to the project are documented and submitted to the primary stakeholder for approval.
- **Project Scope Statement.** The scope statement is updated with changes and approved by the primary stakeholder.
- **Work Breakdown Structure.** The WBS is updated after the scope statement is approved.
- **WBS Dictionary.** If additional items other than quantity changes of current items are added to the scope statement, then those additional items need to be defined in the WBS dictionary.
- **Project Scope Management Plan.** Substantial changes in the scope statement may cause changes in the scope management plan. If the organization is template-based, then it is unlikely that individual projects will drive process changes. However, if the organization is performing highly customized work, changes in the scope statement will have a direct impact on company policies.
- **Performance Reports.** These are as simple as a daily update for the project team members.

Study/Career Tip

When preparing for a job interview, spend some time organizing your employment "stories." If you state that you are good at a particular skill, back that statement up with a story or an example of how you performed in a particular situation.

Critical Thinking Questions

- What other project management processes are affected by scope control changes?
- In what ways?

Change Control

The way changes are controlled is created by the executive team and added to the scope management plan. Here's a case in point.

A CASE IN POINT

Out with the Old, In with the New

During the life cycle of a project, ABC Integrations Project Manager Owen Ginizer learns that a much-improved DVD player is available at a slightly increased price. The company's change policy requires Project Manager Owen to provide the information to the Sales Representative Sal Moore, who notifies the customer. The customer has the opportunity to agree to the change, or to continue with the current plan. If the customer wants the new DVD player, he signs a change order that has been filled out by Sales Representative Sal. The change order includes authorization of the increased price. The change order process also includes testing the newer DVD player for quality and reviewing reports about its dependability before it is offered to the customer.

The change order process helps keep change at a minimum and reminds all concerned parties that a change in price is also needed. The process documents the change, tracks it within the system, and indicates the level of authority needed to approve the change. Because a contract is involved, the change process must comply with the contractual obligations. In other words, the project manager cannot simply change the DVD player without the customer's approval and signature.

Replanning

Replanning is a change in the original plan for accomplishing authorized contractual requirements. There are two types of replanning:

- *Internal replanning.* A change in the original plan that remains within the scope of the authorized contract, caused by a need to compensate for cost, schedule, or technical problems, which have made the original plan unrealistic.
- *External replanning.* Customer-directed changes to the contract in the form of a change order that calls for a modification in the original plan.

To illustrate external replanning, if a customer decides to add the home's security system to the project, it is not necessary to create a new project. Instead, the security system is added by including it in the current project. The material and labor needed for the security system will be added to the existing WBS and WBS dictionary. The project scope will include the information pertaining to the security integration. Schedules will need to be developed to include the additional time needed to interface with the alarm company. The schedule also has to reflect additional time to hook up the security system to the integrated system. Each piece of planning the security integration will be needed, but instead of a separate process, it is integrated into each phase of the existing project.

Study/Career Tips

If you are required to make a class presentation, here are some hints to help you shine in front of your audience.

- *Know your material.* Practice your presentation several times, in front of a mirror if possible.
- *Relax.* Easier said than done, of course, but it will help you sail through your material if you're not overly nervous.
- *Realize that your audience wants you to succeed.* They want you to be captivating, informative, and yes, even entertaining.
- *Don't apologize.* If you're nervous or feel you have made a mistake, don't point it out to your audience. They may not have noticed!

Critical Thinking Question

- Beside the project manager, who do you think at the residential integration company could be responsible for internal replanning?

Configuration Management Issues

The purpose of the configuration management plan is to assure the company and the customer that the approved change is processed with respect to organizational standards. In other words, it is the paper work associated with the change. For example, the customer accepts the change of the DVD player as part of the project. Purchasing will need to receive an up-to-date bill of materials with the old player removed and the new one added. In addition, project management will receive an updated bill of materials for the purpose of invoicing and the creation of a work order.

Schedule Control

Schedule control involves keeping a schedule on track and adjusting it when changes necessitate modifying it. Major changes to a project naturally result in major adjustments to the schedule. Adding the security system to the integrated system is a positive change, but the schedule must be changed. As we noted above, the alarm company needs to be contacted to determine what plans are in place for its installation, the integration company's installation team needs to be included in the alarm schedule, and the completion date of the project may need to be changed. At the same time, the schedule for the original systems needs to be tracked to make sure it continues to move forward on time.

What to Consider in Schedule Control

A number of issues need to be considered in controlling the project schedule. We discuss some of them briefly in the list below.

- **Current status of the project schedule.** Is it on time? Is it ahead, as planned, or behind? Will the original completion date be met?
- **Factors that may change the schedule.** Will requested changes affect the schedule? Are the changes minor, requiring less than an hour to make? What would happen in the event of a storm or personnel accident? How would the company respond?
- **Documentation of project schedule changes.** This is a record of all schedule changes and why they happened. Was the change caused by the company, the customer, an unforeseen emergency, or something else? All of this information helps in planning future projects.
- **Managing the changes when they occur.** Can the company respond as soon as possible to the schedule change? If not, can the company change the schedule as soon as a request requires a change? Typically, a change in the schedule is brought about by an external force, such as a change in the building process. For example, a day prior to installing the cabling in the Smith residence, our Project Manager Owen checks with the builder to confirm the installation date. The builder informs Owen that the other contractors have not finished their work on time, and therefore Owen will need to arrive not tomorrow, but two days from tomorrow. The issue for Owen is to determine if there room in the schedule for the change. If not, when can the team be available to cable the home?

The schedule control process includes:

- **Controlling the Schedule.** How project schedule changes are managed and controlled is explained in the schedule management plan. A project manager allows changes to the project schedule based on a plan devised by the executive team. The plan explains the process that must be followed, including the necessary paperwork, tracking system, and approvals needed to authorize a change.

  Companies can devise a schedule control system in many ways. Changes must be approved at various levels of management, based upon the degree of change, before the project manager adjusts the schedule. The project manager must make sure that his team understands that all changes must be approved. Without approval, the schedule remains the same.
- **Measuring Project Progress.** The project manager should retain the original project schedule even when the schedule is modified. The baseline schedule is used to measure the project team's performance status within that project. Performance measurement techniques are used to determine whether corrective action should be taken because of project variances. The project manager, the lead installers, and the technicians write performance and progress reports for several reasons. The reports track whether the project is on time or not, and explain what has caused the project to go off schedule to prevent the same thing from happening again. The progress reports track the current schedule status and the remaining durations for unfinished schedule activities. Many companies prepare report templates to manage the process so that the reports are easier, more convenient, and consistent throughout the project life cycle.

Study/Career Tips

Hope to do better on tests? Try these strategies:

- Make certain that you fully understand the test directions before attempting to solve any problems or answer any questions.
- Read each question carefully and completely before marking or writing your answer. Re-read if you are at all confused.
- Do not be disturbed about other students finishing before you do. Take your time, don't panic, and you will do much better on the test.

Critical Thinking Question

- What are some possible consequences of not being able to manage schedule control?

Schedule Comparison Bar Charts

One way to easily view a comparison of planned start and finish dates and actual start and finish dates is a comparison bar chart (see Figure 17–1). The bar chart makes it easy to see the differences at a glance. The bar chart includes two bars for each schedule activity or phase of the project. One bar indicates the original schedule, or baseline schedule. The other bar indicates the actual schedule. A project manager or other member of the project team can quickly see the status of the project, and note the areas of the schedule where delays occurred or where float days where needed to get a project back on schedule. Figure 17–1 also displays the baseline versus actual installation hours for a cabling infrastructure project. While the cabling phase was under by four hours, the termination phase hours exceeded estimated hours by four hours. The move-in phase is not finished, but is estimated to be 50 percent complete, and based upon actual hours it can be expected to be four hours over budget.

Cost Control

Cost control is a vital ingredient of any residential integration project. The future of a company depends on making sure that what the customer agrees to pay the company for a project is enough to complete the project and result in a healthy profit. Frequently, the project manager is responsible for monitoring the cost of a project throughout the project life cycle and controlling the costs so that individual phases and the total project stay within the project budget. *A Guide to the Project Management Body of Knowledge* describes cost control as:

- *Influencing the factors that create changes to the cost.* For example, a sudden and dramatic increase in fuel costs causes shippers to levy an excise charge on each shipment received by ABC Integrations. As

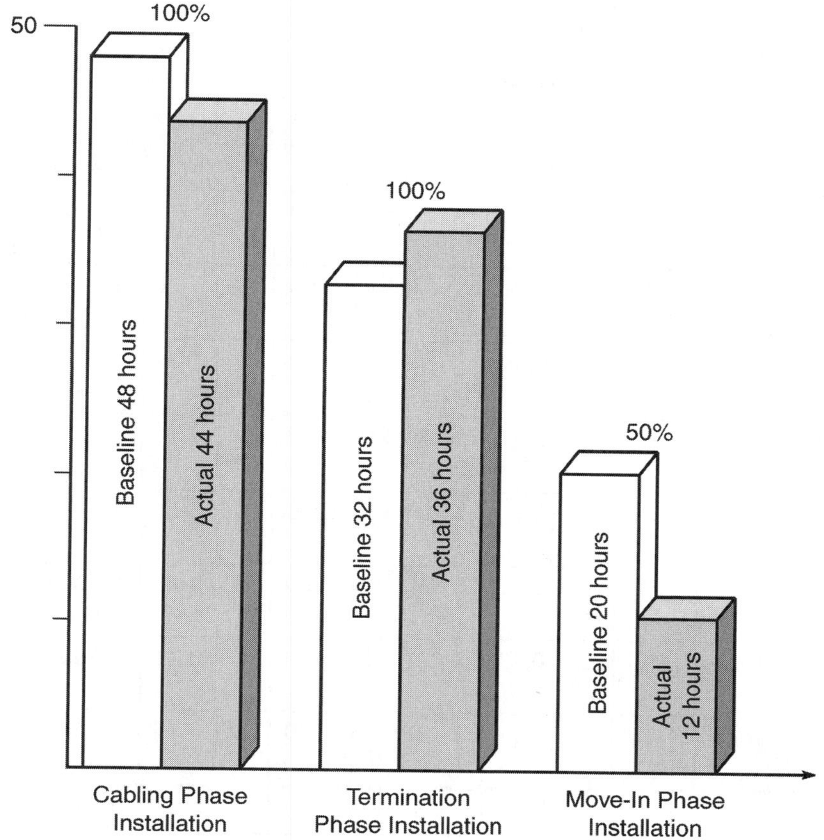

FIGURE 17–1 Schedule Comparison Bar Chart

a result, the cost of shipping is increased across the board. If the contract is a fixed-price contract, ABC Integrations will need to absorb these costs unless there is such a provision included in the contract. This provision will cover ABC Integrations for any increase in cost of materials and cost of transportation.

- *Ensuring requested changes are agreed upon by the customer.* Only approved changes will be carried out and will directly impact the project's bill of materials and project costs.
- *Managing the actual changes when and as they occur includes costs for minor changes in quantities of materials.* For example, ABC Integrations' cost estimate calls for a total of 1,871 feet of Cat5e cable for networking in the customer's home. The actual quantity after cabling the home is 2,000 feet (see Table 17–1). The extra cable increases the cost of the project. However, if the contract is a fixed price, it will not increase the sale price of the contract.
- *Recording all appropriate changes accurately against the cost baseline* is a useful tool in maintaining a record of the changes.

TABLE 17–1
Cost Monitoring for Cabling Phase

	Baseline			Actual		
Description	*Qnty*	*Rate*	*Amount*	*Qnty*	*Rate*	*Amount*
CableScope: CT5eBL Cat5e in box, Blue, Ft, Part No: 1001	1871	$ 0.04	$ 74.84	2000	$ 0.04	$ 80.00
CableScope: CT5eGY Cat5e in box, Grey, Ft, Part No: 1003	220	$ 0.04	$ 8.80	200	$ 0.04	$ 8.00
CableScope: CT5ePK Cat5e in box, Pink, Ft, Part No: 1004	550	$ 0.04	$ 22.00	500	$ 0.04	$ 20.00
CableScope: CT5eWH Cat5e in box, White, Ft, Part No: 1002	1760	$ 0.04	$ 70.40	1800	$ 0.04	$ 72.00
CableScope: RG6QD-WH RG6-QD Coaxial Cable, White, Ft, Part No: 1005	1850	$ 0.11	$ 203.50	1200	$ 0.11	$ 132.00
CableScope: RH6QD-BK RG6-QD Coaxial Cable, Black, Ft, Part No: 1006	1190	$ 0.11	$ 130.90	1200	$ 0.11	$ 132.00
CableScope: S14-4-BL 14 Gauge, 4 Conductor Speaker Cable, Blue, Ft, Part No: 1007	990	$ 0.23	$ 227.70	1000	$ 0.23	$ 230.00
CableScope: S16-2-BL 16 Gauge, 2 Conductor Speaker Cable, White, Ft, Part No: 1008	450	$ 0.11	$ 49.50	500	$ 0.11	$ 55.00
			$ 787.64			$ 729.00

- *Monitoring cost performance to detect and understand variances from the cost baseline.* In Table 17–1 we see the baseline costs of the cabling phase of a project for ABC Integrations. The actual cost of materials for the phase is $58.64 less than estimated in the baseline. The actual quantities can be fed back into the cost estimating system to create a more accurate cost-estimate model for future projects.
- *Preventing incorrect, inappropriate, or unapproved changes from being included* in the reported cost or resource usage. If the costs are dramatically over or under the baseline, serious inquiry must take place to determine the cause. Perhaps additional wallplates were cabled during the phase, or the house size was dramatically different than originally estimated.
- *Informing appropriate stakeholders of approved changes.* When a change has been approved, such as adding security to the integrated system, the project manager is responsible for notifying the builder, architect, designer, customer, and project team of the approved change and when it will take place.

TABLE 17–2
Baseline and Approved Change Order

Approved Change Order			
Description	*Qnty*	*Rate*	*Amount*
CableScope: CT5eBL Cat5e in box, Blue, Ft, Part No: 1001	110	$ 0.04	$ 4.40
CableScope: CT5eGY Cat5e in box, Grey, Ft, Part No: 1003	0	$ 0.04	$ –
CableScope: CT5ePK Cat5e in box, Pink, Ft, Part No: 1004	0	$ 0.04	$ –
CableScope: CT5eWH Cat5e in box, White, Ft, Part No: 1002	110	$ 0.04	$ 4.40
CableScope: RG6QD-WH RG6-QD Coaxial Cable, White, Ft, Part No: 1005	0	$ 0.11	$ –
CableScope: RH6QD-BK RG6-QD Coaxial Cable, Black, Ft, Part No: 1006	0	$ 0.11	$ –
CableScope: S14-4-BL 14 Gauge, 4 Conductor Speaker Cable, Blue, Ft, Part No: 1007	0	$ 0.23	$ –
CableScope: S16-2-BL 16 Gauge, 2 Conductor Speaker Cable, White, Ft, Part No: 1008	0	$ 0.11	$ –
			$ 8.80

- *Acting to bring expected cost overruns within acceptable limits.* Cost overruns do happen, and successful planning builds in a certain amount of cost overruns. The project manager's responsibility is to make sure the cost overruns remain at a reasonable level. One or two extra hours to complete an installation may be satisfactory depending on the size of the project, but doubling the number of hours is way beyond the acceptable limit. That is an indication that the original estimate was based on inaccurate information.

Project cost control looks for causes of both positive and negative variances. The follow items are used in cost control:

- **Cost Baseline.** The cost baseline provides a firm basis for determining whether a cost variance is positive or negative. Table 17–1 shows baseline costs for the cabling phase for a project for ABC Integrations.
- **Performance Reports.** Performance reports provide a regular check of the project status. Figure 17–1 shows a schedule comparison bar chart, which indicates the percent complete of each phase of the installation and the actual versus estimated hours on the project.
- **Approved Change Requests.** Once a change has been approved, a project manager can forecast the time it will take to complete the change. Table 17–2 shows the change in costs for the cabling phase for our example project based upon an approved change order for an additional telephone and networking wallplate.

Study/Career Tips

When employers ask for your salary history, they are referring to the salary you earned for each employer listed on your resume. They are most interested in your salary from your last (or current) position.

When employers request your salary requirements, they are inquiring as to what salary you'd be willing to accept if they were to offer you the job for which you are applying. For many job seekers, it can be difficult to discuss salary history as part of the job hunt. But it's important that you know the difference between these two concepts.

Critical Thinking Question

- Write your own definition for cost control.
- What are its most important aspects?

Quality Control

The quality control process provides an extra measure of protection to the company and the customer by determining whether the project is meeting company standards designed by the executive team and the industry at large. It checks to make sure that project goals and objectives are met and that the processes written by the executive team result in a quality product or service. It also reviews schedule performance, cost control, and other project management functions. It monitors the results of a project and identifies ways to eliminate the causes of unsatisfactory project elements in deliverables and in performance. In companies of sufficient size, quality control is performed and managed by a quality control department. Smaller companies often assign this task to the project manager.

In performing quality control, a project manager should be able to distinguish the following terms clearly:

- **Prevention** is a measure that keeps errors out of the process. For example, ABC Integrations uses industry certification programs, such as CompTIA's HTI + certification and CEDIA's Installer I and II certifications. ABC Integrations requires each of its employees to obtain certification in their respective job roles. Having properly trained and certified employees is the best way to prevent poor quality.
- **Inspection** keeps errors out of the hands of the customer. A project manager is responsible for making sure all system elements are in working condition before turning the system over to the customer. For example, each electronic system at ABC Integrations is built in the shop and thoroughly inspected and tested prior to delivery to the customer's home.
- **Common causes of error.** Common causes of error happen frequently in all companies. To illustrate, a wallplate is cabled with a kinked Cat5e cable. This results from the fact that new installers are placed on the cabling team, and they generally do not understand

how kinking the cable affects system performance. These types of errors are reduced by implementing a training program for new employees.

- **Special or unusual causes of error.** This type of error only happens rarely because of an unusual circumstance. An example of a special or unusual cause of error is having an installed system fail the final electrical inspection because of stricter inspection criteria in a town where the company had never worked before.
- **Tolerances** are the acceptable range in which a result may fall. To illustrate, when Cat5e cable is installed it not only must pass continuity test, it must also be tested across the full bandwidth to ensure it will work with the electronic components connected to it. The cable may pass the continuity test; however, because of kinks or a stretched cable, this may limit the bandwidth the cable can handle. Cables that do not pass bandwidth tests will need to be replaced. Here the tolerance of the cable is expressed as the minimum allowable bandwidth.

Study/Career Tip

The expression goes: It ain't over til it's over. Your paper isn't finished until the last word is typed; the semester isn't over until the last exam is over. See your work through to completion.

Critical Thinking Question

- What is the difference between quality control and quality assurance?

Performance Review

Performance review is the process of collecting status information on the project and reviewing and disseminating the collected information to the project stakeholders. The project manager faces the challenge of staying on track once the project begins, and that can be a greater challenge than building the project plan. The purpose of collecting status information is to provide the best information possible to project stakeholders in order to make day-to-day decisions. Below are some key questions to consider in performing this review process.

- *How often will the project status be collected?* Some information, such as the amount of time spent on the job site, may be collected daily, while information such as the actual quantities of materials used are reported at the end of the phase.
- *How will information be collected?* There are a number of methods for collecting project information, such as time cards, work orders completed with actual data, and written project updates from a project team leader.

- *What information will be collected?* When looking at the actual materials used in a project, should the project manager be concerned with tracking small items such as staples, brackets, and other miscellaneous supplies? Should the travel time to and from be collected? What about the start time and end time at the job site? Each piece of information can be useful; however, the process of collecting the data will have inherent costs associated with it. As a result, a cost-benefit analysis is useful before determining what information is to be collected.
- *What actions will be taken based upon the information?* Once collected, what information will be reported and to whom? How will the data be analyzed? Will it be compared to baseline information on a daily basis, weekly basis, or at the end of each phase?

A project that finishes on time and within budget may seem to be successful. However, there may be problems that a project team experiences that are apart from these two factors. In the performance review process, the project manager faces three issues:

1. **Defining standards of performance.** What is expected of the project team? Yes, a project manager expects the team to complete the project on time and within budget, but what other areas in performance are expected? The project manager needs to know about the team's job quality, cooperation, and project results. The performance review should also include checking for the team's accuracy of installation, how thorough a team has been, how well the team worked together, and how well the team interpreted raw data.
2. **Finding appropriate applications of the standards.** As the project progresses, a Project Manager needs to decide how to test the standards set by the company. He also must understand what the tests reveal. All too frequently this issue is ignored in reviewing a project. In other words, a project manager might be very specific in telling his project team what to do, but fail to convey to the team that company standards need to be are met or exceeded in each project. A project manager is responsible for making sure the team knows that sloppy or below standard work is unacceptable for any reason.
3. **Deciding what actions, if any, a project manager needs to take.** What should a project manager do when she uncovers a problem involving company standards? One of the most difficult issues in project management is assessing the personality of each team member and knowing how each person will react to corrective actions. How does a project manager handle a team member who spends too much time chatting and not enough time working or the team member who is unmotivated or satisfied with less than the company standards? These situations, among others, are the ones that try a project manager's leadership skills.

A project manager also needs to ensure that the project team is working together as a team. Are projects falling behind because the team has issues of authority? Or are personality conflicts undermining team efforts and cooperation? Team problems that are left unresolved can lead to schedule delays, and schedule delays increase labor costs. The result is a project that is neither on time nor within budget—even though everything else is in place.

Study/Career Tip

Attend lectures regularly. Once you miss one, it will be easier to miss more.

Critical Thinking Question

- What are several other ways that a project manager can carry out performance reviews?

In addition to standards that result in a quality project, a project manager applies his own personal standards to the project and the work performed by the project team. As project manager, what is your perception of "excellence?" Since the team's efforts reflect on a project manager's ability to perform his job, it follows that the project manager should be concerned about the impression the team creates with the builder, architect, other trades, and the customer. A team that conveys a sense of professionalism is probably the result of a successful project manager. A team that keeps a customer's home clean leaves a positive impression. A team that is sloppy and throws empty coffee cups and cigarette butts about the property exhibits less than an atmosphere of a professional job or workmanship. One way to get the point across to a project team is to publish a list of project manager expectations on a company Web site or add a copy of the list to each team leader's folder to review with all team members. Another way is by example. A project manager can walk the job site and pick up any unwarranted debris, or take a trash bag and ask the team to bring trash back to the office facility at the end of the day.

Another decision a project manager needs to make is when and how often should a project manager review a team's progress and efforts? Some managers review a team's progress only in a formal employee evaluation, perhaps once or twice a year. Others review their team on an ongoing basis. They make a constant effort to assure quality is maintained in the team's progress. Continual review is the far better choice used by many successful project managers. It is too easy for a project team to slip below quality when there is a six- or twelve-month lapse in the review process. And once quality is decreased, it is much harder to regain the previous level of quality; and it may not be possible to recover at all.

Risk Monitoring and Control

Risk monitoring and control involves examining known risks, reducing future costs for those risks, and evaluating the effectiveness of the risk response. Information is accumulated throughout the project and is used by the executive team and the project manager to help formulate policies to prevent cost overruns in the future.

If managed properly, a project looks easy to outside eyes, when, in fact, it is not. Risk monitoring and control determines if:

- *Assessed risks have changed.* For example, the risk of damaged cabling is high during the termination and testing phase; however, after all the cables have been tested that risk diminishes.
- *Contingency reserves of cost or schedule should be modified to stay current with assessed risks.* The project manager can make these types of adjustments on an as-needed basis as the project progresses.
- *Project assumptions are still valid.* Let's say it is assumed the home will be cabled for all floors during the same installation trip, a total of three continuous days. During the cabling phase, it is learned the basement will not be ready for cabling for an additional two weeks. This assumption must be updated.
- *Company risk management policies are being followed.*

Risk Information Tools

The following tools are used as input to the risk monitoring and control process:

- **Risk Management Plan.** This plan sets policies concerning the assignment of people, including the risk owners, time, and other resources to the project. For example, the plan calls for our project manager to visit the site prior to the cabling phase to ensure the site meets the description provided in the scope statement.
- **Risk Register.** The risk register contains a list of project risks and agreed-upon risk responses. One of the items on the risk register may be an inexperienced electrical inspector. If this is encountered on the project, then the appropriate actions will be taken as called for in the risk register.
- **Approved Change Requests.** These can include modifications to the scope statement, schedule, costs, and the terms of the contract. For example, the customer may request an additional wallplate during the cabling phase.
- **Work Performance Information.** This includes the status of the project, corrective actions, and performance reports. To illustrate, the cabling team may report after the first day of installing cables that they will most likely not finish in the allotted time. The project manager must determine why this has occurred and the best course of action for going forward.
- **Performance Reports.** These reports provide information concerning work completed on the project. For example, the lead installer notifies the project manager when the cabling of the house has been completed, and provides the actual quantity of cables, supplies, and labor used during the phase.

Study/Career Tip

When studying for a test, it's helpful to review your previous tests taken during the semester. Ask yourself: What kinds of questions were asked? Were there factual questions, application questions, definition questions? If they were factual, know the names, dates and places. If they were application-based, study theory. If they were definitions, then be sure to know the terms and their meanings.

Critical Thinking Question

- During the project life cycle, when does risk monitoring and control begin?

Contract Administration

The residential integration company and the customer oversee the project contract for the same reasons. They want to make sure that what has been promised is actually delivered, and they want to make sure their interests are protected. The customer wants to make sure the project deliverables meet her expectations and the legal requirements of the contract. The company wants to make sure the customer meets payment dates on time for the work completed. The contract administration process makes sure that both the buyer and the seller meet the requirements of the contract.

On larger projects, with many deliverables and services, a key aspect of the contract administration is managing interfaces among the many providers. The project manager must work with each subcontractor to ensure the work is completed to company standards. He must then notify the billing or accounting department to authorize payment.

Because of the legal nature of a contract, it is extremely important that the project management team is aware of the legal implications of administering it. In many large companies, a lawyer is retained on a casual basis to administer the contracts for the company. In a small company, this responsibility is given to the project manager. For example, the company lawyer may be used to create a template for the customer contract, while it is up to the project manager to ensure that each project's contract is reviewed. On more casual contracts, such as hiring a satellite installer, the project manager prepares the contract template himself.

Summary

In Chapter 17, we discussed monitoring and controlling the project and verifying the various sections of the scope statement. We discussed scope verification, scope control, how changes are controlled, schedule control, and described a comparison schedule bar chart. We also discussed cost control, quality control, and performance review. We explained how to distinguish between prevention and inspection, discussed common causes of error and unusual causes of error, and explained tolerances. We also discussed contract administration.

The important points in this chapter are:

- Scope verification confirms the scope statement and allows changes to be made before the work begins. One way to verify wallplates defined in a scope statement is to place identifying stickers in the appropriate locations throughout the home and walk through with the customer, allowing her to approve the type of wallplates and the locations.
- Scope control involves how changes are controlled. It is important to make sure the change control process allows only approved change requests.
- Schedule control keeps a project on trim. A comparison bar chart is a useful tool in visualizing the status of a project and whether it is on time, or if it is behind schedule, where the delays occurred.
- Cost control involves making sure that what the customer agrees to pay the company for a project is enough to complete the project and result in a healthy profit. Frequently, the project manager is responsible for monitoring the cost of a project throughout the project life cycle and controlling the costs so that individual phases and the total project stay within the project budget.
- Quality control determines if a project meets company and industry standards throughout the project life cycle.
- A successful project manager can distinguish between prevention and inspection, between common causes of error, and unusual causes of error and understand tolerances.
- Performance review determines how well the project is progressing and whether it is up to the standards set by the company.
- The purpose of performance review is to track the performance of the project manager's accuracy in estimating materials and labor hours. It also is used to measure the progress of the project team. A successful project manager constantly reviews the progress of his project team.
- The purpose of contract administration is to make sure the interests of both the buyer and the seller are protected.

Key Terms

Cable schedule The cable schedule is a table of cables to be installed in the home. It includes the cable type, color, purpose, starting location, ending location, and a unique cable identifier.

Replanning A change in the original plan for accomplishing authorized contractual requirements.

Review Questions

1. Who carries out the scope verification process?
2. What is project scope control?
3. What is the purpose of the project scope control process?
4. What is the purpose of the change control process?
5. What is the purpose of the configuration management plan?
6. What issues should be considered in change control?
7. What are some of the tools used in cost control?
8. What does the quality control process provide?
9. In the performance review process, the project manager faces three issues. What are they?
10. What are some of the tools used in risk management?
11. What is the purpose of the contract administration process?

Closing Procedures

OBJECTIVES

After studying this chapter, you should be able to:

- Explain what is involved in contract closure.
- Review a letter used to close a contract with a subcontractor.
- Discuss early contract termination.
- Explain procurement planning and its relationship to contract closure.
- Discuss the verification process used for each phase of a project.
- Discuss the audit process used in each phase of the project.
- Discuss the process of archiving and distributing contract documents at the close of the project.
- Compose a letter requesting final payment.

OUTLINE

Introduction

The project is completed. The customer has accepted the system and issued the final payment. The project team moves on to the next assignment. Only one step remains. The project manager must close the project, including closing the contract. In this chapter we will look at those procedures.

Contract Closure

In closing the contract, the project manager formally verifies that all the work and deliverables are acceptable to the customer. This can be a phone call to the customer thanking him for his payment and simply asking him if everything is satisfactory. Or it can involve a standard letter of thanks and a request to fill out an evaluation enclosed with the letter, accompanied by a stamped, self-addressed envelope. The project manager can also ask the sales team to talk directly to the customer to ask how he likes the system.

The process to close the contract also includes various administrative chores. The final project outcomes are recorded to help with future projects. Paper files are stored; computer files are archived. Customer names are filed for future reference. In addition to the customer contract for the project, the contracts and agreements with the builder, architect, and subcontractors also are closed for that particular project. Frequently, a project manager closes subcontractor contracts at the end of each phase. See the case in point on page 339.

Study/Career Tip

Time management is critical in our daily lives. Some tasks must be completed at a specific time and in a specific place. "Floating" tasks can be done anywhere, anytime. Carry around a book that you need to read, some cards to review, a letter that you need to write. Maximize your time.

Critical Thinking Question

- What do you think is the leading cause of early contract termination in the residential integration industry?

Early Contract Termination

Sometimes a contract is terminated early **(early contract termination)** as the result of a mutual agreement between the customer and the residential integrator. For example, the customer may default on payments, and the residential integrator may want to close the contract. Or the customer may be dissatisfied with the work and decide to end the contract with the company. Or, the company may fail to fulfill the contract. Contracts include an early termination clause that protects the rights of both the company and the customer when early termination occurs. The clause also describes the responsibilities of both the customer and the company when either wants to terminate the contract. See the case in point on page 340.

Based upon the terms of the contract, both the company and the customer may be allowed to end the contract because of a specific reason or simply because either wants to end it at any time during the contract. The terms of whether this is allowable and how it is allowable

A CASE IN POINT

Closing Our Satellite Subcontractor

Earlier in the project, ABC Integrations Project Manager Owen Ginizer sent a letter, which acted as a contract, to the satellite subcontractor requesting that he install a satellite dish for the owner. Once the subcontractor notifies Project Manager Owen that the work has been completed, this contract needs to be closed out. Here are a few issues that Project Manager Owen faces prior to authorizing payment for the work:

- Has the dish been installed in a location that is consistent with ABC Integrations' standards? In our example, the letter specifies "Install the dish at the home in the area least visible from the front of the home." Either Project Manager Owen or one of the project team members needs to verify that the dish is located properly.
- The contract calls for specific signal strength, which also requires verification. Since this is done as part of the electronics delivery phase, which may be months away, how best can we deal with this issue? Project Manager Owen has a couple of alternatives. He can withhold part of the payment to the subcontractor until verification can be obtained; however, it's unlikely the subcontractor would be happy with that arrangement. If ABC Integrations uses this subcontractor on a regular basis, it is important to keep him happy with the company's work assignments. Therefore, full payment can be issued to the subcontractor with the stipulation that he agrees to return to the site after the electronics are delivered if the signal is not within the specified limits. In that case, the subcontractor would agree to return to adjust the dish. Subcontractors, who expect a steady workflow from a company, are likely to complete the assignment and make any needed signal adjustments. If a subcontractor fails to return to the site for the required adjustment, it's probably time to look for another subcontractor.
- The dish must be grounded. Our contract with the subcontractor calls for the dish grounding, which can be visually checked at the same time the placement of the dish is verified.

Once the project manager (or his designee) verifies the work, the project manager notifies accounting to send payment to the subcontractor. This can happen in different ways. In order to keep paperwork to a minimum for both the subcontractor and ABC Integrations, Project Manager Owen may simply provide a copy of the initial subcontractor letter (the contract) to the office manager. Project Manager Owen writes across the top, "Work Verified 12/5/05, please pay" and then signs the copy of the letter. In a larger organization, the payables person or department may require an invoice from the subcontractor, which has a subcontractor tracking number to indicate which contract it is referencing.

Regardless of the method chosen, receipt of payment indicates the contract has been fulfilled.

are spelled out in the contract. Frequently, there is a financial arrangement in the contract that states who is responsible for any costs associated with ending the contract prematurely. The customer may be responsible for paying for any acceptable work completed up to that time, plus any electronic equipment that already may have been purchased by the company and delivered to the customer, or already assembled for the customer. If the electronics have been ordered but not delivered to the company, then the order can be canceled and the customer is not responsible for those items. However, the contract will probably require payment for installation services up to that point.

A CASE IN POINT
Can't Show Us the Money

ABC Integrations Sales Representative Sal Moore has secured a large contract with his customers, Fred and Wilma Smith. The contract includes a lighting control system, structured cabling system, home-wide music system, and a dedicated home theater system. The cabling and terminate phases are complete and payments are up to date. However, when the time comes to request payment for the electronics and the move-in phase, the payments stop! The customer and Sal sit down and discuss the issue, and unfortunately the customer has run into severe money problems. His company has gone bankrupt and he must sell his dream home before it is even complete.

Fred would like to have the home ready to move in, so it can be sold; but all costs must be dramatically reduced. Project Manager Owen and Sales Representative Sal must make decisions that allow the contract to be closed without losing any short-term profit, while retaining the ability to provide the rest of the contract to the future buyer. Yes, ABC Integrations could chase the current owner for payments and take legal steps to recover the money. However, it is always best to find a way to solve contract issues without legal involvement. This will ensure ABC Integrations has a long-term positive reputation for fairness and integrity within the communities it serves.

Project Manager Owen and Sales Representative Sal decide to not install the lighting control system, and instead, ask the electrician to install standard light switches. They also decide to replace the expensive music speakers with contractor grade speakers, and cover the keypad locations with blank wallplates. The home theater room is to be left completely empty of electronics equipment. The home will look finished and critical services, such as television, telephone, HVAC, and lighting, will be functional, but not connected to an integrated system.

Project Manager Owen and Sales Representative Sal prepare an updated scope statement to give to Engineer Maggie Pi, who prepares a new cost estimate for the customer. This process is sometimes called **value engineering,** which is the analysis of a system to reduce the scope and cost of a project. The cost and scope are negotiated until a fair economic situation can be obtained for the customer and ABC Integrations. The agreed-upon work is completed according to the updated scope statement, and the project is closed the same way that the company closes other contracts.

The Procurement Management Plan

As we learned in Chapter 12, the procurement management plan specifies how to close out purchase orders with vendors and subcontractors. In our satellite subcontractor example, the policy of how the subcontractor will be paid is covered within this plan. For significant purchases, a plan to administer the contract is prepared based upon the specific buyer-specified items within the contract, such as documentation and requirements for delivery and performance that the buyer and seller must meet. See the case in point on page 341.

Study/Career Tip

Many of us struggle with becoming and staying organized. If you practice staying organized, there's an instant gratification. It feels so good you'll keep doing it. You can't be organized if you don't want to be. But as soon as you are organized, you'll begin to want to be all the time.

A CASE IN POINT
The Theater Room

ABC Integrations has hired an outside company to design and install all of the interiors for a dedicated home theater room. This company, called Theater Designs Inc., has a standard contract and payment terms that are used in all of their projects. Theater Designs procedures call for a walkthrough with ABC Integrations and the customer on the final day of installation. This walkthrough is used as the verification process, and if defects are found, Theater Designs Inc. fixes them on the spot to ensure the system has been completed to everyone's satisfaction.

Theater Designs Inc. requires the final payment at the time of the final walkthrough as well as an authorized signature from ABC Integrations and the homeowner to verify that the work is complete and satisfactory.

The final payment is substantial: $15,000. The customer is to pay ABC Integrations for the work and ABC Integrations is to pay Theater Designs, Inc. ABC must have the funds available on that day, even though the payment from the customer has not been deposited. In this case, Project Manager Owen gets special permission from the company owner to maintain a negative balance of payments versus invoicing for the project for this short period while the customer's check clears. While the policy of ABC Integrations is never to have a negative customer balance, cases like these do arise which require special attention during the project close. This special attention includes the company owner authorizing the negative balance, and our bookkeeper, Becka Books, tracking the payments closely to ensure ABC Integrations does not stay negative for very long.

Critical Thinking Questions

- Do you think it's prudent to have a negative cash flow on a project as described in the *A Case In Point?*
- Why or why not?

Contract Closing Procedures

Contract closing procedures is the process that explains the details of what should be done each time a contract is closed. A checklist is helpful in following this procedure. The closing procedure is applied to each phase of the project. Each of the following tasks is performed as part of the closing procedures.

- Verification
- Audit
- Contract Documents
- Payment

Verification

As we discussed earlier in this chapter, every project deliverable requires verification. Verification is the process of ensuring the product or service works as described within the scope statement and meets quality standards as specified in the quality management plan. Below are some examples of how the verification procedure may be carried out at the completion of each phase.

- **Cabling Phase.** The verification for this phase is accomplished visually. The cables are free of nicks, sharp bends, and crimps from staples and other fastening devices. The cables are left neatly bundled in the electronics closet. The cable installation looks clean and professional, and has no tangled cables. The quality management plan should include a list of checks for the cabling phase.
- **Terminate Phase.** The verification in this phase is accomplished using various test equipment. In this phase each cable is terminated, tested, and labeled. If a cable does not pass the test, new terminations are applied, and possibly a new cable is installed.
- **Move-In Phase.** In this phase the home is made ready for the owner to take occupancy. The telephone and television services are tested at each wallplate, using a common television and telephone owned by the company. In addition, speakers are tested using a simple CD player and amplifier to ensure the speakers are free of distortion and rattling.
- **Electronics Build.** In this phase the electronics are assembled, programmed, and tested. This process is always performed in the shop. The reason this process is performed in the shop is to remove the house cabling as a variable in the system during the test and to allow the electronics and programming to be tested as one functional unit. Each feature of the system is tested, and the system is presented to the customer for verification that customer preferences have been properly programmed.
- **Electronics Delivery.** Once the system is delivered and connected to the house cabling, all should work without test, right? After all, the cables and electronics have been previously tested. Guess again. Since all electronic systems have tolerances, the combination of the house and the system can cause issues within the system. In this phase, the system is thoroughly tested in the home, prior to presentation to the customer when any final bugs are worked out.

Study/Career Tip

Technological innovations have made it easy to search and apply for jobs using a few mouse clicks and your keyboard. You must also realize that these same innovations have made it almost as easy for scammers to take advantage of you. Be mindful of the many scams out there that involve providing personal information over the Internet or jobs that seem too good to be true.

Critical Thinking Question

- How does a residential integration firm use its audit information going forward?

Audit

While the verification process ensures quality, the audit process ensures that the right quantity of products and services have been delivered. This process checks the scope of the project against what was actually accomplished in the field.

- **Cabling Phase.** Each wallplate location is counted and compared to the scope statement. The location of the wallplate is compared to the scope statement. If the location or quantity is different, the project manager follows through with the installation team and sales representative, which triggers a change order or corrective action on part of the cabling team.
- **Terminate Phase.** The wallplates are counted yet again in this phase, and the distribution panels are compared to the scope statement. Each distribution panel has a capacity and is checked. For example, the RF distribution panel may only handle eight wallplates per the scope statement; however, nine cable TV wallplates have been installed. The project manager needs to consider two issues. Was this a change order created as part of the cabling phase? How will the extra wallplate be accounted for with respect to the distribution panel?
- **Move-In Phase.** In this phase the home is made ready for the owner to take occupancy. A count of speakers and models is compared against the scope statement.
- **Electronics Build.** In this phase the electronics are assembled, programmed, and tested. As part of the audit each component is compared to the scope statement and the serial numbers are recorded. This information is used in the creation of the warranty document.
- **Electronics Delivery.** Once the system is delivered and connected to the house cabling, each electronic component is checked against the scope statement. A simple method for this is to initial and date next to each product description in the scope statement. Any discrepancies can be noted, such as a television that is installed in a different location from the one called for in the scope statement. A quick phone call to the sales representative may find that the customer requested the change and the change is acceptable. Or the opposite may happen. The project manager could find out a mistake occurred and that it needs to be corrected.

Contract Documents

Contract documents include the contract, any approved changes to the contract, the scope statement, and project acceptance criteria. For example, ABC Integrations includes block diagrams of all systems with the contract. This practice allows for future upgrades. In addition, owner's manuals and small parts, such as equipment remotes and assorted materials that vendors provide with the equipment but are not used in the installation, are placed in a box and delivered with final payment. Table 18–1 shows a list of documents and their destination at the end of a project for ABC Integrations.

- **Final Scope Statement.** This is the final version of the scope statement. The scope statement may be updated during the project.

TABLE 18–1
Closing Project Documents

Document	*Accounting Archive*	*Project Archive*	*Provided to the Customer*
Final Scope Statement		X	X
As-Built Block Diagrams		X	X
Payment Request Letters	X		
Project Invoices	X		
Change Orders	X	X	
Product Owner's Manuals			X
Final System Programming		X	X
Programming Preference Worksheets		X	X
Statement of Warranty		X	X

When updating the original scope statement all the changes are dated and added to the scope statement without removing any previous information.

- **As-Built Block Diagrams.** Block diagrams show how each system is connected and contain information such as room names, cabling labels, color codes, test points, and other information required to service the system.
- **Payment Request Letters.** Each time payment is requested, a letter is sent summarizing the original contract value, all change orders, and payments made to date.
- **Project Invoices.** Each time products and services are delivered to the site, an invoice is entered into the accounting system.
- **Change Orders.** All change orders must be accepted by the customer. These include the scope and cost of the change.
- **Product Owner's Manuals.** These are the manuals that come with the system electronics.
- **Final System Programming.** This is the programming as well as any specialized software required to upload the programming. For example, our customer receives a touchscreen-based remote that not only has a computer-based file that specifies the remote's programming, but also the software required to program the remote. This may also include graphic files and user interface layouts.
- **Programming Preference Worksheets.** These preference worksheets include items such as favorite radio stations, thermostat settings, and time-of-day lighting sequences.
- **Statement of Warranty.** This document specifies the warranty period of each electronic component, as well as the terms of the service provided by ABC Integrations.

Study/Career Tip

Many study and career tips have been offered throughout this textbook. But these are only the "tip" of the proverbial iceberg. Conduct an Internet search or contact your school's career center for more tricks of the studying and career trades.

Payment

Typically, residential integrators require payments prior to installation work. This results in a positive customer balance at all times. A payment is made prior to each phase, which allows the residential integration company to draw from the funds as products and services are invoiced. The final step in each phase is to create the payment request for the next phase. Figure 18–1 shows a sample letter that is sent to each ABC Integrations customer upon completion of the cabling phase.

Dear Fred Smith:

We would like to thank you for your continued support throughout our project. All of us at ABC Integrations want to ensure that you are 100% satisfied with our services. At this time the cabling of your home is complete and, based upon our proposal and contract, the next payment is now due.

For your reference, please find the following reconciliation of your account:

Contract Amounts	
Contract No.	*Amount*
0123FS	$12,653.00
123FS-CO1	$ 826.00
Total of Contracts	$13,479.00

Payment Amounts	
Payment Date	*Amount*
Paid 11/5/05	$ 4,000.00
Due 12/15/05, Prior to Termination	**$ 4,000.00**
Due 2/1/06, Prior to Move-In	$ 4,000.00
Balance due upon completion	$ 13,479.00

Again, thank you for your support. We look forward to hearing from you should you have any comments or questions.

FIGURE 18–1 Payment Request Letter

Summary

In Chapter 18, we discussed contract closure, considered issues in writing letters to close contracts with subcontractors, and discussed early contract termination. We explained the contract closure issues in procurement planning. We discussed four parts of the contract closure procedure—verification, audit, contract documents and final payment.

Important points in this chapter are:

- In closing the contract, the project manager formally verifies that all the work and deliverables are acceptable to the customer. Contract closure also involves various administrative chores. Final project outcomes are recorded, paper and computer files are archived, and customer names are filed.
- Contracts should include a clause that governs responsibilities and protects both the customer and the company in the case of early contract termination.
- The four issues in contract termination are: verification, audit, contract documents, and final payment.
- Verification is the process of ensuring the product or service works as described within the scope statement and meets quality standards as specified in the quality management plan. Verification is performed in each phase of a project.
- The audit process ensures that the right quantity of products and services have been delivered. The audit checks the scope of the project against what was actually accomplished in the field.
- Contract documents include the contract, any approved changes to the contract, the scope statement, and project acceptance criteria.
- Residential integrators require payments prior to installation work. This results in a positive customer balance at all times. The request for payment letter is sent to the customer at the completion of each phase.

Key Terms

As-built block diagrams Block diagrams that show how each system is connected. They contain information such as room names, cabling labels, color codes, test points, and other information required to service the system.

Early contract termination The event that occurs when either the buyer or the seller decides to pull out of the contract before it is completed.

Final scope statement The final version of the scope statement, including the original text and all updates.

Statement of warranty This document specifies the warranty period of each electronic component, as well as the terms of the service provided by the company.

Value engineering Value engineering is a creative approach used to optimize life cycle costs, save time, increase profits, improve quality, expand market share, solve problems, and/or use resources more effectively.

Review Questions

1. What are the administrative tasks associated with closing a contract?
2. How can contracts legally be terminated early?
3. What are the four steps in contract closing procedures?
4. What are considered contract documents?

Pulling It All Together—ABC Integrations

Introduction

This Appendix addresses the case study company, ABC Integrations, that we followed throughout the textbook, and a set of processes that we developed for this company. The processes are presented on a CD that accompanies this textbook. The names of the computer files on the CD are listed under the various sections in the Appendix. These sections contain each of the processes that the *A Guide to the Project Management Body of Knowledge* (PMBOK Guide) defines as necessary elements in project management. We have adapted and developed these elements for ABC Integrations. The processes developed for ABC Integrations provide an example of how you can develop these processes at your organization. It is the process itself that is important for you to learn. Learn the process and you can develop it, alter it, and use it, as appropriate, in your role as project manager.

In this Appendix, we have attempted to pull together all the elements you'll need to fulfill your role as project manager successfully. If ABC Integrations were a real company, these elements would be needed to establish the company and its overall project management approach. Although most residential integration companies develop processes and procedures for every aspect of their company, we present only a sample of each type. For example, there is only one mission statement; however, a company can develop as many core beliefs as appropriate for the company and its environment. The same is true for the section on the company products. We provide files for a structured cabling system. You can use the same process to develop procedures for all your products. We also present a number of templates that can be used as informational tools to develop the scope statement and other documents at your company. We have also included flowcharts, where appropriate, to help you visualize the process.

The Case Study—ABC Integrations

Throughout this textbook we build on a single case study. Each chapter expands on prior chapters to build a complete suite of project management templates and forms for our case study. One of the purposes of the textbook is to teach the fundamentals required for you, as project manager, to build your own suite of documents for your organization. The following section identifies and describes the various components used to build the fictional residential integration company, ABC Integrations. Here you will get a complete accounting of how the company is structured and organized.

Mission Statement

The company mission statement is used to measure day-to-day business decisions, and it provides a clear view of what the company does for its customers. A detailed explanation of the mission statement is covered in Chapter 3, The Project Team. ABC Integrations' mission statement is:

> *ABC Integrations provides only high quality turnkey systems for entertainment, comfort, and convenience in the modern digital home.*

Core Beliefs

Some companies go beyond the mission statement and add additional concepts that are known as core beliefs, which further spell out the company's operating philosophy. A detailed explanation of core beliefs is covered in Chapter 3, The Project Team. ABC Integrations' core beliefs are:

- *ABC Integrations provides standardized, cost-effective, and high-performance systems that exceed our customers' expectations.*
- *We agree to requirements and deliverables with the customer before work has begun, establish milestones, and monitor progress throughout the project.*
- *We strive to minimize on-site installation time.*

Organizational Chart

ABC Integrations has 15 employees. This represents a typically sized company in the residential integration industry. ABC Integrations installs two different types of systems—structured cabling and music.

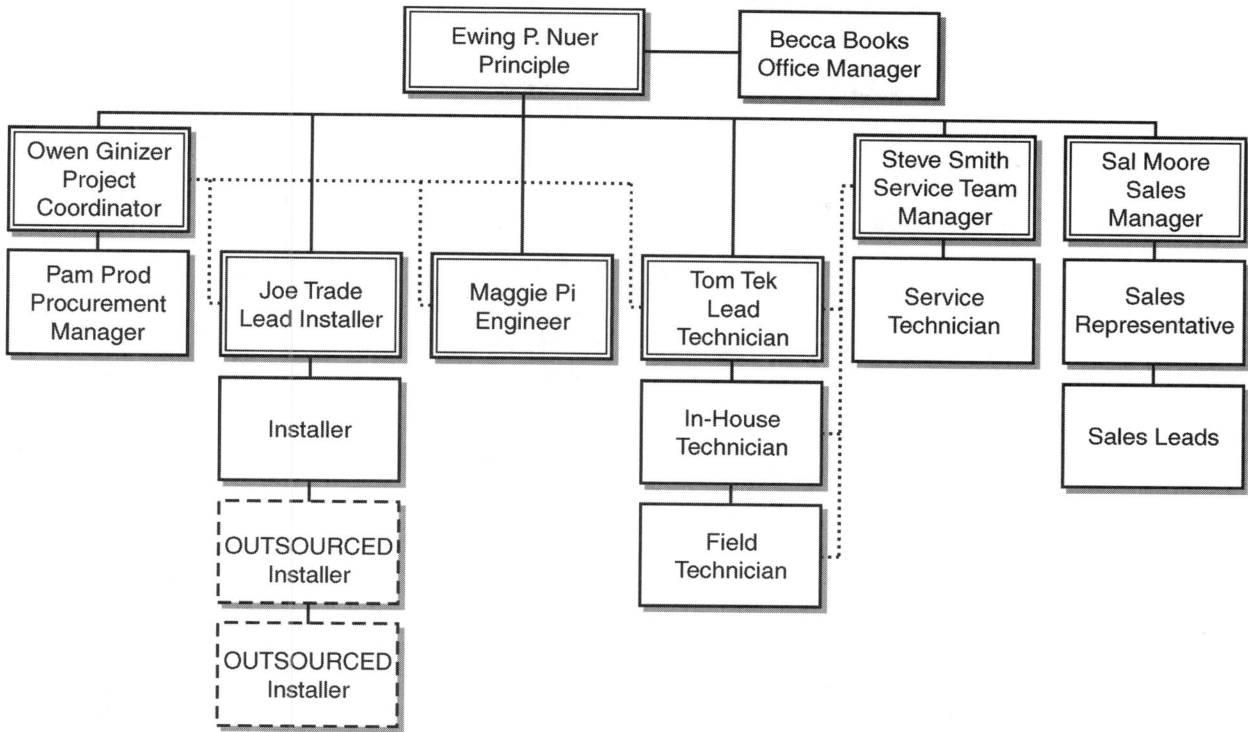

FIGURE A–1 ABC Integrations Organizational Chart

While the typical residential integration company installs more systems, our examples from these two systems clearly illustrate the learning points of this textbook.

Figure A–1 represents the organizational chart for our case study company. ABC Integrations' organizational structure is based upon a strong matrix organization and it places much of the day-to-day responsibilities on the project manager. Organizational structures are covered in detail in Chapter 3, The Project Team.

Let's look at our case study to review the flow of the ABC Integrations organizational chart and how the company's structure is a good fit for the residential integration industry.

Customer Profile

The customer profile is developed by the executive team, which looks at the company's typical customer. The profile includes information that generally describes the demographic that the company services. It includes general information about the customer. Is the customer married or single? Does the customer have children? What is their financial bracket? What is the size of their home? What type of home do they have? Are they a professional, skilled laborer, or general office employee? What level education have they attained? What level is their knowledge of technology? The executive team uses this information to develop the products that it sells. The team must know the customer it wants to reach before it can assess the products and services it will offer.

Once the customer profile is created, the executive team develops the company products and services to fit that customer profile. A company that sells high-end, complex, and custom integration systems with lighting systems that control up to 200 light switches is not going to do well selling to customers who live in 1300-square-foot homes. Neither will the company that sells only family room home theaters do very well selling to customers, who want extensive home integration systems. The company's products must meet the customer's needs and requirements. The customer profile is stored in the case study file "PM01 - Project Charter - Structured Cabling.doc".

Accounting

ABC Integrations has adapted the following accounting policies. For more on this, see Chapter 4, Understanding Cost Accounting.

- **Accounting Method.** Accrual. This was chosen due to the inventory nature of the company. ABC Integrations regularly holds inventory for customers. The value is carried in the inventory asset account.
- **Customer Payments.** All payments received are entered into a liability account for customer deposits. These funds are taken against invoices as products and services are delivered.
- **Invoicing.** All products are invoiced as they leave the company shop. Services are invoiced on a monthly basis using a "percent complete" for each of the installation phases. This information is provided by Project Manager Owen Ginizer.
- **Customer Contracts.** These contracts are typically fixed price contracts. This type of contract is chosen to clearly fix the price of the products and services to be provided to the customer.

Enterprise Environmental Factors

These factors include ABC Integrations' culture, its project management information system, and a human resource pool. The culture of the company can be best described as a small, family-oriented business.

Project Management Information System

Eight of the fifteen employees have laptops, and they connect to an in-house network. In addition, there is a desktop computer in the company for the rest of the employees to check e-mail upon returning from the field at the end of each day. Since the company works from laptops and has a need to work remotely, a series of MS Word documents and MS Excel spreadsheets make up the bulk of the information systems. In addition, the project manager utilizes MS project for managing projects. He issues his reports in Adobe PDF format for the rest of the company.

The president, office manager, and the procurement manager all interface with the company accounting program, QuickBooks, which is located on a server in the office. Each has access to the server remotely via a VPN connection.

The Human Resource Pool

The human resource pool consists of not only the company organization chart (see Figure A–1), but includes the list of subcontractors utilized by the company to perform a wide range of specialized services for its customers. Below is a list of those subcontractors:

- **Satellite.** This contractor installs satellite dishes.
- **HVAC.** This contractor provides technical assistance with respect to HVAC integration. He helps interface with the HVAC contractor and installs the thermostats when the project HVAC contractor is unable to do it.
- **Video Calibration.** This contractor calibrates plasma monitors, projectors, and other video displays.
- **Audio Calibration.** This contractor calibrates the home theater systems, as well as provides technical assistance in designing complex home theater systems.
- **Information Technology.** This contractor not only works on ABC Integrations' internal computer system, he also provides technical assistance with customer networks.

Organizational Process Assets

The organizational process assets include the policies, procedures, standards, and guidelines of the organization. In addition, historical information and lessons learned from previous projects are considered organizational process assets. The case study files referenced within Appendix A are considered ABC Integrations' organizational process assets.

The Project Management Knowledge Areas

The project management knowledge areas used in this case study are derived from *A Guide to the Project Management Body of Knowledge*, as published by the Project Management Institute (**www.pmi.org**). The knowledge areas of project management can be divided into two categories: project constraints and facility constraints. Project constraints include cost, time, scope, and quality. Facility constraints include communications, human resources, risk, and procurement. The last project management knowledge area is project integration, which is the "glue" that holds each of the knowledge areas together. Figure A–2 shows how these areas interact.

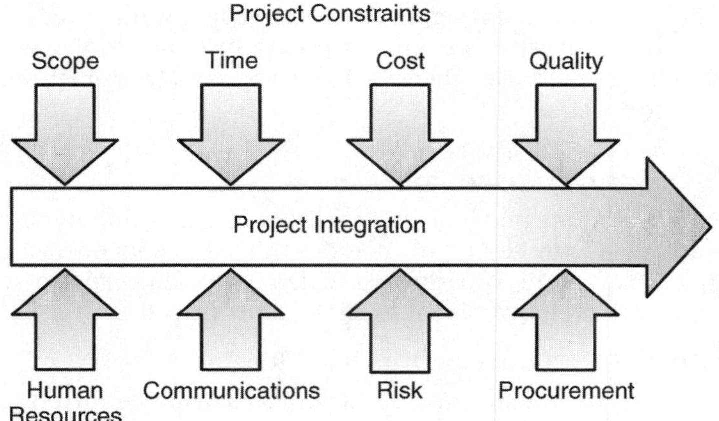

FIGURE A–2 Project Management Knowledge Areas

The following sections describe each of the project management knowledge areas in greater detail.

Project Integration Management

Project integration management includes the process and activities needed to identify, define, combine, unify, and coordinate the various processes and project management activities within the project management process groups. In the project management context, integration includes characteristics of unification, consolidating, articulation, and integrative actions that are crucial to project completion, successfully meeting customer and stakeholder requirements, and managing expectations. Figure A–3 shows the work process flow of ABC Integrations' project integration management. These processes include:

Develop Project Charter

The project charter is the document that formally authorizes the project, and it is the first step in defining the project. For ABC Integrations, the executive team writes the charter, which becomes the foundation for every job the company undertakes. To that end, a customer demographic is written to summarize the typical ABC Integrations customer. This is used to focus the project charter in a specific direction and to ensure that the company will not accept work that is not in keeping with the project charter.

ABC Integrations' project charter is stored in the case study file "PM01 - Project Charter - Structured Cabling.doc".

Develop Scope Statement Template

The scope statement template is a document that provides an in-depth look at the scope of a project. At ABC Integrations, the executive team develops the scope statement template, which is then filled in by the

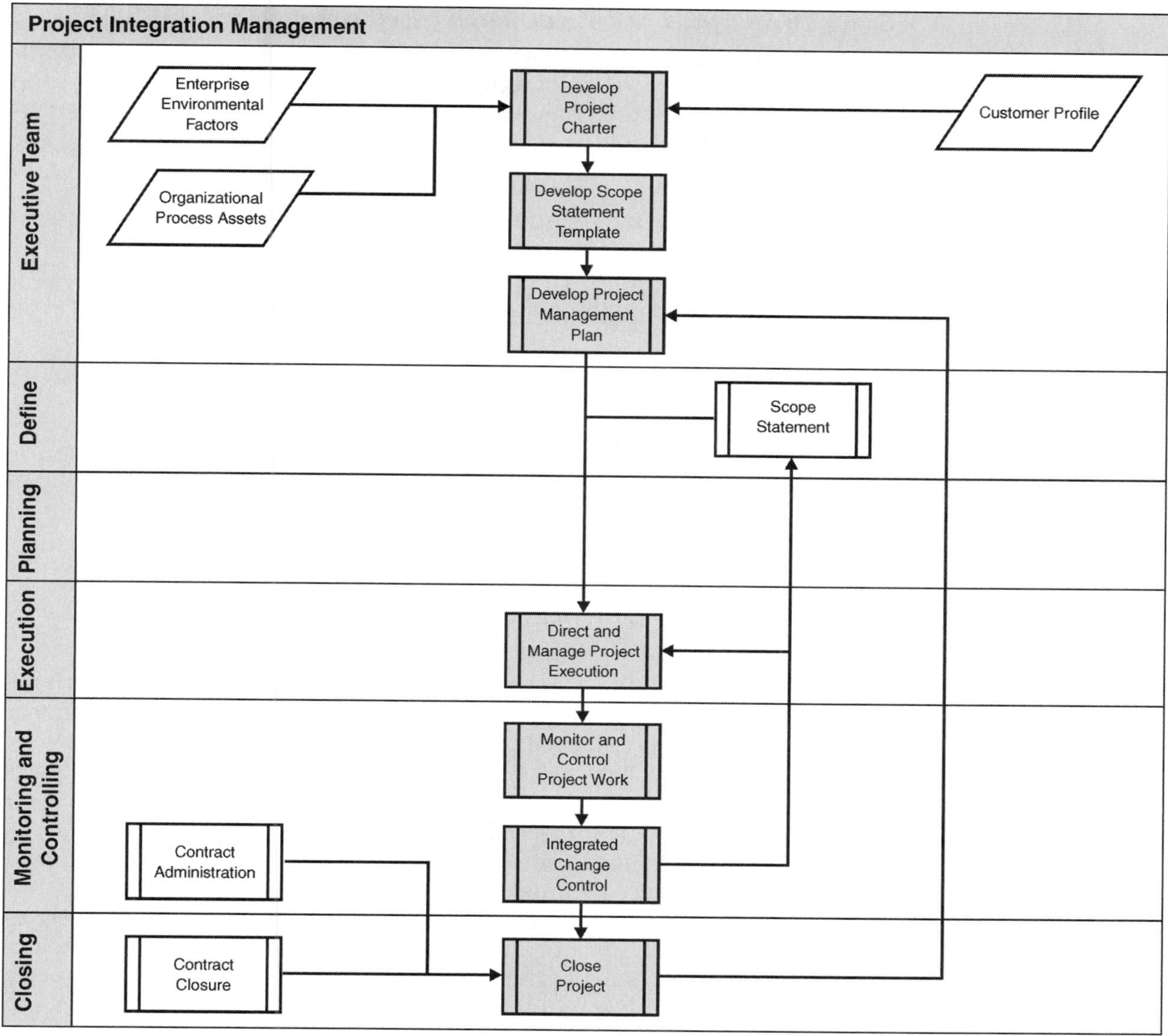

FIGURE A–3 Project Integration Management

sales representative for each customer. This helps decrease the time required to create a scope statement for each job, as well as mitigating potential risks due to vague scope definition. The case study file can be found in "PM05 - Template - Project Scope Statement - Structured Cabling.doc".

Develop Project Management Plan

The project management plan documents actions required to define, prepare, integrate, and coordinate all subsidiary plans. ABC Integrations has created this plan in several forms; a sample is "PM08 - Policy - Project Scope Management Plan.doc".

Direct and Manage Project Execution

This is the process of executing the work defined in the project management plan to achieve the project's requirements defined in the project scope statement. ABC Integrations uses several forms for managing and controlling the project execution. These files include:

- "PM11 - Form - Work Order.doc". A generic work order form.
- "PM12 - Memo - Customer Project Open.doc". A letter to the customer upon opening the project.
- "PM13 - Memo - GC Project Open.doc". A letter to the general contractor upon opening the project.

Monitor and Control Project Work

Monitor and control project work is the process created and used to monitor and control a project to meet the performance objectives defined in the project management plan.

Integrated Change Control

This process reviews all change requests, approves changes, and controls changes to the deliverables and organizational process assets.

Close Project

Close project is the finalizing of all activities across all of the project process groups to formally close the project.

Project Scope Management

Project scope management includes the processes required to ensure that the project includes all the work required, and only the work required, to complete the project successfully. Project scope management is primarily concerned with defining and controlling what is and is not included in the project. Figure A–4 shows the work process flow of ABC Integrations' scope management.

Project Scope Management processes include:

Scope Management Planning

This process involves creating a project scope management plan that documents how the project scope will be defined, verified, and controlled, and how the work breakdown structure (WBS) will be created and defined. Scope planning is covered in Chapter 9, The Scope Management Plan, and Chapter 7, The Work Breakdown Structure. The scope management plan for ABC Integrations can be found in "PM08 - Policy - Project Scope Management Plan.doc".

Scope Definition

The project scope statement explains the project details in easy-to-understand terms and leaves out the technology jargon. The purpose

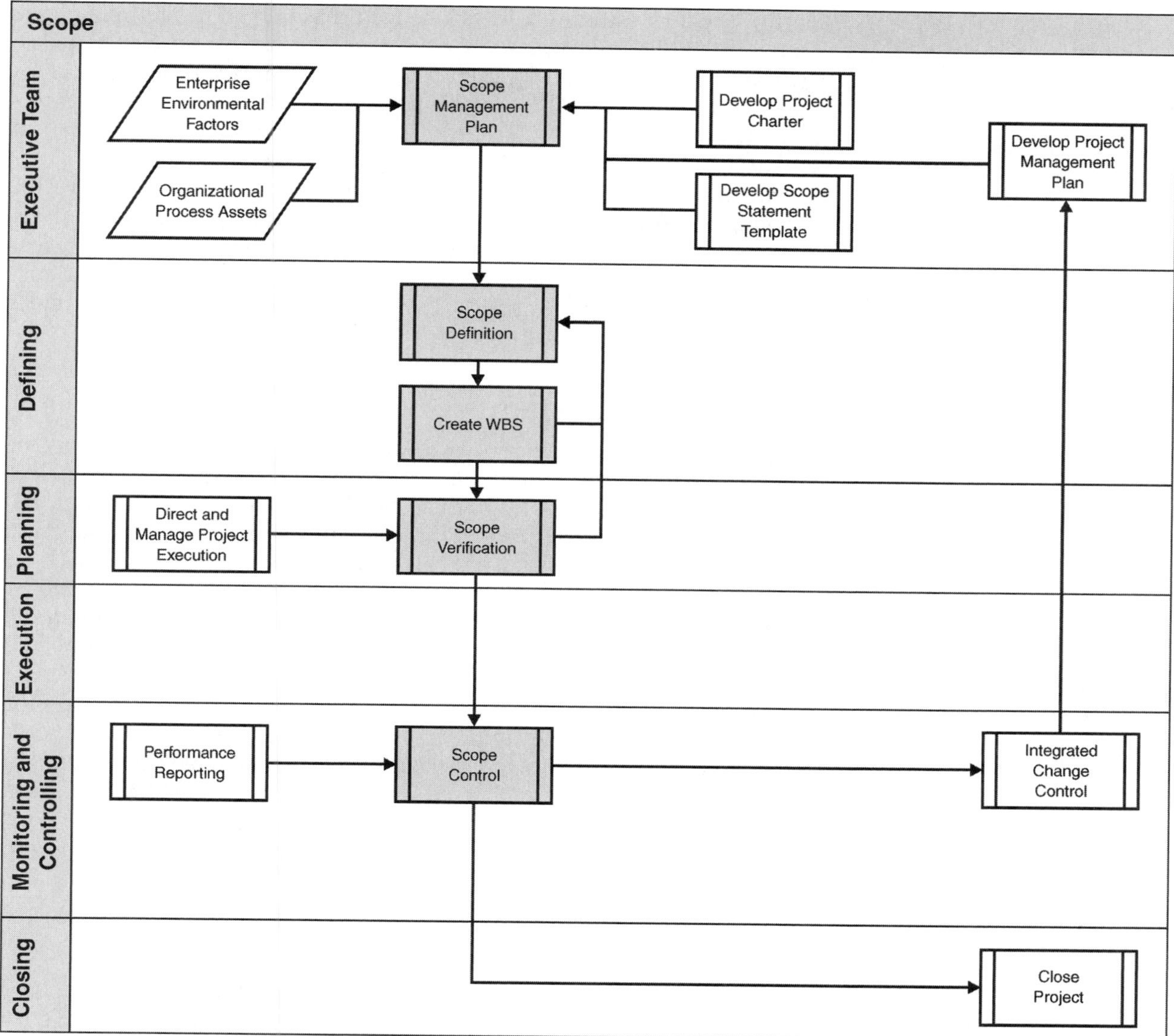

FIGURE A–4 Scope Management Processes

of the scope statement is to let both the customer and the project team know exactly what is expected of both parties during the installation and when major events are scheduled to occur. The case study file can be found in "PM14 – Smith Fred and Wilma - Project Scope Statement - Structured Cabling.doc". It contains a scope statement developed from the template for our customers Fred and Wilma Smith.

Create WBS

The WBS subdivides the major project deliverables and project work into smaller, more manageable components. At ABC Integrations, the WBS template is prepared by the executive team. The WBS is created as part of the cost estimate and can be found as part of the template case study file "PM07 - Template - Cost Estimate - Structured Cabling.xls", as well as a copy created for our customers Fred and Wilma Smith. This case study file is "PM15 - Smith Fred and Wilma - Cost Estimate - Structured Cabling.xls". The cost estimate is covered in more detail in Chapter 8, Cost Estimating.

In addition to breaking down the project to analyze cost, ABC Integrations also breaks the project into a WBS that aids in scheduling. The template for this can be found in "PM10 - Template - Project Tracking - Structured Cabling.mpp", and the specific file for our customers Fred and Wilma Smith is "PM16 - Smith Fred and Wilma - Project tracking - Structured Cabling.mpp".

Scope Verification

Scope verification is the process designed to formalize acceptance of the completed project deliverables. The project manager carries out the verification process, which is covered in detail in Chapter 17, Managing and Controlling. Our case study uses the "stickering" method described in Chapter 17. A sample of a sticker is shown in Figure 17–1, and a template to create the sticker has been included in "PM17 - Template - Cabling Sticker - Telephone and Networking.doc". The stickers mark the actual locations in the home of all the wallplates, keypads, and touchscreens. The scope document, "PM14 - Smith Fred and Wilma - Project Scope Statement - Structured Cabling.doc", calls for a telephone wallplate in the kitchen. The sales representative and project manager walk the site and place a sticker for the telephone wallplate in the kitchen at the exact location where it is to be installed. This process provides an opportunity for changes to be made before the project begins and the installations are completed. It also verifies the specific locations for the wallplates.

Scope Control

Scope control is the process designed to control changes to the project scope, including the processes for change orders, variance analysis, replanning, and configuration management. This process is covered in detail in Chapter 17, Managing and Controlling. The scope management plan for ABC Integrations includes the scope control process and can be found in "PM08 - Policy - Project Scope Management Plan.doc".

Project Time Management

Project time management includes the processes required to accomplish timely completion of the project. Figure A–5 shows the work process flow of ABC Integrations' time management. These processes include:

Activity Definition

Activity definition identifies the specific schedule activities that need to be performed to produce the various project deliverables.

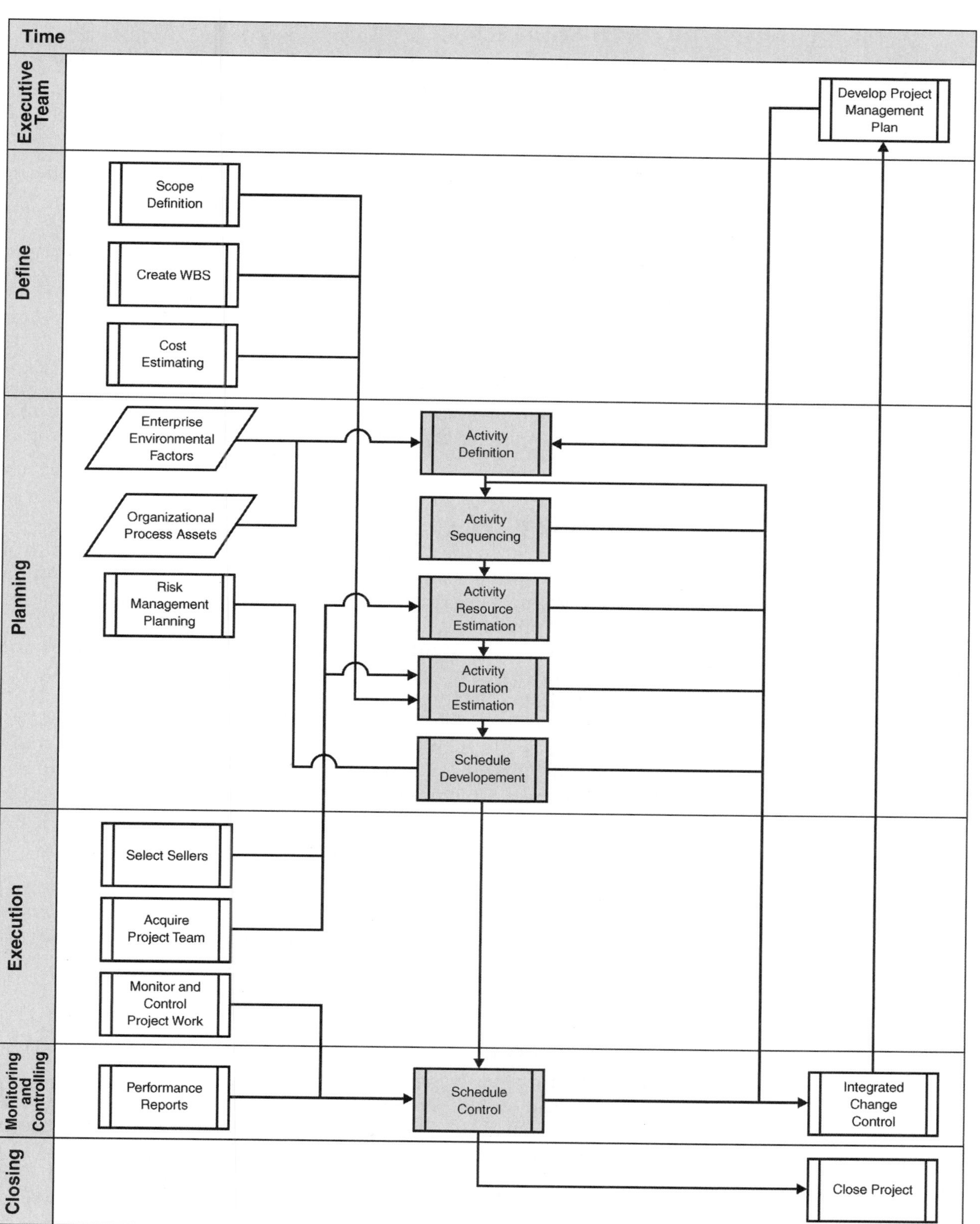

FIGURE A–5 Time Management Process

Activity Sequencing

Activity sequencing is the process created and used to identify and document dependencies among schedule activities.

Activity Resource Estimating

Activity resource estimating is estimating the types and quantities of resources required to perform each schedule activity.

Activity Duration Estimating

Activity duration is the process used to estimate the number of work periods that will be needed to complete individual schedule activities.

Schedule Development

Schedule development is analyzing activities sequences, durations, resource requirements, and schedule constraints to create the project schedule.

Schedule Control

Schedule control is the design and use of the processes to control changes to the project schedule. It involves schedule management, a schedule baseline, performance reports, and approved change requests.

All of the project time management processes are covered in detail in Chapter 10, Time Management Planning, with the exception of schedule control, which is covered in detail in Chapter 17, Managing and Controlling. ABC Integrations has created a template in Microsoft Project, which includes this process; the case study file is "PM10 - Template - Project Tracking - Structured Cabling.mpp". Our case study customer information is included in "PM16 - Smith Fred and Wilma - Project Tracking - Structured Cabling.mpp".

Project Cost Management

Project cost management includes the processes involved in planning, estimating, budgeting, and controlling costs so that the project can be completed within the approved budget. Figure A–6 shows the work process flow of ABC Integrations' cost management.

Project Cost Management processes include:

Cost Estimating

Cost estimating involves developing an approximation or estimate of the cost of the resources required to complete a project. This process is covered in detail in Chapter 8, Cost Estimating. ABC Integrations utilizes a cost estimate template found in case study file "PM07 - Template - Cost Estimate - Structured Cabling 122605.xls". The cost estimate for our customers, Fred and Wilma Smith, is in "PM15 - Smith Fred and Wilma - Cost Estimate - Structured Cabling.xls".

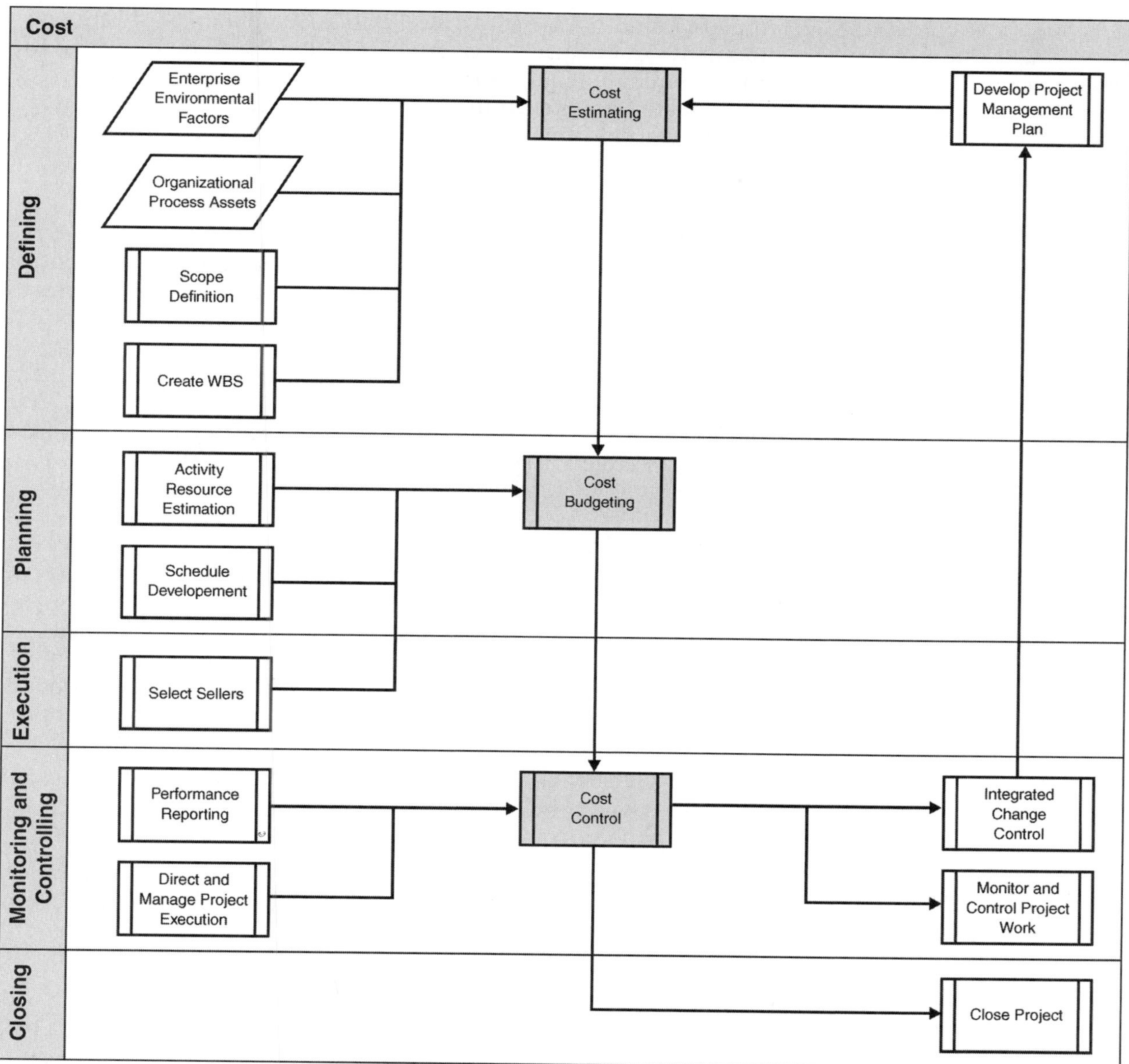

FIGURE A–6 Cost Management Processes

Cost Budgeting

Cost budgeting is aggregating the estimated costs of individual activities or work packages to establish a cost baseline. Cost budgeting is represented as the bill-of-materials (BOM) for the project. A critical step in the cost management plan is generating the bill-of-materials, which is derived from the cost estimate. By presenting and viewing the bill-of-materials in different ways, it is possible to accomplish several purposes. This process is covered in detail in Chapter 15, Cost Management Planning. ABC Integrations utilizes a cost estimate template found in case study file "PM07 - Template - Cost Estimate - Structured Cabling.xls". This template includes cost budgeting in the form of purchase requirements, sales invoices, and work orders. The cost estimate for our customers, Fred and Wilma Smith, is in "PM15 - Smith Fred and Wilma - Cost Estimate - Structured Cabling.xls".

Cost Control

Cost control includes the process designed to control costs so that the project can be completed within the approved budget. Cost control influences the factors that create cost variances and control changes to the project budget. This process is covered in detail in Chapter 17, Managing and Controlling. ABC Integrations has created a template in Microsoft Project, which includes this process; the case study file is "PM10 - Template - Project Tracking - Structured Cabling.mpp". Our case study customer information is included in "PM16 - Smith Fred and Wilma - Project Tracking - Structured Cabling.mpp".

Project Quality Management

Project quality management involves the processes and activities of the performing organization that determine quality polices, objectives, and responsibilities so that the project will satisfy the needs for which it was undertaken. It implements the quality management system through policy and procedures, with continuous process improvement activities conducted throughout the project, as appropriate. Figure A–7 shows the work process flow of ABC Integrations' quality management. These processes include:

Quality Planning

Quality planning is the process created and used to identify which quality standards are relevant to the project and to determine how to meet those standards.

Perform Quality Assurance

Performing quality assurance is the process designed and used to apply the planned, systematic quality activities to ensure that the project employs all processes needed to meet the requirements of the project. This process is covered in detail in Chapter 16, Project Execution.

At ABC Integrations, in-house technicians build the equipment racks that will hold all the electronics. As a quality assurance activity, another in-house technician tests all the equipment in the rack to make sure it is in working order before it is disassembled and packed up for delivery to the customer's home. By continually using this quality assurance process, technicians figured out a way to organize the electronics in the rack in a more efficient way by always putting them in the same order and marking each piece in for the same order when it is prepared for delivery.

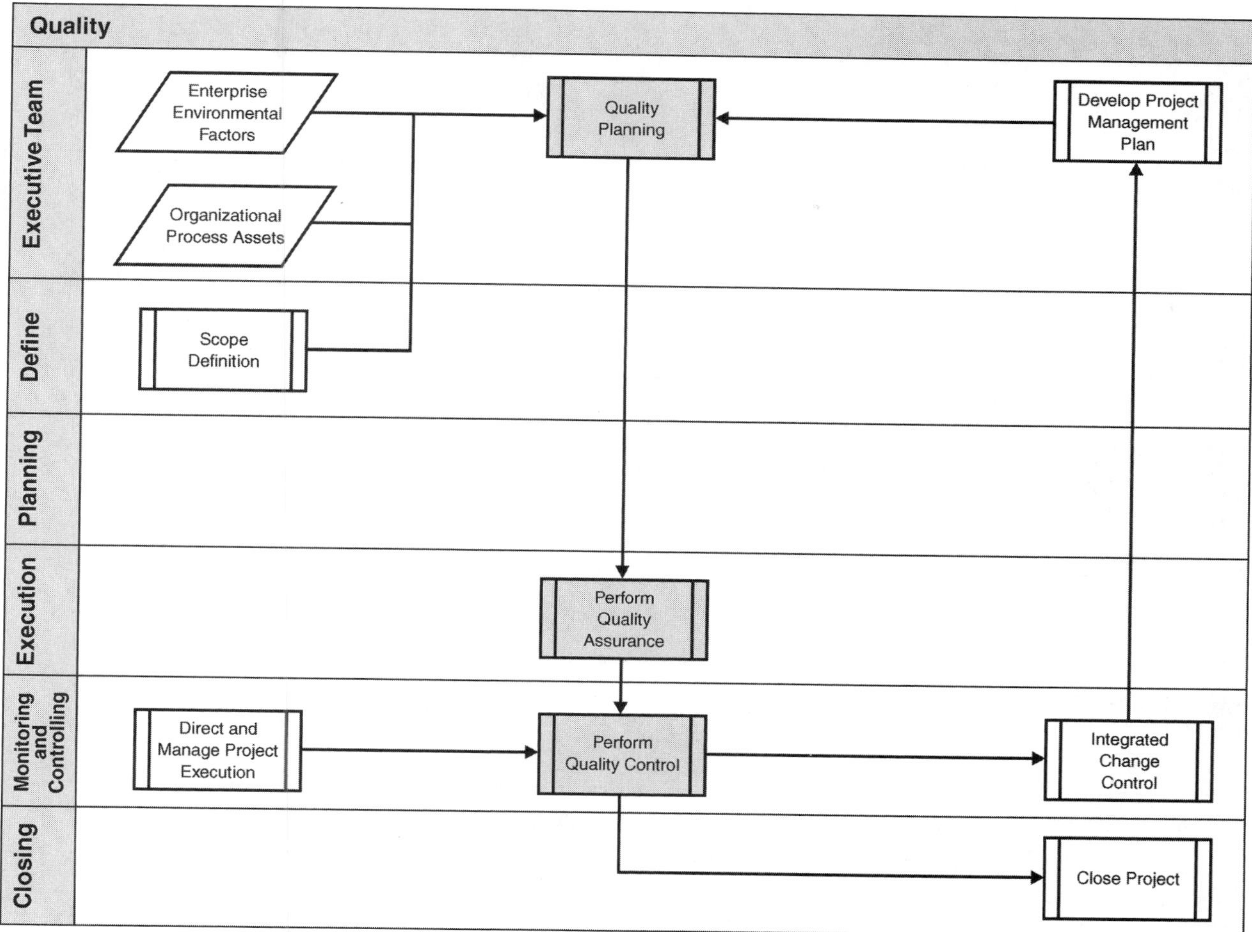

FIGURE A–7 Quality Management Processes

Perform Quality Control

Quality control is the process designed to monitor specific project results to determine whether they comply with relevant quality standards and to identify ways to eliminate causes of unsatisfactory performance. This process is covered in detail in Chapter 17, Monitoring and Controlling.

ABC Integrations uses industry certification programs to maintain quality. In addition, each electronic system at ABC Integrations is built in the shop and thoroughly tested prior to delivery to the customer's home.

Project Human Resource Management

Project human resource management includes the processes that organize and manage the project team. The project team is comprised of the people who have assigned roles and responsibilities for completing the project. While it is common to speak of rolls and responsibilities being

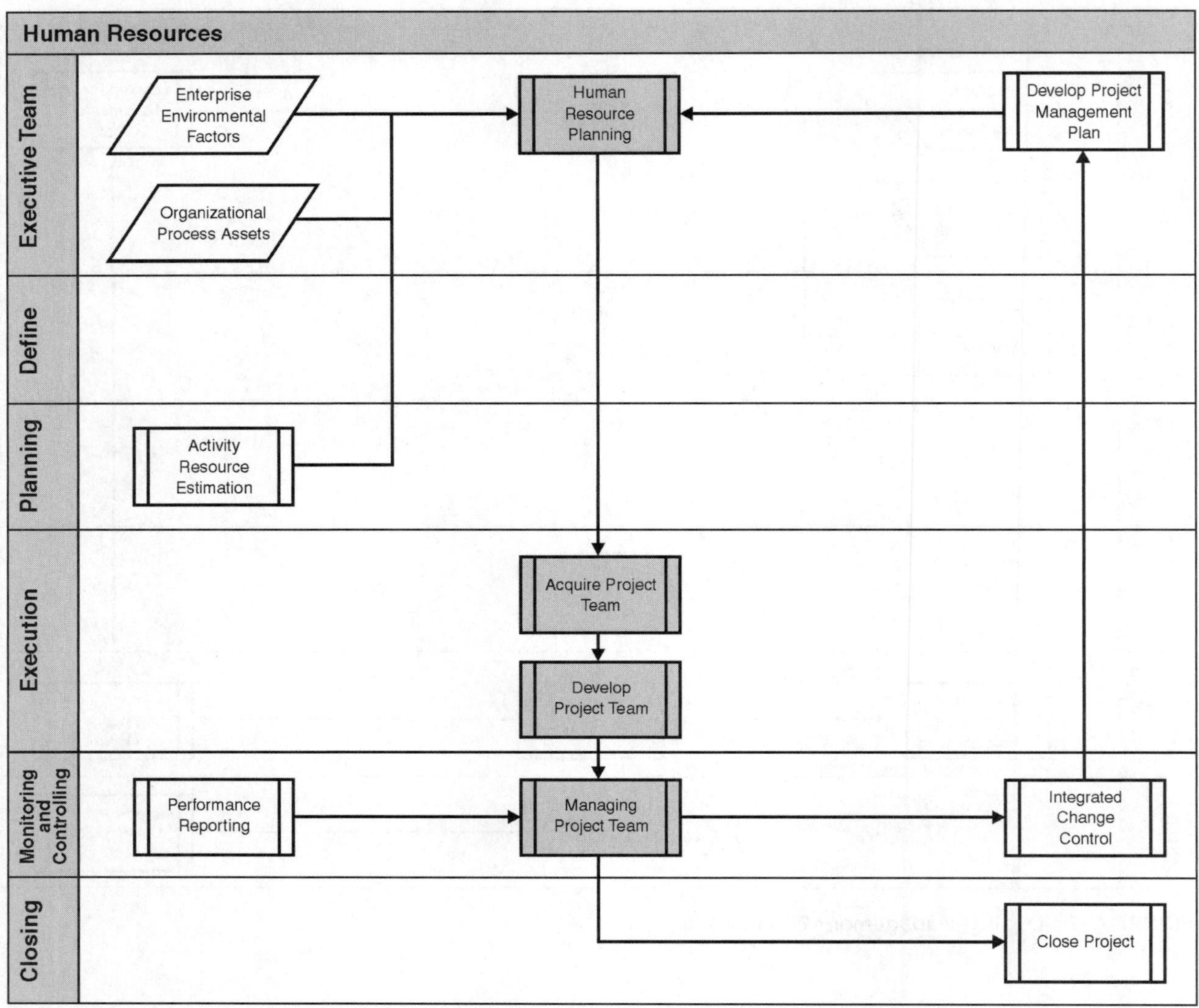

FIGURE A–8 Human Resource Processes

assigned, team members should be involved in much of the project's planning and decision-making processes. Early involvement of team members adds expertise during the planning process and strengthens commitment to the project. The type and number of project team members can often change as the project progresses. Project team members can be referred to as the project's staff. Figure A–8 shows the work process flow of ABC Integrations' human resource management.

Project Human Resource Management processes include:

Human Resource Planning

Human resource planning includes identifying and documenting project roles, responsibilities, and reporting relationships, as well as creating the staffing management plan.

Acquire Project Team

Acquiring the project team is the process created and used to obtain the human resources needed to complete the project. This process is covered in detail in Chapter 16, Project Execution.

Develop Project Team

Developing the project team is the process created and used to improve the competencies and interaction of team members to enhance project performance. This process is covered in detail in Chapter 16, Project Execution.

Manage Project Team

Managing the project team is the process designed to track team member performance, provide feedback, resolve issues, and coordinate changes to enhance project performance. This process is covered in detail in Chapter 17, Managing and Controlling.

Project Communications Management

Project communications management is the design and use of the processes required to create, collect, distribute, store, retrieve, and ultimately dispose of project information in a way that is timely and appropriate. The project communications management processes afford the critical connections among the project team and stakeholders necessary for successful communications. Project managers can spend an excessive amount of time communicating with the project team, stakeholders, customer, and sponsor. The purpose of communications management is to assure that everyone connected to the project understands how communications affect the project as a whole. Figure A–9 shows the work process flow of ABC Integrations' communications management.

Process Communications Management processes include:

Communications Planning

Communications planning is the process created and used to determine the information and communications needs of the project stakeholders.

Information Distribution

Information distribution is the process created and used to make information needed for the project readily accessible to project stakeholders within an appropriate length of time. This process is covered in detail in Chapter 16, Project Execution. ABC Integrations has developed a number of template-based letters for the distribution of information. These include:

- "PM11 - Form - WorkOrder.doc". A generic work order form.
- "PM12 - Memo - Customer Project Open.doc". A letter to the customer upon opening the project.

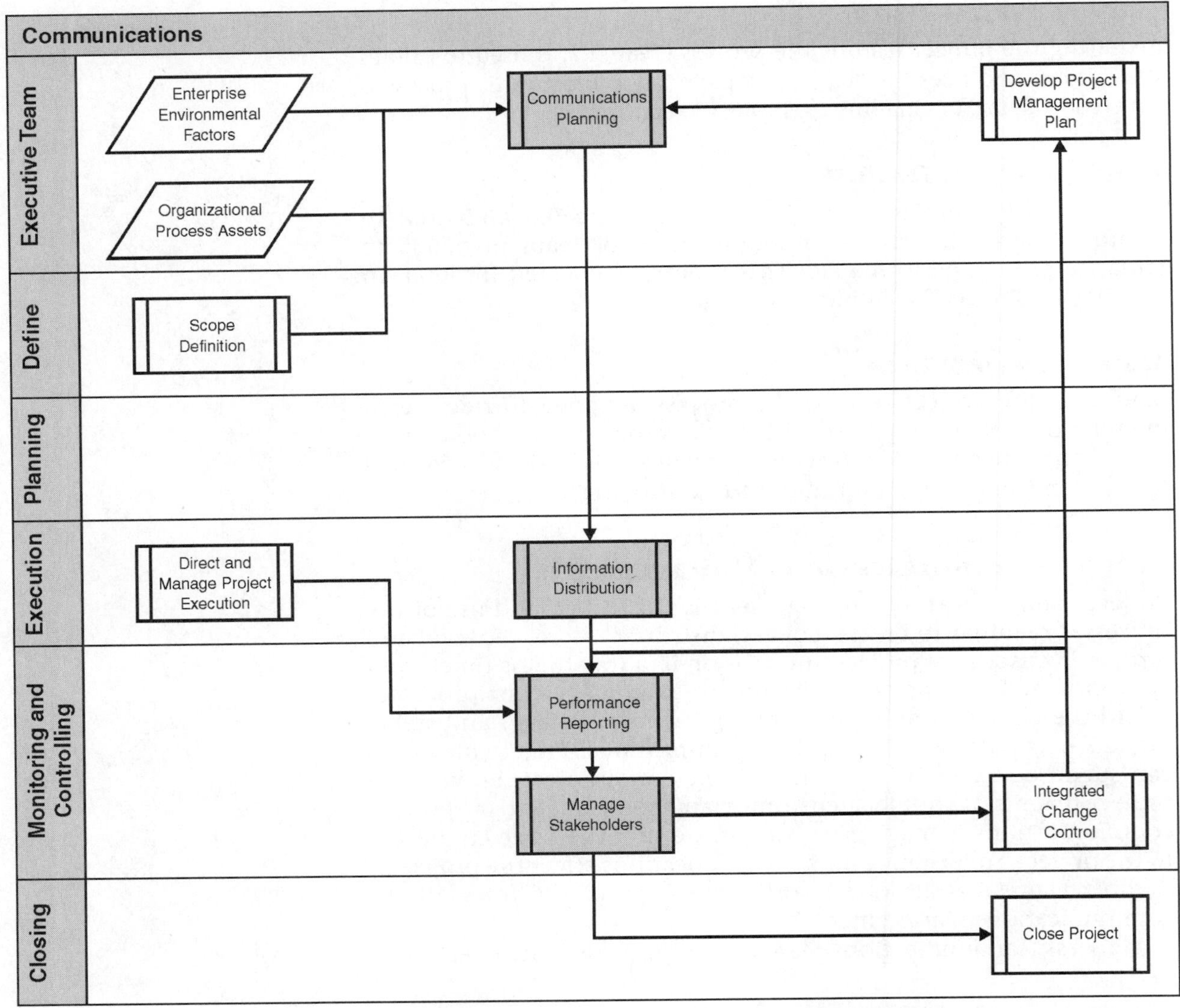

FIGURE A–9 Communications Management Processes

- "PM13 - Memo - GC Project Open.doc". A letter to the general contractor upon opening the project.

Performance Reporting

Performance reporting is the process used to collect and distribute project performance information, which includes status reporting, progress measurement, and forecasting. This process is covered in detail in Chapter 17, Managing and Controlling. ABC Integrations uses as an internal reporting tool in the case study file "PM19 - Template - Lessons Learned.doc". This form allows the project team member to detail what was correctly accomplished with the project phase, and what was not. This information is then used to improve the project management as well as to formulate reports to the project stakeholders.

Manage Stakeholders

Managing stakeholders is the process used to manage communications and interactions with project stakeholders—builder, architect, subcontractors, customer, and project team. It provides a way to meet their requirements and resolve issues that might arise with stakeholders. This process is covered in detail in Chapter 17, Managing and Controlling.

Project Risk Management

Project risk management is designing and using the processes needed to carry out a risk management plan. It identifies, analyzes, responds, and monitors and controls all risk aspects of a project. The purpose of project risk management is to increase the probability and affect of positive events, also known as opportunities, and to decrease the probability and effect of negative events or problems that arise on project goals. Figure A–10 shows the work process flow of ABC Integrations' risk management. These processes include:

Risk Management Planning

Risk management planning is the process used to choose how to approach, plan, and execute the risk management activities for a project. This process is covered in detail in Chapter 11, Risk Management Planning.

Risk Identification

Risk identification is the process used to decide which risks could affect the project and document the characteristics of those risks. This process is covered in detail in Chapter 11, Risk Management Planning.

Qualitative Risk Analysis

Qualitative risk analysis is the process used to set priorities for identified risks so that the probability of whether or not they'll occur can be assessed as well as their potential impact on a project. The analysis also determines what action, if any, should be taken. This process is covered in detail in Chapter 11, Risk Management Planning.

Quantitative Risk Analysis

Quantitative risk analysis is the process used to analyze numerically the effect on the overall project objectives of the identified risks. This process is covered in detail in Chapter 11, Risk Management Planning.

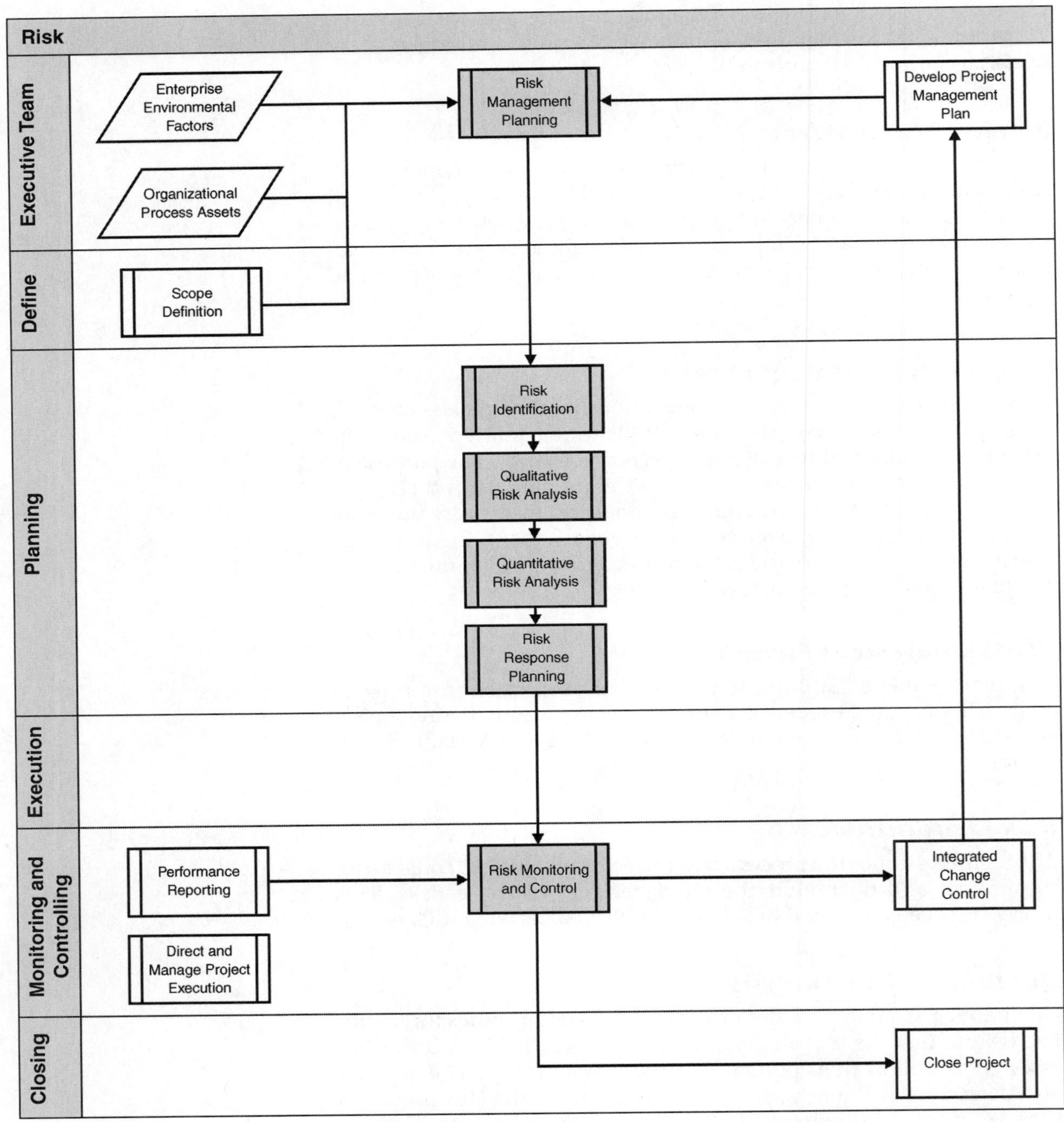

FIGURE A–10 Risk Management Processes

Risk Response Planning

Risk response planning is the process used to develop options and actions to enhance opportunities (positive events) and to reduce threats (negative events) to project objectives. There are five ways to respond to risk: avoidance, mitigation, assumption, transference, and prevention. This process is covered in detail in Chapter 11, Risk Management Planning.

Risk Monitoring and Control

Risk monitoring and control is the process used to track identified risks, monitor residual risks, identify new risks, execute risk response plans, and evaluate success throughout the project life cycle. This process is covered in detail in Chapter 17, Managing and Controlling.

Project Procurement Management

Project procurement management involves the processes used to purchase or acquire the products, services, or outcomes needed from outside the company to complete a project. Procurement can be viewed in two ways. The organization can be either the buyer or the seller of the product, service, or outcome within a contract.

Project procurement management is the process used to manage a contract and control change processes that are necessary to administer contracts or purchase orders written by authorized project team members. Project procurement management also includes the administration of any contract written by an outside organization (the buyer) that is purchasing the project from the performing organization (the seller) and administering contractual obligations placed on the project team by the contract.

Figure A–11 shows the work process flow of ABC Integrations' procurement management.

Project Procurement Management processes include:

Plan Purchases and Acquisitions

Planning purchases and acquisitions is the process used to decide what to purchase or acquire. It also determines when and how to execute those purchases and acquisitions. We cover this process in detail in Chapter 12, Procurement Management Planning.

Plan Contracting

Plan contracting is the process used to document products, services, and outcomes that are required. It also identifies potential sellers. We cover this process in detail in Chapter 12, Procurement Management Planning.

Request Seller Responses

Requesting seller responses is the process created and used to obtain information, quotations, bids, offers, or proposals, as appropriate during a project. We cover this process in detail in Chapter 16, Project Execution. The case study file, "PM07 - Template - Cost Estimate - Structured Cabling.xls", contains a purchase requirements sheet, which is used by ABC Integrations to request seller responses for product purchases.

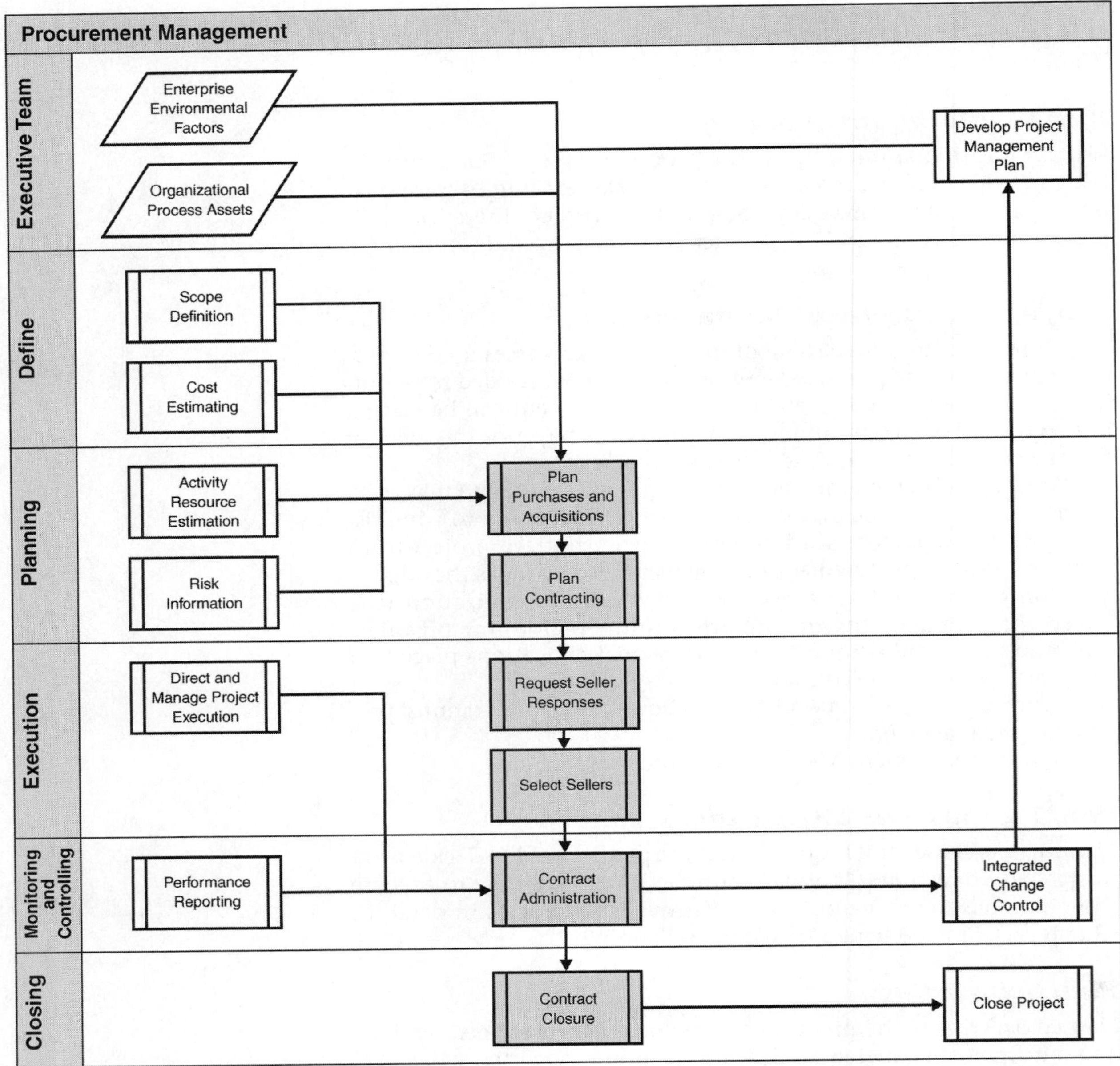

FIGURE A–11 Procurement Management Processes

Select Sellers

Selecting sellers is the process used for reviewing offers, choosing from among potential sellers, and negotiating a written contract with a seller. We cover this process in detail in Chapter 16, Project Execution.

Contract Administration

Contract administration is the process used to manage the contract and the relationship between the buyer and the seller. It involves reviewing and documenting how a seller is performing or has performed to establish required corrective actions and provide a basis for future relationships with the seller, managing contract related changes and, when appropriate, managing the contractual relationship with the outside buyer of the project. This process is covered in detail in Chapter 17, Monitoring and Controlling.

Contract Closure

Contract closure is the process created and used to complete and settle each contract, which includes the resolution of any open or incomplete items, and the process used to close each contract. This process is covered in detail in Chapter 18, Closing Procedures.

PMP Certification

What Is PMP Certification?

Project Management Professional (PMP) certification is governed by the Project Management Institute, (PMI). To be eligible for the PMP credential, you must first meet specific education and experience requirements and agree to adhere to a code of professional conduct. The final step in becoming a PMP is passing a multiple-choice examination designed to objectively assess and measure your project management knowledge. This computer-based examination is administered globally. In addition, those who have been granted the PMP credential must demonstrate an ongoing professional commitment to the field of project management by satisfying PMI's Continuing Certification Requirements.

The PMP is by far the project management credential of choice across a vast array of industries and companies. By attaining the PMP credential, your name will be included in the largest and most prestigious group of certified professionals in the project management community.

PMP Certification Prep Materials

The PMI recommends that candidates begin exam preparation by purchasing the *PMP® Suggested Resource Pack* available through the PMI bookstore, **http://www.pmi.org.**

The following publications provide information on how the exam was developed, project management knowledge domains, and PMP application requirements.

- *PMP® Suggested Resource Pack*
- *A Guide to the Project Management Body of Knowledge (PMBOK® Guide), 2004 Edition*
- *Project Management Professional (PMP®) Role Delineation Study*
- *PMP Certification Handbook*
- *Code of Professional Conduct*

Other Suggested Readings

The following is a list of suggested materials, which, depending on your needs, may compliment an overall study plan.

- *Project Management: A Systems Approach to Planning, Scheduling, and Controlling, 8th Edition.* Harold Kerzner, PhD (John Wiley & Sons)
- *PMP® Exam Practice Test and Study Guide, 5th Edition.* Edited by J. LeRoy Ward, PMP (ESI International)
- *PMP® Exam Prep, 4th Edition.* Rita Mulcahy, PMP (RMC Publications)
- *Project Management: A Managerial Approach, 5th Edition.* Jack Meredith and Samuel Mantel (John Wiley & Sons)
- *The Fast Forward MBA in Project Management, 2nd Edition.* Eric Verzuh (John Wiley & Sons)
- *The New Project Management, 2nd Edition.* J. Davidson Frame (Jossey-Bass)

Professional Responsibility Reference List

This is a list of suggested materials, which, depending on your needs, may compliment an overall study plan.

- *Doing Business Internationally: The Guide to Cross-Cultural Success, 2nd Edition.* Terence Brake, Danielle Walker, and Thomas Walker (McGraw-Hill Professional Book Group)
- *The Cultural Dimension of International Business.* Gary P. Ferraro (Prentice Hall)
- *Global Literacies: Lessons on Business Leadership and National Cultures.* Robert Rosen (Editor), Patricia Digh, Marshall Singer, and Carl Phillips (Simon & Schuster)

Trade Organizations

Association for the Advancement of Cost Engineering (AACE)

With about 5,500 members worldwide, AACE International serves cost management professionals: cost managers and engineers, project managers, planners and schedulers, estimators and bidders, and value engineers. **http://www.aacei.org.**

Building Industry Consulting Services International (BICSI)

BICSI offers the ITS (information transport systems) professional standards-based, vendor-neutral training, technical publications, prestigious registration, and educational conferences to aid in furthering their careers.

Members are people involved in the design and installation of cabling infrastructures. These members design and install the cabling infrastructure that feeds communications, automation, and electronic security systems in commercial buildings, apartment complexes and condominiums, multi-building campuses, and multi- and single-family residences. **http://www.bicsi.org.**

Custom Electronics Design and Installation Association (CEDIA)

CEDIA is an international trade association of companies that specialize in designing and installing electronic systems for the home. CEDIA members include residential electronics systems professionals who have emerged as the "fourth contractor" in the building and remodeling industries alongside electrical, plumbing, and HVAC professionals. **http://www.cedia.net.**

Computing Technology Industry Association (CompTIA)

CompTIA is the leading association representing the international technology community. Its goal is to provide a unified voice, global advocacy and leadership, and to advance industry growth through standards, professional competence, and education and business solutions. **http://www.comptia.org.**

Electronics Industry Association (EIA)

EIA is a national trade organization that includes the full spectrum of United States manufacturers. The alliance is a partnership of electronic and high-tech associations and companies whose mission is promoting the market development and competitiveness of the United States high-tech industry through domestic and international policy efforts. **http://www.eia.org.**

National Electrical Manufacturers Association (NEMA)

NEMA provides a forum for the standardization of electrical equipment. The organization has also made numerous contributions to the electrical industry by shaping public policy development and operating as a central confidential agency for gathering, compiling, and analyzing market statistics and economics data. **http://www.nema.org.**

National Fire Protection Association (NFPA)

The mission of NFPA is providing and advocating scientifically-based consensus codes and standards, research, training, and education. NFPA serves as the world's leading advocate of fire prevention and is an authoritative source on public safety. In fact, NFPA's 300 codes and standards influence every building, process, service, design, and installation in the United States, as well as many of those used in other countries. **http://www.nfpa.org.**

National Systems Contractors Association (NSCA)

The NSCA is the leading not-for-profit association representing the commercial electronic systems industry. The National Systems Contractors Association is a powerful advocate of all who work within the low-voltage industry, including systems contractors/integrators, product manufacturers, consultants, sales representatives, a growing number of architects, specifying engineers, and others. **http://www.nsca.org.**

Project Management Institute (PMI)

PMI is focused on the needs of project management professionals worldwide. PMI professionals come from virtually every major industry including, aerospace, automotive, business management, construction, engineering, financial services, information technology, pharmaceuticals, healthcare, and telecommunications. **http://www.pmi.org.**

Telecommunications Industry Association (TIA)

TIA is the leading U.S. non-profit trade association serving the communications and information technology industry, with proven strengths in market development, trade shows, domestic and international advocacy, standards development, and enabling e-business. Through its worldwide activities, the association facilitates business development opportunities and a competitive market environment. TIA provides a market-focused forum for its member companies, which manufacture or supply the products and services used in global communications. **http://www.tiaonline.org.**

Glossary

Acceptance criteria The conditions that a company sets before beginning a project to gain the customer's acceptance of the products upon completion of installation.

Accrual-basis accounting A method of accounting in which each item is entered as it is earned or incurred regardless of when actual payments are received or made.

Activity* A component or work performed during the course of a project.

Activity definition* The process of identifying the specific schedule activities that need to be performed to produce the various project deliverables.

Activity description* A short phrase or label for each schedule activity used in conjunction with an activity identifier to differentiate that project schedule activity from other schedule activities. The activity description normally describes the scope of work of the schedule activity.

Activity duration* The time in calendar units between the start and finish of a schedule activity.

Activity identifier* A short, unique numeric or text identification assigned to each scheduled activity to differentiate that project activity from other activities. Typically, unique within any one project schedule network diagram.

Activity list* A documented tabulation of schedule activities that show the activity description, activity identifier, and a sufficiently detailed scope of work description so the project team members understand what work is to be performed.

Activity sequencing* The process of identifying and documenting dependencies among schedule activities.

Approval requirements Conditions established in the scope statement that are applied to project objectives, deliverables, documentation, and work that when signed by the customer gives approval for the project. In addition, the conditions required prior to any changes in the project scope.

As-built block diagrams Block diagrams that show how each system is connected. They contain information such as room names, cabling labels, color codes, test points, and other information required to service the system.

As-built documentation Drawings and diagrams that provide an accurate representation of how the product or facility is actually built.

Assumptions Physical conditions in a project that are accepted or assumed without proof. Within the project scope statement, project assumptions describe the potential impact of such assumptions, if they prove to be false.

Authorizing party In residential integration, the authorizing party typically is the homeowner or homebuyer, in effect, the customer, or the person who initiates the project.

Authority* The right to apply project resources, expend funds, make decisions, or give approvals.

Backward pass* The calculation of the late finish dates and late start dates for the uncompleted portions of all schedule activities. Determined by working backward through the schedule network logic from the project's end date. The end date may be calculated in a forward pass or set by the customer or sponsor.

Benchmarking Benchmarking is used to create a standard for quality improvement.

Bilaterally A contract is originated bilaterally when the buyer (residential integrator) requests a quotation (RFQ) for the product, a written proposal (RFP), or an invitation to bid (ITB).

Bill-of-materials* A document that is a formal hierarchical tabulation of the physical assemblies, subassemblies, and components needed to fabricate a product.

Block diagrams Simple drawings of how cables in an integrated system are to be connected.

Bookkeeping The process of maintaining, auditing, and processing financial information for business purposes.

Bottom-up estimating Bottom-up estimating starts at the bottom (separate individual items in a project) and works up to the finished project (everything added together).

Bottom-up WBS Once project team members identify specific tasks related to a project, the tasks are combined to create the next higher level of assembly. Smaller and simpler tasks are combined to create larger and more complex tasks.

Boundaries Conditions that are set before the work begins to identify what is included and what is not included in the project.

Cable schedule The cable schedule is a table of cables to be installed in the home. It includes the cable type, color, purpose, starting location, ending location, and a unique cable identifier.

Cabling documentation The documentation that explains any prior cabling that has been competed in a home.

Cash-basis accounting A method of accounting in which each item is entered as payments are received or made.

Cash flow The movement of money in and out of an organization.

Cash flow gap An excessive outflow of cash that may not be covered by a cash inflow for weeks, months, or even years.

Cash inflows The movement of money into an organization.

Cash outflows The movement of money out of an organization.

COGS (Cost of Goods Sold) In accounting, the cost of goods sold describes the direct expenses incurred in producing a particular good for sale, including the actual cost of materials that comprise the good, and direct labor expense in putting the good in salable condition. Cost of goods sold does not include indirect expenses such as office expenses, accounting, shipping, advertising, and other expenses that can not be attributed to a particular item for sale.

Commercial database Contains information pertaining to cost, installation time, and other key product information that is available through third-party resources.

Communications management plan* The document that describes the communications needs and expectations for the project, how and in what format information will be communicated, when and where each communication will be made, and who is responsible for providing each type of communication. A communication management plan can be formal or informal, highly detailed or broadly framed, based on the requirements of the project stakeholders. The communications management plan is contained in, or is a subsidiary plan of the project management plan.

Communications A process through which information is exchanged among persons using a common system of symbols, signs, and behaviors.

Communications planning* Determining the information and communications needs of the project stakeholder—who needs what information, when they will need it, and how it will be given to them.

Constraint* The state, quality, or sense of being restricted to a given course of action or inaction. An applicable restriction or limitation, either internal or external to the project, that will affect the performance of the project or a process. For example, a schedule constraint is any limitation or restraint placed on the project schedule that affects when an activity can be scheduled and is usually in the form of fixed imposed dates. A cost constraint is any limitation or restraint placed on the project budget such as funds available over time. A project resource constraint is any limitation or restraint placed on resource usage, such as what resource skills or disciplines are available and the amount of a given resource available during a specified period.

Contingency plan A contingency plan is designed to deal with a particular risk event, or problem.

Continuous improvement The process of continually striving to improve. It is a by-product of a quality assurance plan.

Contract* A contract is a mutual binding agreement that obligates the seller to provide the specified product or service or result and obligates the buyer to pay for it.

Contract value The agreed upon amount for goods and services that are expected to be delivered.

Cost* The monetary value or price of a project activity or component that includes the monetary work of the resources required to perform and complete the activity or component, or to produce the component. A specific cost can be composed of a combination of cost components including direct labor hours, other direct costs, indirect labor hours, other indirect costs, and purchased price. However, in the earned value management methodology, in some instances, the term cost can represent only labor hours with conversion to monetary worth.

Cost baseline A time-phased budget used to measure and monitor cost performance.

Cost-benefit analysis Cost-benefit analysis compares the costs of benefits of a change to determine whether to implement the change.

Cost budgeting* The process of aggregating the estimated costs of individual activities or work packages to establish the cost baseline.

Cost differential The difference between the estimated or bid cost and the actual cost. It can either be a credit or a debit.

Cost estimating* The process of developing an approximation of the costs of the resources needed to complete project activities.

Cost management plan* The document that sets out the format and establishes the activities and criteria for planning, structuring, and controlling the project costs. A cost management plan can be formal or informal, highly detailed or broadly framed, based on the requirements of the project stakeholders. The cost management plan is contained in, or is a subsidiary plan of, the project management plan.

Cost plus fixed fee* A type of contract in which the buyer reimburses the seller for the seller's allowable costs (allowable costs are defined by the contract), plus a fixed amount of profit (fee).

Cost plus incentive fee* A type of contract in which the buyer reimburses the seller for the seller's allowable costs (allowable costs are defined in the contract), and the seller earns its profit if it meets defined performance criteria.

Cost plus percentage fee Provides reimbursement of allowable costs of services performed plus an agreed upon percentage of the estimated costs as profit.

Costs Price paid, or otherwise linked with economic transaction.

Critical activity* Any schedule activity on a critical path in a project schedule. Most commonly determined by using the critical path method. Although some activities are "critical," in the dictionary sense, without being on the critical path, this meaning is seldom used in the project context.

Critical path* Generally, but not always, the sequence of schedule activities that determines the duration of the project. Generally, it is the longest path through the project. However, a critical path can end, as an example, on a schedule milestone that is in the middle of the project schedule and that has a finish-no-later-than imposed date schedule constraint.

Critical path method* A schedule network analysis technique used to determine the amount of scheduling flexibility (the amount of float) on various logical network paths in the project schedule network, and to determine the minimum total project duration. Early start and finish dates are calculated by means of a forward pass, using specified start date. Late start and finish dates are calculated by means of a backward pass, starting from a specified completion date, which sometimes is the project early finish date determined during the forward pass calculation.

Customer experience The customer's awareness and expectations of the quality of work performed and the level of service provided throughout the entire project life cycle.

Customer information Information about the customer such as contact information, family members, and personal habits that help define the project.

Decomposition The subdividing of the major project deliverables or subdeliverables into smaller, more manageable components until the deliverables are defined in sufficient detail to support the development of project tasks.

Deliverables* Any unique and verifiable product, result, or capability to perform a service that must be produced to complete a process, phase, or project. Often used more narrowly in reference to an external deliverable, which is a deliverable that is subject to approval by the project sponsor or customer.

Deliverables The products or services defined by the product scope definitions and provided to the customer, also including ancillary or non-technical results, such as accounting reports or written how-to instructions.

Differential quantity The difference quantity between the baseline and actual quantities.

Direct costs Costs that are directly related to the project life cycle.

Duration* The total number of work periods, not including holidays or other nonworking periods, required to complete a schedule activity or Work Breakdown Structure component. Usually expressed in workdays or workweeks. Sometimes incorrectly equated with elapsed time.

Early contract termination The event that occurs when either the buyer or the seller decides to pull out of the contract before it is completed.

Early finish date* In the critical path method, the earliest possible point in time on which the uncompleted portions of a schedule activity or the project can finish, based upon the schedule network logic, the data date, and any schedule constraints. Early finish dates can change as the project progresses and changes are made to the project management plan.

Early start date* In the critical path method, the earliest possible point in time on which the uncompleted portions of a schedule activity or the project can start, based upon the schedule network logic, the data date, and any schedule constraints. Early start dates can change as the project progresses and changes are made to the project management plan.

Effort* The number of labor units required to complete a schedule activity or Work Breakdown Structure component. Usually expressed as staff hours, staff days, or staff weeks.

Enterprise environmental factors* Any or all external environmental factors and internal organizational environmental factors that surround or influence the project's success. These factors are from any or all of the enterprises involved in the project and include organizational culture and structure, infrastructure, existing resources, commercial databases, market conditions, and project management software.

Equipment location The location in the home where the integration equipment such as the television and telephone panels are located.

Event* Something that happens, an occurrence, an outcome.

Executive team The executive team is the governing branch of the organization. In a small company, this may be the owner, and in larger companies, it may be the director of operations in conjunction with the heads of each of the departments—engineering, sales, project management, installation, and administration.

Existing construction A project that is carried out in an existing structure, such as in a home remodel.

Expected monetary value (EMV) The product of an event's probability of occurrence and the gain or loss that will result.

Expense Expense is a general term for an outgoing payment made by a business or individual.

Fallback plans Plans developed for risks that have a high impact on the project objectives.

Fast tracking* A specific project schedule compression technique that changes network logic to overlap phases that would normally be done in sequence, such as the design phase and construction phase, or to perform schedule activities in parallel.

Final scope statement The final version of the scope statement, including the original text and all updates.

Floor plan A view of the layout of the floor structure from an overhead perspective is called a floor plan. The floor plan contains information such as location and size of windows and doors, room size, location of exterior and partition walls, location of electrical fixtures, outlets and switches and plumbing fixture locations and other items. Each floor has its own floor plan.

Forward pass* The calculation of the early start and early finish dates for uncompleted portions of all network activities.

Free float* The amount of time that a schedule activity can be delayed without delaying the early start of any immediately following schedule activities.

Functional organization A functional organization is prevalent in companies that do not take on project-based work as part of their regular business. A company with a functional-oriented structure is based on the theory that staff divisions of responsibility approximately have equal delegations of authority.

GAAP Generally Accepted Accounting Principles (GAAP) are the accounting rules used to prepare financial statements for publicly traded companies and many private companies in the United States.

Gross profit The earnings from an ongoing business after direct costs of goods sold have been deducted from sales revenue for a given period.

High flux projects High flux projects change in scope through the project life.

House layout A general floor plan of the home, not to scale, designed to show an overall layout of the home.

Indirect costs Costs that are shared over many projects.

Inflows *See* Cash inflows.

Initial project organization The individual members of the teams that work on a project.

Initial defined risks Known project risks.

Initial project organization The individual members of the teams who work on a project.

Installation documentation Documents that show any prior installations in the home.

Intangible costs Costs that cannot be quantitatively measured or valued.

Invitation to Bid (ITB) A document that requests suppliers to submit a bid in order to secure work or material.

Job costing report A report that summarizes the costs of the project by phase and is broken out by products and services.

Late end date The latest date an activity can end without impacting the schedule.

Late start date The latest date an activity can start without impacting the schedule.

Learning curve The length of time it takes to learn successfully any given task. When many items are produced repeatedly, the unit cost of those items normally decreases in a regular pattern as more units are produced.

Logical relationship* A dependency between two project schedule activities, or between a project schedule activity and a schedule milestone. The four possible types of logical relationships are finish-to-start, finish-to-finish, start-to-start, and start-to-finish.

Lowball The practice of estimating a project below the cost of delivering that project, in hopes of creating profit with additional project changes.

Make-or-buy analysis A decision process in which it is determined whether to manufacture internally or buy from external sources some component, article, or item of equipment.

Market conditions Refers to the strength of the market or a market segment.

Matrix-based organization The most appropriate structure for the residential integration company, it is a compromise between the functional and projectized organization. Where a particular organization lies along the spectrum depends on the products and services that the company offers to its customers.

Methodology* A system of practices, techniques, procedures, and rules used by those who work in a discipline.

Milestone* A significant point or event in the project.

Net profit The earnings or income after subtracting miscellaneous income and expenses (patent royalties, interest, capital gains) and federal income tax from operating profit.

Non-standard systems Systems that an organization has never installed.

Objective* Something toward which work is to be directed, a strategic position to be attained, or a purpose to be achieved, a result to be obtained, a product to be produced, or a service to be performed.

Organization* A group of persons organized for some purpose or to perform some type of work within an enterprise.

Organizational chart* A method for depicting interrelationships among a group of persons working together toward a common objective.

Operating profit The earnings or income after all expenses (selling, administrative, depreciation) have been deducted from gross profit.

Operations turnover The process of shifting the responsibility for a project from the defining phase (sales) to the operations phases (project management).

Opportunity The cumulative effect of the chances of uncertain occurrences that will affect project objectives positively.

Outflows *See* Cash outflow.

Outsourcing The procuring of services or products from an outside supplier or manufacturer in order to cut costs and/or to speed up project deliverables.

Overhead An average amount used for indirect costs that is treated like a direct cost.

Payable Accounts payable is one of a series of accounting transactions covering payments to suppliers owed money for goods and services.

Performance reports* Documents and presentations that provide organized and summarized work performance information, earned value management parameters and calculations, and the analyses of project work progress and status. Common formats for performance reports include bar charts, S-curves, histograms, tables, and project schedule network diagrams showing current schedule status.

Performing organization* The enterprise whose employees are most directly involved in doing the work of the project.

Predecessor activity* The schedule activity that determines when the logical successor activity begins.

Procurement management plan* A type of procurement document used to request proposals from prospective sellers of products or services. In some application areas, it may have a narrow or more specific meaning.

Product scope definition A definition that explains the details and parameters of a product and is included in the scope statement.

Product strategy A product roadmap used to determine what direction to take, how to get there, and why that direction is expected to be successful.

Profit margin The ratio between revenue and profits.

Progressive elaboration* Continuously improving and detailing a plan as more details and specific information and more accurate estimates become available as the project progresses, and thereby producing more accurate and complete plans that result from the successive iterations of the planning process.

Project* A temporary endeavor undertaken to create a unique product, service, or result.

Project assumptions *See* assumptions.

Project boundaries *See* boundaries.

Project charter* A document issued by the project initiator or sponsor that formally authorizes the existence of a project, and provides the project manager with the authority to apply organizational resources to project activities.

Project configuration management requirements The administrative constraints placed on the project, including but not limited to, payment schedule, change order policy, and termination of the contract.

Project constraints Specific conditions that limit the installation team's options on a project.

Project deliverables *See* deliverables.

Project life cycle* A collection of generally sequential project phases whose name and number are determined by the control needs of the organization or organizations involved in the project. A life cycle can be documented with a methodology.

Project management* The application of knowledge, skills, tools, and techniques to project activities in order to meet or exceed stakeholders' needs and expectations from a project.

Project manager* The person assigned by the performing organization to achieve the project objectives.

Project objectives A project's success criteria that can be measured in terms of time, cost, schedule, or quality.

Project plan* A formal, approved document used to guide both project execution and project control. The primary uses of the project plan are to document planning ssumptions and decisions, facilitate communication among stakeholders, and document approved scope, cost, and schedule baselines. A project plan may be summary or detailed.

Project requirements The conditions or capabilities that must be met or possessed by the deliverables of the project.

Project schedule* The planned dates for performing schedule activities and the planned dates for meeting schedule milestones.

Project scope statement* The narrative description of the project scope, including major deliverables, project objectives, project assumptions, project constraints, and a statement of work, that provides a documented basis for making future project decisions and for confirming or developing a common understanding of the project scope among the stakeholders. The definition of the project scope—what needs to be accomplished.

Project scope management plan* The document that describes how the project scope will be defined, developed, and verified and how the work breakdown structure will be created and defined. It provides guidance on how the project scope will be managed and controlled by the project management team. It is contained in or is a subsidiary plan of the project management plan. The project scope management plan can be informal or broadly framed, or formal and highly detailed, based on the needs of the project.

Project specifications Any external documents, such as floor plans, and other third party specifications that affect the project.

Project team* All the project team members, including the project management team, the project manager, and for some projects, the project sponsor.

Project time management* Includes the process required to accomplish timely completion of the project.

Projectized organization The project manager has complete authority over the project team. The team is created and assigned on a full-time basis to the project manager, and the parent organization remains functionally organized.

Progressive elaboration* Continuously improving and detailing a plan as more details and specific information and more accurate estimates become available as the project progresses, and thereby producing more accurate and complete plans that result from the successive iterations of the planning process.

Purchase order A contract between the buyer and seller to purchase goods and services.

Qualitative risk analysis* The process of prioritizing risks for subsequent further analysis or action by assessing and combining higher probability of occurrence and impact.

Quality* The degree to which a set of inherent characteristics fulfills requirements.

Quality assurance A proactive method to ensure quality. It creates quality policies and procedures that ensure that project standards are correctly and verifiably met during the project.

Quality management planning* A subset of project management that includes the processes required to ensure that the project will satisfy the needs for which it was undertaken.

Quantitative risk analysis* The process of numerically analyzing the effect on overall project objectives of identified risks.

Receivable Accounts receivable is one of a series of accounting transactions dealing with the billing of customers who owe money to a person, company or organization for goods and services that have been provided to the customer.

Replanning A change in the original plan for accomplishing authorized contractual requirements.

Request for Proposal (RFP) A document generated when a contract is originated bilaterally. The contract is developed by releasing an RFP for a high dollar contract. The RFP method commonly is used by builders and homeowners when soliciting proposals for custom residential integration systems in the home.

Request for Quotation (RFQ) A document that lists products with no pricing that is sent to suppliers asking them to fill in the prices. Most commonly used in low-cost supplies and materials.

Resources* Skilled human resources (specific disciplines either individually or in crews or teams), equipment, services, supplies, commodities, materials, budgets, or funds.

Retrofit A project in which a device or system is installed for use in or on an existing structure.

Revenue Money received for goods and services that have been delivered.

Riser diagrams A set of diagrams that show the interconnections between equipment installed in a residential integration project.

Risk* The cumulative effect of the chances of uncertain occurrences which will adversely affect project objectives. It is the degree of exposure to negative events and their probable consequences. Project risk is characterized by three factors: risk event, risk probability, and the amount at stake.

Risk management plan* Documents how the risk processes will be carried out during the project.

Risk matrix The presentation of information about risks in a matrix format, enabling each risk to be presented as the cell of a matrix whose rows are usually the stages in the project life cycle and whose columns are different causes of risk. A risk matrix is useful as a checklist of different types of risk that might arise over the life of a project but it must always be supplemented by other ways of discovering risks.

Risk response plan* A document detailing all identified risks, including description, cause, probability of occurring, impact(s) on objectives, proposed responses, owners, and current status.

Scope* The sum of all products, services, and results to be provided as a project.

Schedule milestone A significant event in the project schedule, such as an event restraining future work or marking the completion of a major deliverable. A schedule milestone has zero duration. Sometimes called a milestone activity.

Scope planning* The process of progressively elaborating the work of the project, which includes developing a written scope statement that includes the project justification, the major deliverables, and the project objectives.

Service policy An organization policy or procedure for servicing projects after completion. It includes what services and products are included in the warrantee period as well as charges outside the limits of the product warranty.

Site conditions The conditions of the worksite.

Site location The physical location of the worksite.

Soft skills A set of nontechnical skills that enhance the ability of the project manager to successfully accomplish his work.

Sponsor* The person or group that provides the financial resources, in cash or in kind, for the project.

Stakeholders* Persons and organizations such as customers, sponsors, performing organization and the public, that are actively involved in the project, or whose interest may be positively or negatively affected by execution or completion of the project. They may also exert influence over the project and its deliverables.

Standard systems Systems that an organization regularly offers to customers.

Statement of warranty This document specifies the warranty period of each electronic component, as well as the terms of the service provided by the company.

Sunk costs Money that has been spent in the past and is nonrecoverable.

System information Information about any one of the integrated electronic systems, such as options, placement, and quantities of products desired.

Tangible costs Costs that can be quantitatively measured or valued.

Task* A term for work whose meaning and placement within a structured plan for project work varies from the application area, industry, and brand of project management software.

Template* A partially complete document in a predefined format that provides a defined structure for collecting, organizing, and presenting information and data. Templates are often based upon documents created during prior projects. Templates can reduce the effort needed to perform work and increase the consistency of the results.

Threats Threats are project risks that jeopardize the project's objectives and project success.

Top-down estimating Top-down estimating uses the cost of the major components and the time required to complete the system to determine an overall estimate.

Top-down WBS The largest items of the project are divided into subordinate items. This method organizes the project to provide the basis for the project schedule.

Total float* The total amount of time that a schedule activity may be delayed from its early start date without delaying the project finish date, or violating a schedule constraint. Calculated using the critical path method technique and determining the difference between the early finish dates and late finish dates.

Uncertainty The possibility that events may occur which will impact the project either favorably or unfavorably. Uncertainty gives rise to both opportunity and risk.

Unilaterally A contract is originated unilaterally when a purchase order is provided to the vendor. Multiple copies of the purchase order are created.

Unusual circumstances Any items that are not included on the information gathering forms, but are required for the purpose of defining the project.

Validation* The technique of evaluating a component or product during or at the end of a phase or project to ensure it complies with the specified requirements.

Value engineering* Value engineering is a creative approach used to optimize life cycle costs, save time, increase profits, improve quality, expand market share, solve problems, and/or use resources more effectively.

Verification Confirmation that the decomposition of a project into deliverables and subdeliverables is correct, leading to the formation of work packages.

WBS dictionary* A document that describes each component in the work breakdown structure (WBS). For each WBS component, the WBS dictionary includes a brief definition of the scope or statement of work, defined deliverables(s), a list of associated activities, and a list of milestones. The information may include responsible organization, start and end dates, resources required, an estimate of cost, charge number, contract information, quality requirements, and technical reference to facilitate performance of the work.

Work Breakdown Structure* A deliverable-oriented hierarchical decomposition of the work to be executed by the project team to accomplish the project objectives and create the required deliverables. It organizes and defines the total scope of the project. Each descending level represents and increasingly detailed definition of the project work. The WBS is decomposed into work packages. The deliverable orientation of the hierarchy includes both internal and external deliverables.

Work Package* A deliverable or project work component at the lowest level of each branch of the work breakdown structure. The work package includes the schedule activities and the schedule milestones required to complete the work package deliverable or project work component.

---

* Project Management Institute, *A Guide to the Project Management Body of Knowledge (PMBOK® Guide), Third Edition,* Project Management Institute, Inc., 2004.

Index

Q

R

W

X